Power Systems: Analysis, Control and Protection

Power Systems: Analysis, Control and Protection

Editor: Linda Morand

NY RESEARCH
P R E S S

New York

Published by NY Research Press
118-35 Queens Blvd., Suite 400,
Forest Hills, NY 11375, USA
www.nyresearchpress.com

Power Systems: Analysis, Control and Protection
Edited by Linda Morand

International Standard Book Number: 978-1-63238-657-1 (Hardback)

Cataloging-in-Publication Data

Power systems : analysis, control and protection / edited by Linda Morand.
    p. cm.
Includes bibliographical references and index.
ISBN 978-1-63238-657-1
1. Electric power systems. 2. Electric power systems--Control. 3. Electric power systems--Protection. 4. Power (Mechanics). 5. Power-plants. I. Morand, Linda.
TK1001 .P69 2019
621.31--dc23

Contents

Preface

Every book is initially just a concept; it takes months of research and hard work to give it the final shape in which the readers receive it. In its early stages, this book also went through rigorous reviewing. The notable contributions made by experts from across the globe were first molded into patterned chapters and then arranged in a sensibly sequential manner to bring out the best results.

A power system combines the diverse aspects of generation, transmission and distribution of electrical energy to supply energy for a variety of household and industrial applications. The study of power systems is an inter-disciplinary subject that integrates electrical and electronic engineering for the design and operation of grids and other power systems. One of the major difficulties in power systems is in maintaining the frequency value. Even minor fluctuations in the frequency can damage appliances and synchronous machines. Power systems have one or more sources of power, such as batteries, fuel cells or photovoltaic cells. Some of the components of power systems are conductors, capacitors, reactors, etc. Protective devices such as circuit breakers and protective relays are also crucial to power systems. This book attempts to understand the multiple branches that fall under the discipline of power systems and how such concepts have practical applications. The various advancements in the field are glanced at and their applications as well as ramifications are looked in detail. Power systems engineers, students and researchers will find this book full of crucial and unexplored concepts.

It has been my immense pleasure to be a part of this project and to contribute my years of learning in such a meaningful form. I would like to take this opportunity to thank all the people who have been associated with the completion of this book at any step.

Editor

Optimal scheduling strategy for virtual power plants based on credibility theory

Qian Ai[*], Songli Fan and Longjian Piao

Abstract

The virtual power plant (VPP) is a new and efficient solution to manage the integration of distributed energy resources (DERs) into the power system. Considering the unpredictable output of stochastic DERs, conventional scheduling strategies always set plenty of reserve aside in order to guarantee the reliability of operation, which is too conservative to gain more benefits. Thus, it is significant to research the scheduling strategies of VPPs, which can coordinate the risks and benefits of VPP operation. This paper presents a fuzzy chance-constrained scheduling model which utilizes fuzzy variables to describe uncertain features of distributed generators (DGs). Based on credibility theory, the concept of the confidence level is introduced to quantify the feasibility of the conditions, which reflects the risk tolerance of VPP operation. By transforming the fuzzy chance constraints into their equivalent forms, traditional optimization algorithms can be used to solve the optimal scheduling problem. An IEEE 6-node system is employed to prove the feasibility of the proposed scheduling model. Case studies demonstrate that the fuzzy chance strategy is superior to conservative scheduling strategies in realizing the right balance between risks and benefits.

Keywords: Credibility theory, Distributed energy resource (DER), Scheduling strategy, Uncertain factors, Virtual power plant (VPP)

Introduction

Nowadays, due to the rising prices of fossil fuels and the threat of climate change caused by greenhouse gases, distributed energy resources (DERs) have drawn widespread attention because of their clean and renewable characteristics [1–3]. However, the output of DERs is fluctuating and unpredictable. As the penetration of intermittent renewables in the grids is increasing gradually, more technical challenges need to be addressed in the schedule and control of their operation [4–7]. Meanwhile, the liberalization of the electricity market makes DERs inevitable. However, the small capacity, intermittent output and the lack of appropriate interaction with the system operator are the biggest barriers ahead of DERs for participation in the electricity market. To solve these problems mentioned above, the DERs could be aggregated as an entity which can behave like a conventional generator, naming the virtual power plant (VPP) [8–10].

In [11, 12], a VPP is defined as a coalition of DERs including distributed generations (DGs), storage devices, and interruptible loads. Considering the characteristics of each DER and the impact of network, VPPs generate one unit portfolio which can be utilized to offer services to the system operator, and even to make contracts in the wholesale market. Therefore, by introducing the concept of the VPP, the visibility and controllability of DERs for system operators will be improved considerably, just the same as the conventional transmission–connected power plants.

To realize the concept of the VPP, scholars around the world have done lots of studies on VPPs in many aspects [13–17], including VPP modeling methods, negotiating behaviors in the market, bidding strategies, reliability evaluation, management systems, and so on. Among them, the optimal scheduling strategy of DERs in the VPP is a hot topic. Reference [18] proposed a non-equilibrium model which takes into account the supply and demand balancing constraint and security constraint of VPPs. On this basis, the strategy proposed in [19] considered the effect of reliability and determined the optimum hourly operating strategy of DERs by applying

* Correspondence: aiqian@sjtu.edu.cn
Department of Electrical Engineering, Shanghai Jiao Tong University, Shanghai 200240, China

the Monte Carlo simulation method. However, these scheduling strategies are based on the deterministic market prices. Considering the uncertainties in prices, [20] developed a new risk constrained scheduling for VPPs to help the aggregator bid in the day-ahead market.

However, the optimal bidding strategy of those researches mainly focused on maximizing the VPP's profit in different types of markets. With regard to reserve level, these methods spare large capacity as recovery to balance the supply and demand. Although the system stability can be guaranteed in this deterministic way, the obtained scheduling results are always inevitably conservative. Actually, more benefits can be gained via reducing the reserve capacity appropriately.

This paper proposes a fuzzy chance-constrained scheduling model to provide solutions for the optimal scheduling strategy for VPPs based on credibility theory. In this model, as the prediction errors of renewables are characterized as fuzzy parameters, the related constraints correspondingly contain fuzzy variables which need to be tackled properly. By introducing the concept of the confidence level, the feasibility of the conditions where the fuzzy chance constraints can be satisfied is quantified, so as to characterize the risk tolerance of VPPs. The proposed fuzzy chance model is difficult to solve due to the incorporation of fuzzy variables. By transforming the constraints containing fuzzy variables into their equivalent forms, the chance constraints are converted into deterministic constraints which consider the fuzzy risks (or reliability) as well. Then traditional optimization algorithms can be utilized to solve the optimal scheduling problem, in order to achieve a compromise between risks and benefits.

The rest of the paper is organized as follows: the theoretical basis of this paper is presented in Credibility theory and fuzzy chance-constrained programming, including credibility theory and fuzzy chance constrained programming; the proposed VPP scheduling strategy based on credibility theory is demonstrated in VPP scheduling strategy based on credibility theory; solution methods used in this paper are stated in Methods; case studies and discussion are given in Case study; and finally, the conclusion is drawn in Conclusions.

Credibility theory and fuzzy chance-constrained programming

Distinguished from the conventional power plants, the scheduling of VPPs is required to deal with a number of uncertain variables from renewables and loads. The common ways to process those uncertain variables are categorized into determinate and indeterminate methods. Determinate method is to set plenty of reserve aside in order to guarantee the reliability of operation. However, this approach is too conservative to coordinate the risks

and benefits of VPP operation. More benefits can be obtained by ignoring the small probability events and making decisions within the scope of risk tolerance [21]. The fuzzy chance-constrained programming method proposed in this paper is one kind of indeterminate scheduling strategy.

In this approach, the decision result is allowed to violate the constraints which contain fuzzy variables, but the feasibility of satisfying the constraints should be no less than the preset confidence level. The confidence level is a concept which reflects the risk tolerance of the system, or the reliability requirements of the system when facing uncertain variables. In conclusion, the optimal fuzzy chance-constrained scheduling strategy is a compromise between risks and economic profits. Credibility theory provides theoretical support to solve the confidence level problem of the fuzzy chance strategy decision, which greatly contributes to the development and improvement of the fuzzy chance programming theory.

Credibility theory
The concept of the fuzzy set theory was initiated by Zadeh via the membership function [22]. In order to measure a fuzzy event, the possibility measure is proposed. Since then the possibility theory has been studied by many researchers. Although the possibility measure has been widely used, it does not possess the self-duality property. Thus, if the possibility measure equals one it does not mean the event will happen definitely, while the event may happen even though the possibility measure is zero. So the solution to the fuzzy decision problem has been a mathematical conundrum for a long time.

Credibility theory was propounded by Liu in 2004 as a branch of mathematics for studying the fuzzy behavior [23]. It establishes a complete axiomatic system which is parallel with probability theory in dealing with uncertainty. Based on five axioms mentioned in [23], a credibility measure is defined.

According to Liu [23], the following five axioms should hold to ensure that the number Cr(A) has certain mathematical properties.

Definition 2.1:

Axiom 1: $\mathrm{Cr}\{\Theta\} = 1$;
Axiom 2: Cr is non-decreasing, i.e., if A ⊆ B, there is always $\mathrm{Cr}\{A\} \le \mathrm{Cr}\{B\}$;
Axiom 3: Cr is self-dual: $\forall A \in P(\Theta)$, $\mathrm{Cr}\{A\} + \mathrm{Cr}\{A^c\} = 1$;
Axiom 4: $\forall A_i \in \{A_i|\ \mathrm{Cr}\{A_{ij}\} \le 0.5\}$, there is $\mathrm{Cr}\{\bigcup_i A_i\} \wedge 0.5 = \sup_i \mathrm{Cr}\{A_i\}$.
Axiom 5: Let $\Theta_1, \Theta_2, ..., \Theta_n$ be the nonempty sets corresponding to that $\mathrm{Cr}_1, \mathrm{Cr}_2, ..., \mathrm{Cr}_n$ satisfy the axioms as respectively defined above, and let $\Theta = \Theta_1 \times \Theta_2 \times ... \times \Theta_n$.

Then, we have $\mathrm{Cr}\{(\theta_1, \theta_2, ..., \theta_n)\} = \mathrm{Cr}_1\{\theta_1\} \wedge \mathrm{Cr}_2\{\theta_2\} \wedge ... \wedge \mathrm{Cr}_n\{\theta_n\}$ for each $(\theta_1, \theta_2, ..., \theta_n) \in \Theta$.
where Θ is a nonempty set; $P(\Theta)$ is the possibility set of Θ, the element of $P(\Theta)$ is a fuzzy event; A is the subset of Θ; $\mathrm{Cr}(A)$ is a non-negative number indicating the credibility of Event A which will occur; and $\wedge$ represents the minimal operator.

Accordingly, the credibility measure describes the credibility level of a fuzzy event; parallel with the confidence level in probability theory, it satisfies the self-duality. It means that the event whose credibility measure is 1 will definitely occur, and similarly, the event whose credibility measure is 0 will never happen. So credibility theory solves the confusion caused by a subordinate degree, and provides theoretical foundations for the fuzzy decision.

In the fuzzy decision of VPP scheduling, some definitions and theorems of credibility theory are utilized, as listed below:

Definition 2.2 (Membership Function): The membership function $\mu(x)$ of the fuzzy parameter ξ is defined as:

$$\mu(x) = \mathrm{Pos}\{\theta \in \Theta | \xi(\theta) \leq x\} \tag{1}$$

Definition 2.3 (Credibility Measure): The credibility measure can be obtained from the membership function, which is named as the credibility inversion theorem:

$$\mathrm{Cr}\{\xi \in A\} = \frac{\sup \mu(x)_{x \in A} + 1 - \sup \mu(x)_{x \in A^c}}{2} \tag{2}$$

Fuzzy chance-constrained programming
Fuzzy chance-constrained programming refers to a type of decision-making problem which contains fuzzy parameters. The forward-looking decision results of this problem may not satisfy the constraints with fuzzy parameters to some extent, but the possibility of satisfying the constraints should be no less than the pre-specified confidence level. The general form of the fuzzy chance-constrained programming problem is:

$$\begin{cases} \max \bar{f} \\ \mathrm{Cr}\{f(\boldsymbol{x}, \boldsymbol{\xi}) \geq \bar{f}\} \geq \alpha \\ \mathrm{Cr}\{g_i(\boldsymbol{x}, \boldsymbol{\xi}) \leq 0, i = 1, 2, \cdots, n\} \geq \beta \end{cases} \tag{3}$$

where $\boldsymbol{x}$ is the decision vector; $\boldsymbol{\xi}$ is the fuzzy parameter vector; $g_i(\boldsymbol{x}, \boldsymbol{\xi}) \leq 0$ is the ith constraint; α is the confidence level of the object function; β is the confidence level of the fuzzy constraint. The decision vector $\boldsymbol{x}$ is feasible only when the credibility of $g_i(\boldsymbol{x}, \boldsymbol{\xi}) \leq 0$ is no less than β. Additionally, the optimal solution can maximize the object function f at the confidence level of α.

VPP scheduling strategy based on credibility theory

Traditional VPP scheduling strategy

The primary objectives of the VPP optimal scheduling problem are various, such as minimizing the cost of producing energy, or maximizing the profits of VPPs. In this paper, the priority of VPP operation is to make use of the renewable energy resources (RES), and based on this, the objective function is to minimize the costs of conventional unit generation and the interruptible load.

$$\min f = \sum_{t=1}^{T} \left\{ \sum_{i=1}^{N_G} [J_{Gi}(t) + u_i(t-1)u_i(t)SU_i(t)] + \sum_{j=1}^{N_{DR}} J_{DRj}(t) \right\} \tag{4}$$

$$J_{Gi}(t) = u_i(t)\left(a_i P_{Gi}^2 + b_i P_{Gi} + c_i\right) \tag{5}$$

$$J_{DRj}(t) = d_j P_{DRj} u_j(t) \tag{6}$$

where $J_{Gi}(t)$ is the operational cost of unit i at time t; $SU_i(t)$ is the start-up cost of the unit i at time t; a_i, b_i, c_i are the coefficients of the operational cost of the unit i; d_j is the coefficient of the cost by curtailing the load j; P_{Gi} is the output power of the unit i; P_{DRj} is the virtual generation via curtailing the load; $u_i(t)$ is the binary state variable of the unit i at time t: it equals "1" if the unit i is on at time t, and equals "0" if the unit i is off at time t. Similarly, $u_j(t)$ is the binary state variable of the load j at time t. N_G and N_{DR} are the numbers of conventional units and loads in the demand response respectively.

The constraints considered in this model are presented as follows:

1) Supply–demand balancing constraint

$$\sum_{i=1}^{N_G} P_{Gi}(t) + \sum_{j=1}^{N_{DR}} P_{DRj}(t) + P_{RES}(t) = P_L(t) \tag{7}$$

2) System reserve constraint

$$\sum_{i=1}^{N_G} P_{Gi}^{\max} + \sum_{j=1}^{N_{DR}} P_{DRj}^{\max} + P_{RES}(t) \geq P_L(t) + R(t) \tag{8}$$

3) DG constraints

$$u_i(t)P_{Gi}^{\min} \leq P_{Gi}(t) \leq u_i(t)P_{Gi}^{\max} \tag{9}$$

$$-RD_i \leq P_{Gi}(t) - P_{Gi}(t-1) \leq RU_i \tag{10}$$

4) Interruptible load constraint

$$u_j(t)P_{DRj}^{\min} \leq P_{DRj}(t) \leq u_j(t)P_{DRj}^{\max} \tag{11}$$

where $P_{RES}(t)$ is the power generated by renewables at time t; $P_L(t)$ is the total load at time t; $R(t)$ is the reserve

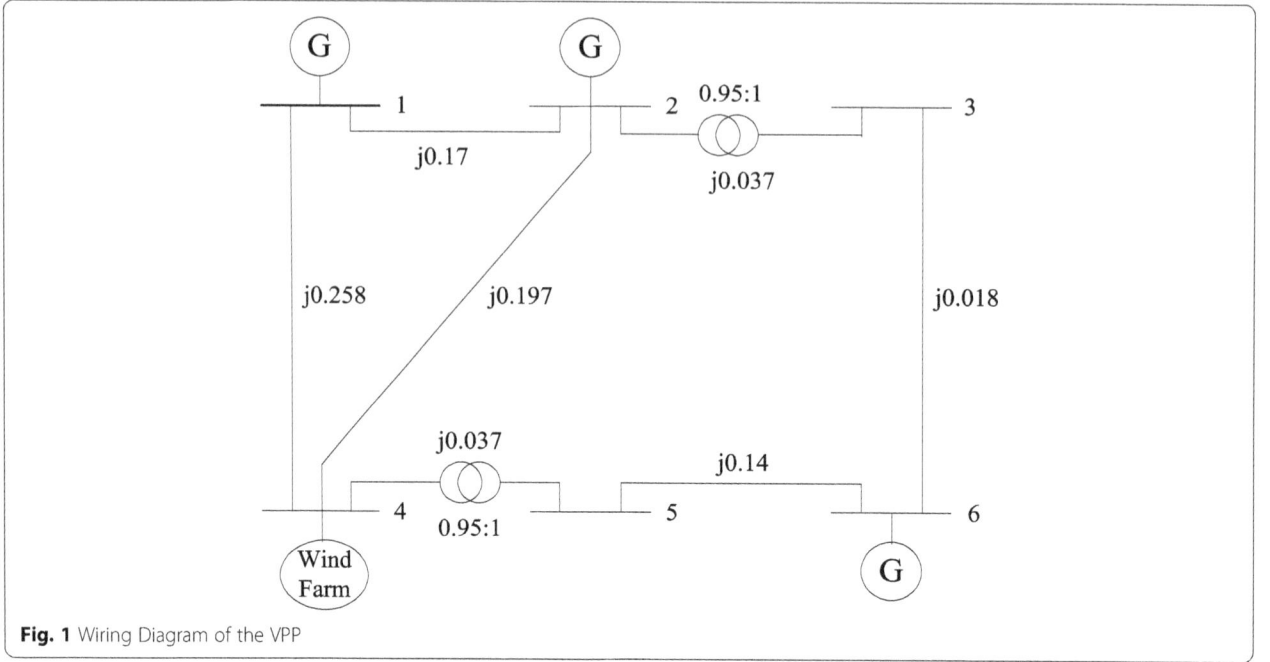

Fig. 1 Wiring Diagram of the VPP

capacity for the fluctuation of load and renewables at time t; $P_{Gi}^{\min}$ and $P_{Gi}^{\max}$ are the minimum and the maximum output limit of the unit i respectively; RU_i and RD_i are the ramp-up rate and the ramp-down rate of the unit i; $P_{DRj}^{\min}$ and $P_{DRj}^{\max}$ are the minimum and maximum allowed curtailed value of the load j respectively.

Fuzzy chance-constrained VPP scheduling

The fuzzy parameters of VPP scheduling strategy come from the unpredictable output of renewables in VPPs [24]. Accurate mathematical expressions of those uncertain characteristics should be established, which are the foundations for the uncertain scheduling strategy. The main characterizing methods include modeling output of renewables and modeling their predicted errors. In this paper, the modeling of the forecast errors is considered to describe the fuzziness of renewables instead of modeling the output. The modeling of renewables output will make three constraints containing fuzzy variables, which are the supply–demand balance constraint, the spinning reserve constraint, and the static security constraint. If the predicted value is regarded as a deterministic value and the prediction error is utilized to describe the fuzziness of renewables, only the spinning reserve constraint should be considered.

The percentage of renewables prediction error is defined as

$$\varepsilon\% = \frac{P_{RES}-P_{RES}^F}{P_{RES}^F} \times 100\% \tag{12}$$

where P_{RES} and P_{RES}^F are the real value and the forecasted value of the output respectively.

The membership function of the prediction error is:

$$\mu = \begin{cases} \dfrac{1}{1+\sigma\left(\dfrac{\varepsilon}{E_+}\right)^2}, & \varepsilon > 0 \\ \dfrac{1}{1+\sigma\left(\dfrac{\varepsilon}{E_-}\right)^2}, & \varepsilon \leq 0 \end{cases} \tag{13}$$

where E_+ and E_- are the average values of positive and negative prediction errors obtained from statistics, respectively; σ is the weighting factor.

According to credibility theory, the credibility measure of the fuzzy parameter ξ can be expressed as:

$$\mathrm{Cr}(\xi \leq \varepsilon) = \begin{cases} 1-\dfrac{1}{2[1+\sigma\left(\dfrac{\varepsilon}{E_+}\right)^2]}, & \varepsilon > 0 \\ \dfrac{1}{2[1+\sigma\left(\dfrac{\varepsilon}{E_-}\right)]}, & \varepsilon \leq 0 \end{cases} \tag{14}$$

Thus, according to credibility theory, the spinning reserve constraint of the traditional scheduling model

Table 1 Parameters of conventional generators

Node	a	b	c	$P_G^{\max}$	$P_G^{\min}$	RU(RD)
1	0.1	13.5	176.9	220	100	55
2	0.1	32.6	129.9	100	10	50
6	0.1	17.6	137.4	20	10	20

Table 2 Load data of the VPP in 24 h

Time	Load	Time	Load	Time	Load	Time	Load
1	175.2	7	173.4	13	242.2	19	246.0
2	165.2	8	177.6	14	243.6	20	237.4
3	158.7	9	186.8	15	248.9	21	237.3
4	154.7	10	206.9	16	255.8	22	232.7
5	155.1	11	228.6	17	256.0	23	195.9
6	160.5	12	236.1	18	246.7	24	195.6

should be changed into chance constraints containing fuzzy variables.

$$\mathrm{Cr}\left\{\sum_{i=1}^{N_G} u_i(t)P_{Gi}^{\max} + \sum_{j=1}^{N_{DR}} u_j(t)P_{DRj}^{\max} + P_{RES}^F(t)(1+\xi) \geq P_L(t) + R(t)\right\} \geq \alpha \tag{15}$$

where α is the credibility level, which represents the reliability level of the reserve constraint. In the realistic problem, α should be more than 0.5.

According to the equivalent theorem of credibility theory, (15) can be transformed into an equivalent form as indicated in (16):

$$\frac{P_L(t) + R(t) - P_{RES}^F(t) - \sum_{i=1}^{N_G} u_i(t)P_{Gi}^{\max} - \sum_{j=1}^{N_{DR}} u_j(t)P_{DRj}^{\max}}{P_{RES}^F(t)} \leq K_\alpha \tag{16}$$

$$K_\alpha = \inf\left\{K|K = \mu^{-1}(2(1-\alpha))\right\}, \forall \alpha \geq 0.5 \tag{17}$$

The equivalent form of the fuzzy chance constraint can be further deduced to the following forms:

$$\sum_{i=1}^{N_G} u_i(t)P_{Gi}^{\max} + \sum_{j=1}^{N_{DR}} u_j(t)P_{DRj}^{\max} + (1+K_\alpha)P_{RES}^F(t) \geq P_L(t) + R(t) \tag{18}$$

$$K_\alpha = |E_-\%|\sqrt{\frac{2\alpha-1}{2\sigma(1-\alpha)}} \tag{19}$$

It can be proved that K_α is a monotonously increasing function. Equation (18) illustrates the links between the confidence level of the fuzzy constraint and spinning reserve allocation when taking the fuzziness of RES predicted error into account. It shows that to improve the confidence level of the spinning reserve constraint, a greater spinning reserve should be allocated. The form of the formula is similar to the traditional spinning reserve constraint. Thus, the model can be easily solved by correcting the reserve levels using the regular optimization algorithm with the coefficient K_α.

Methods

As introduced above, once chance constraint is transformed into its crisp equivalent, traditional optimization methods can be employed to solve the scheduling problem. Based on the natural selection and genetic manipulation, genetic algorithms (GAs) are heuristic random searching methods, and they possess excellent robustness and commonality. However, there are some shortfalls in the original GA, e.g. the slow convergence speed, falling easily into the locally optimal solution, and so on. A self-adaptive GA method was introduced to improve the performance. The probabilities of crossover and mutation for each generation are adaptively determined, which can overcome the premature convergence and the slow convergence speed in later evolutionary processes.

The crossover and mutation operators are two important genetic operators. In this paper, the crossover

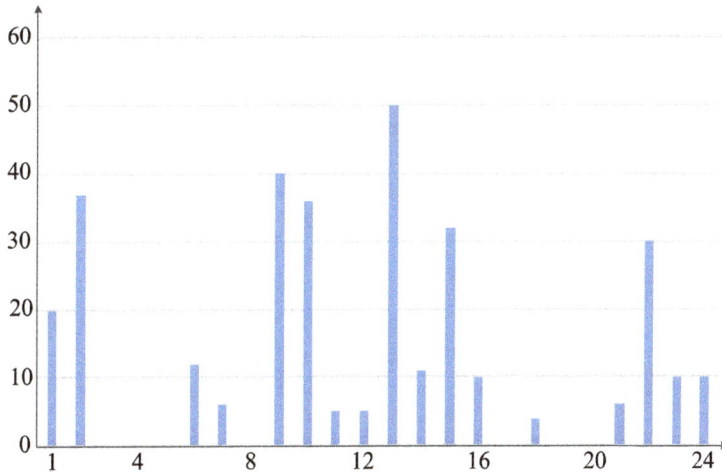

Fig. 2 Forecasted Output of the Wind Farm in 24 h

operator adopts double-cut-point crossover and the mutation operator is implemented through power disturbance. The crossover and mutation probabilities adopt self-adaptation mechanism and they will adjust adaptively according to the values of the fitness function [25], as shown in (20) and (21):

$$
P_c = \begin{cases} P_{c2} - \dfrac{(P_{c2}-P_{c3})\left(f'-f_{avg}\right)}{f_{max}-f_{avg}}, & f' \geq f_{avg} \\[2ex] P_{c1} - \dfrac{(P_{c1}-P_{c2})\left(f'-f_{min}\right)}{f_{avg}-f_{min}}, & f' < f_{avg} \end{cases}
$$

$$(20)$$

$$
P_m = \begin{cases} P_{m2} - \dfrac{(P_{m2}-P_{m3})\left(f'-f_{avg}\right)}{f_{max}-f_{avg}}, & f' \geq f_{avg} \\[2ex] P_{m1} - \dfrac{(P_{m1}-P_{m2})\left(f'-f_{min}\right)}{f_{avg}-f_{min}}, & f' < f_{avg} \end{cases}
$$

$$(21)$$

where P_c and P_m are the crossover and mutation probabilities respectively; f_{max}, f_{min}, f_{avg} are the maximum, minimum, and average fitness; f' is the greater fitness of the two individuals for genetic operation. Therefore, the superior individual with greater fitness is more likely to be reserved and the inferior individual tends to be transformed into a new one. The improved method could guarantee the population diversity and accelerate its convergence; furthermore, it could avoid the phenomena of premature and slow convergent speed in the later stage of evolution.

Case study

To evaluate the proposed approach, two test systems are studied. At first, a small VPP consisting of three conventional generators, one 50 MW wind farm and loads (conventional and interruptible) is employed. Then a system consisting of 10 generators and one wind farm is implemented to show the effectiveness of the proposed model.

Table 3 Optimal solutions compared between deterministic and fuzzy chance-constrained model

Cost/$	$a = 0.6$	$a = 0.7$	$a = 0.8$	$a = 0.9$	Deterministic model
Start-up cost of units	400	400	200	161	387
Operation cost	143,380	144,681	144,280	144,341	144,300
Cost of curtailing load	2253	934	1644	1620	2333
Total cost	146,033	146,015	146,124	146,122	147,020

Table 4 Optimal solutions without curtailing loads

Cost/$	$a = 0.6$	$a = 0.7$	$a = 0.8$	$a = 0.9$	Deterministic model
Start-up cost of units	400	400	200	200	200
Operation cost	145,854	145,865	146,080	146,096	146,287
Total cost	146,254	146,265	146,280	146,296	146,487

Test systems

The structure of the proposed VPP is shown in Fig. 1. The parameters of the generators and loads in 24 h are indicated in Tables 1 and 2. The wind farm is integrated into the system at Node 4, and its forecasted output in 24 h is shown in Fig. 2. The load at Node 1 is the interruptible load which can be curtailed if necessary. The curtailed load can be regarded as virtual power which can be adjusted continuously and provide the spinning reserve. Its virtual generation capacity ranges from 10 MW to 40 MW. The cost paid for the curtailed load is 45 $/MW.

The parameters of the self-adaptive GA are set as follows: the population size $Np = 30$; maximum iterations $N_{ite} = 300$; $P_{c1} = 0.85$, $P_{c2} = 0.5$, $P_{c3} = 0.2$, $P_{m1} = 0.09$, $P_{m2} = 0.05$, $P_{m3} = 0.01$. Considering the stochastic nature of GA, 20 test trials were conducted for each case.

Results and discussion

The scheduling model, consisting of (4)–(7), (9)–(11), and (18)–(19), is solved by the self-adaptive GA method in Methods. Apart from the data described in Test systems, different confidence levels are also applied, ranging from 0.6 to 0.9, in steps of 0.1. The corresponding costs for satisfying the fuzzy chance spinning reserve constraint are listed in Table 3, including start-up cost, operation cost, cost of curtailing load, and total cost. In addition, the solutions of the fuzzy chance model are compared with those of the traditional deterministic model.

From Table 3, the total costs vary under different confidence levels; especially when $\alpha = 0.6$, $\alpha = 0.7$, $\alpha = 0.8$, $\alpha = 0.9$, the total cost are 146,033 $, 146,015 $, 146,124 $, 146,122 $, respectively, which are all less than the cost in deterministic model. It can be inferred that the proper selection of confidence levels will help achieve a tradeoff between economy and reliability. However, as the credibility level is increasing gradually from 0.6 to 0.9, the optimized results (i.e., the total costs) in Table 3 do not show

Table 5 Wind power and the load of the VPP in 6 h

Hour	1	2	3	4	5	6
Wind Power/MW	42	63	70	60	58	40
Load/MW	1036	1110	1258	1406	1480	1628

Table 6 Parameters of conventional generators

unit	a	b	c	P_G^{max}	P_G^{min}	RU(RD)
1	0.00043	21.60	958.20	470	150	80
2	0.00063	21.05	1313.6	460	135	80
3	0.00039	20.81	604.7	340	73	80
4	0.00070	23.90	471.60	300	60	50
5	0.00079	21.62	480.29	243	73	50
6	0.00056	17.87	601.75	160	57	50
7	0.00211	16.51	502.71	130	20	50
8	0.00480	23.23	639.40	120	47	30
9	0.10908	19.58	455.60	800	20	30
10	0.00951	22.54	692.40	55	55	30

a clear monotonic feature. It should be noted that this does not mean that the results of the fuzzy chance model based on credibility theory are confusing. It is because the results are interfered by the demand-side resources.

If the loads are not allowed to be curtailed, the effect of the interruptible load on the solutions will be excluded. Then the total cost of VPP operation in different confidence levels will be 146,254 \$, 146,265 \$, 146,280 \$, 146,296 \$ and 146,487 \$ respectively, showing a clear monotonously increasing trend, just as shown in Table 4. This is because, as the credibility level increases, VPP operator needs to spare more reserve capacity to guarantee the growing reliability requirement. Undoubtedly, that would lead to the rise of the total cost. Meanwhile, it is obvious that the total costs in model without curtailing loads are higher than those in original model through comparing the values in Tables 3 and 4. Therefore, the introduction of the demand response (or interruptible loads) helps to improve the economy of the whole system, which is consistent with the current research results.

Furthermore, these results stimulate VPP to incorporate demand response and chance constraint programming into its operation strategy.

To further confirm the effectiveness of the proposed fuzzy chance model, the operation strategy of a larger VPP system in 6 h is tested. In order to exclude the impact of the interruptible load cost, loads in this system do not participate in the operation. The forecasted output of the wind farm and loads is shown in Table 5, and the parameters of the 10 generators are listed in Table 6.

The operational costs of VPPs under different confidence levels are depicted in Fig. 3, which are the same as the results in [24], except that confidence levels are set as $1 - \alpha$ in this paper. Figure 3 shows the corresponding cost is increasing when the credibility level rises from 0.6 to 0.9. It is because the confidence level represents the reliability of VPP operation. To reduce the risk resulting from the unpredictable output of renewables, more spinning reserves are required, which will definitely increase the operational cost. The two case studies prove that the proposed model can put forward an optimal operation strategy which can reach the balance between economy and risks/reliability.

Conclusions

VPP is a promising way to integrate the DERs into the power system. As the output of DERs is usually unpredictable, the scheduling of VPPs has to deal with a number of uncertain variables, which is a fuzzy programming problem. To realize the proper balance between benefits and risks, credibility theory is introduced in the optimal scheduling strategy for VPPs in this paper, thereby proposing a fuzzy chance scheduling model. The concept of the confidence level can quantify the possibility of satisfying the fuzzy chance constraints, which represents the

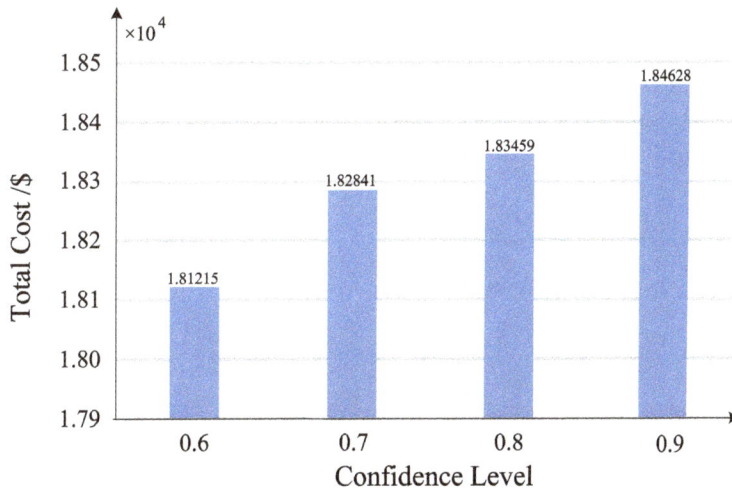

Fig. 3 Total Cost of the VPP under Various Confidence Levels

risk tolerance of VPPs. Further, through transforming the fuzzy chance constraints into their equivalent forms, the conventional optimization algorithms can be utilized to solve the optimal scheduling problem. The case study proves that the operational cost of VPPs will increase when the confidence level increases. That is to say, unnecessary costs can be reduced when the risk of VPP operation is within tolerance.

Acknowledgment
This work is supported by the National Natural Science Foundation of China (No. 51577115). Meanwhile, the authors would like to thank the editor and reviewers for their sincere suggestions on improving the quality of this paper.

Authors' contribution
QA and SF carried out the study of virtual power plants,and drafted the manuscript. All authors read and approved the final manuscript.

About the authors
Q. Ai received the B.S. degree in electrical engineering and automation from Shanghai Jiao Tong University, Shanghai, China, in 1991 and M.S. degree in electrical engineering from Wuhan University, Wuhan, China, in 1994 and Ph.D. degree in electrical engineering from Tsinghua University, Beijing, China, in 1997. After spending one year at Nanyang Technological University, Singapore and two years at University of Bath, UK, he is currently a professor at Shanghai Jiao Tong University, Shanghai, China. His research interests include load modeling, smart grid, and intelligent algorithms.
S. L Fan received the B.S. degree in electrical engineering and automation from Sichuan University, Chengdu, China, in 2013. She is currently pursuing the Ph.D. degree in electrical engineering at Shanghai Jiao Tong University, Shanghai, China. Her research interest include virtual power plant operation and optimization.
L. J Piao received the B.S. and M.S. degree in electrical engineering from Shanghai Jiao Tong University, Shanghai, China, in 2012 and 2015, respectively. He is currently pursuing the M.S. degree in electrical engineering at Shanghai Jiao Tong University, Shanghai, China. His current research interests include multi-agent system in smart grids, and electric vehicle coordinated charging using game theory.

Competing interests
The authors declare that they have no competing interests.

References
1. Bamberger Y, Baptista J, Belmans R. et al. (2006). *Vision and Strategy for Europe's Electricity Networks of the Future: European Technology Platform Smart Grids[R]*. Office for Official Publications of the European Communities.
2. Coll-Mayor D, Paget M & Lightner E. (2007). Future intelligent power grids: analysis of the vision in the European Union and the United States[J]. *Energy Policy, 35*(4), 2453–2465.
3. Wang C, & Li P. (2010). Development and challenges of distributed generation, the micro-grid and smart distribution system [J]. *Automation of Electric Power Systems, 2*, 004.
4. McDermott TE, Dugan RC. (2002). Distributed generation impact on reliability and power quality indices[C]//Rural Electric Power Conference, 2002. IEEE: D3-D3_7.
5. Newman DE, Carreras BA, Kirchner M. et al. (2011). The impact of distributed generation on power transmission grid dynamics[C]//System Sciences (HICSS), 2011 44th Hawaii International Conference on. IEEE: 1–8.
6. Gomez JC, Vaschetti J, Coyos C. et al. (2013). Distributed generation: impact on protections and power quality[J]. *Latin America Transactions, IEEE (Revista IEEE America Latina), 11*(1), 460–465.
7. Fahim SR, Helmy W. (2012). Optimal study of distributed generation impact on electrical distribution networks using GA and generalized reduced gradient[C]//Engineering and Technology (ICET), 2012 International Conference on. IEEE: 1–6.
8. Braun M, & Strauss P. (2008). A review on aggregation approaches of controllable distributed energy units in electrical power systems[J]. *International Journal of Distributed Energy Resources, 4*(4), 297–319.
9. Mohammadi J, Rahimi-Kian A, Ghazizadeh MS. (2011). Joint operation of wind power and flexible load as virtual power plant[C]//Environment and Electrical Engineering (EEEIC), 2011 10th International Conference on. IEEE: 1–4.
10. Pudjianto D, Ramsay C, & Strbac G. (2007). Virtual power plant and system integration of distributed energy resources[J]. *Renewable Power Generation, IET, 1*(1), 10–16.
11. Pudjianto D, Ramsay C, Strbac G. et al. (2008). The virtual power plant: Enabling integration of distributed generation and demand[J]. *FENIX Bulletin, 2*, 10–16.
12. Saboori H, Mohammadi M, Taghe R. (2011). Virtual power plant (VPP), definition, concept, components and types[C]//Power and Energy Engineering Conference (APPEEC), 2011 Asia-Pacific. IEEE: 1–4.
13. El Bakari K, Kling WL. (2011). Development and operation of virtual power plant system[C]//Innovative Smart Grid Technologies (ISGT Europe), 2011 2nd IEEE PES International Conference and Exhibition on. IEEE: 1–5.
14. Zdrilić M, Pandžić H, Kuzle I. (2011). The mixed-integer linear optimization model of virtual power plant operation[C]//Energy Market (EEM), 2011 8th International Conference on the European. IEEE: 467–471.
15. Salmani MA, Tafreshi SMM, Salmani H. (2009). Operation optimization for a virtual power plant[C]//Sustainable Alternative Energy (SAE), 2009 IEEE PES/ IAS Conference on. IEEE: 1–6.
16. Ruiz N, Cobelo I, & Oyarzabal J. (2009). A direct load control model for virtual power plant management[J]. *Power Systems, IEEE Transactions on, 24*(2), 959–966.
17. Mashhour E, Moghaddas-Tafreshi SM. (2009). Trading models for aggregating distributed energy resources into virtual power plant[C]// Adaptive Science & Technology, 2009. ICAST 2009. 2nd International Conference on. IEEE: 418–421.
18. Mashhour E, & Moghaddas-Tafreshi SM. (2011). Bidding strategy of virtual power plant for participating in energy and spinning reserve markets—Part I: Problem formulation[J]. *Power Systems, IEEE Transactions on, 26*(2), 949–956.
19. Soltani M, Raoofat M, Rostami MA. (2012). Optimal reliable strategy of virtual power plant in energy and frequency control markets[C]//Electrical Power Distribution Networks (EPDC), 2012 Proceedings of 17th Conference on. IEEE: 1–6.
20. Taheri H, Ghasemi H, Rahimi-Kian A. et al. (2011). Modeling and optimized scheduling of virtual power plant[C]//Electrical Engineering (ICEE), 2011 19th Iranian Conference on. IEEE: 1–6.
21. Xue Z, Li G, Zhou M. (2011). *Credibility theory applied for estimating operating reserve considering wind power uncertainty[C]//PowerTech, 2011 IEEE Trondheim*. Trondheim, Norway: IEEE, 1–8.
22. Zadeh LA. (1965). Fuzzy sets[J]. *Information and Control, 8*(3), 338–353.
23. Liu B. (2006). A survey of credibility theory[J]. *Fuzzy Optimization and Decision Making, 5*(4), 387–408.
24. Xin A, & Xiao L. (2011). Dynamic economic dispatch for wind farms integrated power system based on credibility theory[J]. *Proceedings of the CSEE, 31*(S1), 12–18.
25. Kuang H, Jin J, & Su Y. (2006). Improving crossover and mutation for adaptive genetic algorithm[J]. *Computer Engineering and Applications, 12*, 93–99.

2

ANN based directional relaying scheme for protection of Korba-Bhilai transmission line of Chhattisgarh state

Anamika Yadav*, Yajnaseni Dash and V. Ashok

Abstract

As it is crucial to protect the transmission line from inevitable faults consequences, intelligent scheme must be employed for immediate fault detection and classification. The application of Artificial Neural Network (ANN) to detect the fault, identify it's section, and classify the fault on transmission lines with improved zone reach setting is presented in this article. The fundamental voltage and current magnitudes obtained through Discrete Fourier Transform (DFT) are specified as the inputs to the ANN. The relay is placed at section-2 which is the prime section to be protected. The ANN was trained and tested using diverse fault datasets; obtained from the simulation of different fault scenarios like different types of fault at varying fault inception angles, fault locations and fault resistances in a 400 kV, 216 km power transmission network of CSEB between Korba-Bhilai of Chhattisgarh state using MATLAB. The simulation outcomes illustrated that the entire shunt faults including forward and reverse fault, it's section and phase can be accurately identified within a half cycle time. The advantage of this scheme is to provide a major protection up to 99.5% of total line length using single end data and furthermore backup protection to the forward and reverse line sections. This routine protection system is properly discriminatory, rapid, robust, enormously reliable and incredibly responsive to isolate targeted fault.

Keywords: Artificial neural network, Fault classification, Fault detection, Fault direction estimation, Section identification

Introduction

As electric power system encompasses comprehensive interacting elements, there is a chance of occurring faults or unwanted short circuit conditions for all time. The flow of heavy currents due to short circuits causes damage to the equipments and other elements of power system [1]. Precise differentiation of faults and exact indication of fault type are the two key aspects for protecting the transmission line from various disturbances, faults and their consequences. Distance relay is one of the universal short circuit protection schemes in a transmission line which performs both primary and backup protection. Primary protection is as quick as possible with no time delay whereas back up protection is operational merely if primary relay fails. Various protecting zones are provided through distance relay based on the percentages of line impedances. The conventional distance relaying scheme usually provide protection to only 80% of the line in Zone 1, Zone 2 covers 120% of line and Zone 3 is set to cover the longest remote line. It is set up to operate with a twice time delay of Zone 1. Performance degradation occurs due to several circumstances like fault-path resistance, remote in-feed currents, and shunt capacitance [2, 3].

Easy identification of the fault which might have been occurred in the transmission line can be achieved by an intelligent expert like ANN [4]. Diverse protecting mechanisms of transmission lines have been proposed earlier to detect and classify fault utilizing high frequency noise generated by fault and NNs [5], initial current travelling wave technique [6], wavelet transform [7], wavelet fuzzy combined approach [8], high speed protective relaying using ANN architecture and digital signal processing concepts. [9], modular yet integrated approach using modified Kohonen-type neural network [10], combined supervised and unsupervised neural network with ISODATA clustering algorithm [11], RBF NN with OLS learning method [12] and Combined fuzzy neural network [13–16] wavelet

* Correspondence: ayadav.ele@nitrr.ac.in
National Institute of Technology Raipur, Raipur, Chhattisgarh, India

analysis and ANN [17–19], ANN Approach [20–25]. However these techniques did not identify the fault direction and section. Besides there are other techniques developed for fault detection and location using ANNs [26–29], Thevenin equivalent impedance and compensation factor [30], Clarke Concordia transformation, eigen value approach and NN [31], Kohonen network approach [32], Radial basis neural network [33, 34] and synchronized phasor measurement units (PMU) [35, 36]. However these techniques did not identify the fault type, fault direction and the faulty section. In some of the ANN based directional relaying techniques the fault types and the fault phases haven't been classified [37–39].

Presence of variety of intelligent techniques available for protecting the transmission line is still having certain issues which limit their applicability. This paper primarily aims to develop an efficient protection technique based on ANN with improved first zone reach setting and detection of fault is accomplished within half cycle of time. Thus, this paper has two key objectives: firstly to detects and classifies the fault and secondly to identify the zone of fault and the direction (whether it is forward or reverse fault). Effect of variation in different fault parameter like fault types, locations, inception angles, resistances were also exemplified here. As in this scheme, only single end data used, so there is no need of communicating link for remote end data.

Power system under study

An existing 400 kV transmission line network of Chhattisgarh linking two utilities NTPC and CSEB has been considered. The power system network has two line sections of total length 216 km (NTPC, Korba – 17 km- CSEB, Korba West –199 km- Khedamara, Bhilai). The power system network of Chhattisgarh State Electricity Board (CSEB) under study consist of 400 kV, 50Hz single circuit three phase transmission line of length 216 km as shown in single line diagram in Fig. 1. At Bus-1, four generating station of 500 MW and three generating station of 210 MW of NTPC Korba are connected which transfers power to Bhilai-Khedamara 400 kV grid at bus B4. At bus-2, two generating station of CSEB Korba West of 210 MW are connected through a 17 km transmission line. Thereafter the power is transferred through 199 km transmission line to Bhilai-Khedamara 400 kV grid at bus-3. Another three phase source of 400 kV, 50Hz is connected to bus-3 for simulating the remote end infeed which also represents the thevenin equivalent source of the interconnected grid. The proposed relay is situated at bus-2, so that directional relaying principle can be employed by considering the section-2 of 199 km length as primary section to be protected and section-1 of 17 km length as reverse line section as seen by the relay. Various types of shunt faults were simulated using three phase fault breaker.

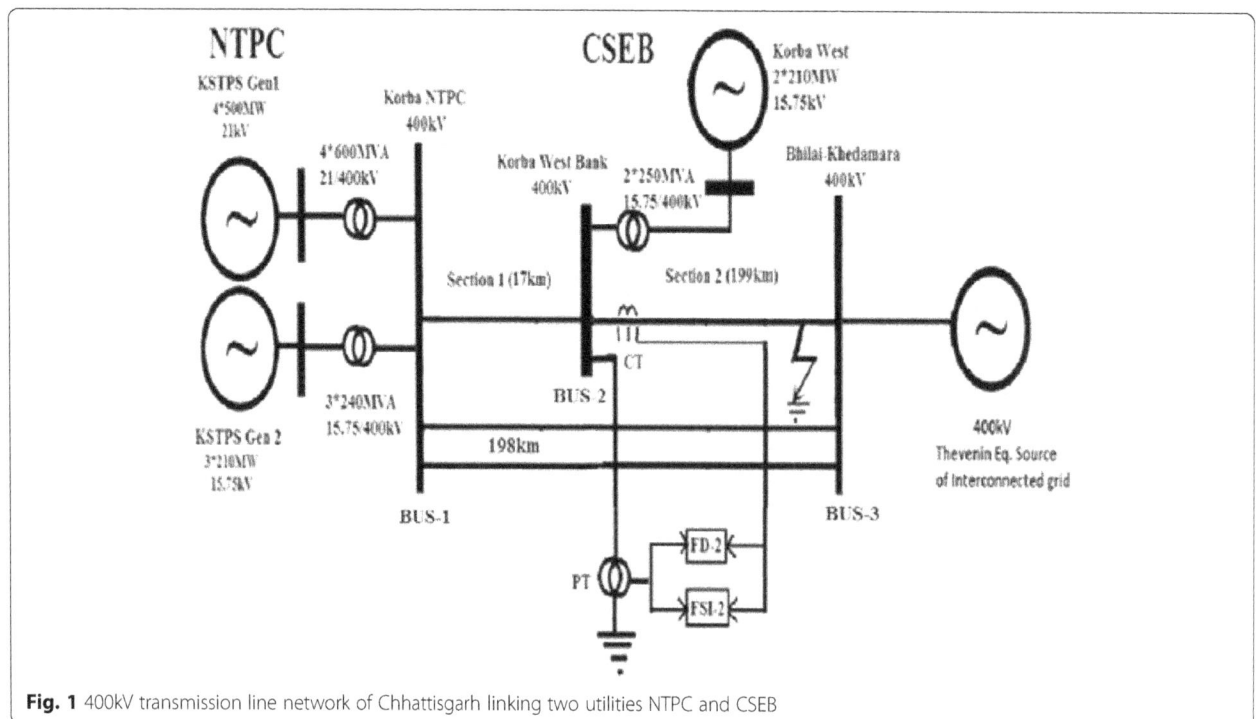

Fig. 1 400kV transmission line network of Chhattisgarh linking two utilities NTPC and CSEB

Methods

Proposed ANN based fault detector, section estimator and fault classifier

The proposed scheme deals with protective relaying tasks including fault section estimation and classification. Two ANNs have been designed, one for classification and other for section estimation. With the help of these networks, detection and classification of all 10 shunt faults, its section/zone and direction can be assessed by the using merely one terminal data. The simulations were conducted concerning different power systems and fault conditions for representing the robustness of the proposed scheme. Performance evaluation of the current study has been carried out using the MATLAB software. A transmission line (400 kV) of length 216 km sectionalized in two zones fed from both the ends as described in the previous section has been chosen for this work.

Various kinds of shunt faults which may occur in transmission line in each section are LG, LL, LLG and LLL. These faults were simulated in a preferred power system model developed by using MATLAB software. The 3-phase voltage and current were measured at bus-2. These signals were sampled at 1 kHz and then passed through analog filter. The analog filter is a Butterworth low pass filter of order 2 with pass band edge frequency (480 Hz) for removing higher order harmonics from the signal. Later, one full cycle recursive DFT was used for calculating the fundamental components of voltage and current. Then these signals were normalized in range -1 to +1 and fed to ANN for training, testing and validation. The flow chart for proposed absolute protecting scheme based on ANN was presented in Fig. 2.

The design process of the ANN based relay goes through the following steps:

a) Selection of inputs and outputs for ANN.
b) Simulation of different fault cases to form training and test data sets.
c) Architecture and training of ANN with appropriate training data sets.

Selection inputs and outputs for ANN

The deviations of the computed voltages and currents signals in the time domain are extremely perceptible and specific under verity of fault circumstances. Designing a fast and reliable ANN based relay concerns about the variations of current and voltage signals before and after the incidence of fault. When faults occur, diverse frequency components of signals appear, and the DC magnitude is attenuated according to the progression of time. In addition, a number of non-fundamental

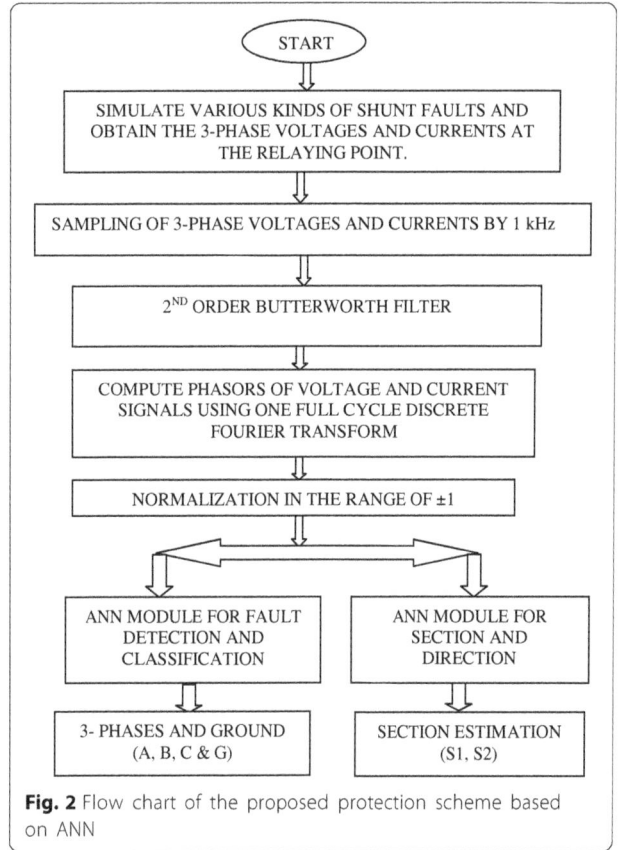

Fig. 2 Flow chart of the proposed protection scheme based on ANN

frequency components altered for different fault locations. As the performance of ANN depends upon the input and output characteristics, so, it is essential to preprocess and extract the useful features from the input data to train the ANN. The three phase instantaneous voltage and current signals were presented in Fig. 3 during AG fault at 5 km in Section 2 at 61 ms time with fault resistance 0.001Ω. After 61 ms time, the faulty phase current starts increasing and voltage signal starts decreasing in magnitude. Figure 4 shows the fundamental components of voltage and current signals after preprocessing with DFT.

The inputs to the network were selected as the magnitudes of the fundamental components (50 Hz) of 3-phase voltages and currents assessed at the relay location at single end of the line. Therefore, the total inputs to ANN to detect & classify fault, and estimate fault sections are six as presented in Eq. (1). Post fault samples (10) of fundamental components of 3-phase voltage and current signals were extracted for forming the input matrix of ANN training as shown in Eq. (2).

$$X = \begin{bmatrix} V_{af}, & V_{bf}, & V_{cf}, & I_{af}, & I_{bf}, & I_{cf} \end{bmatrix} \qquad (1)$$

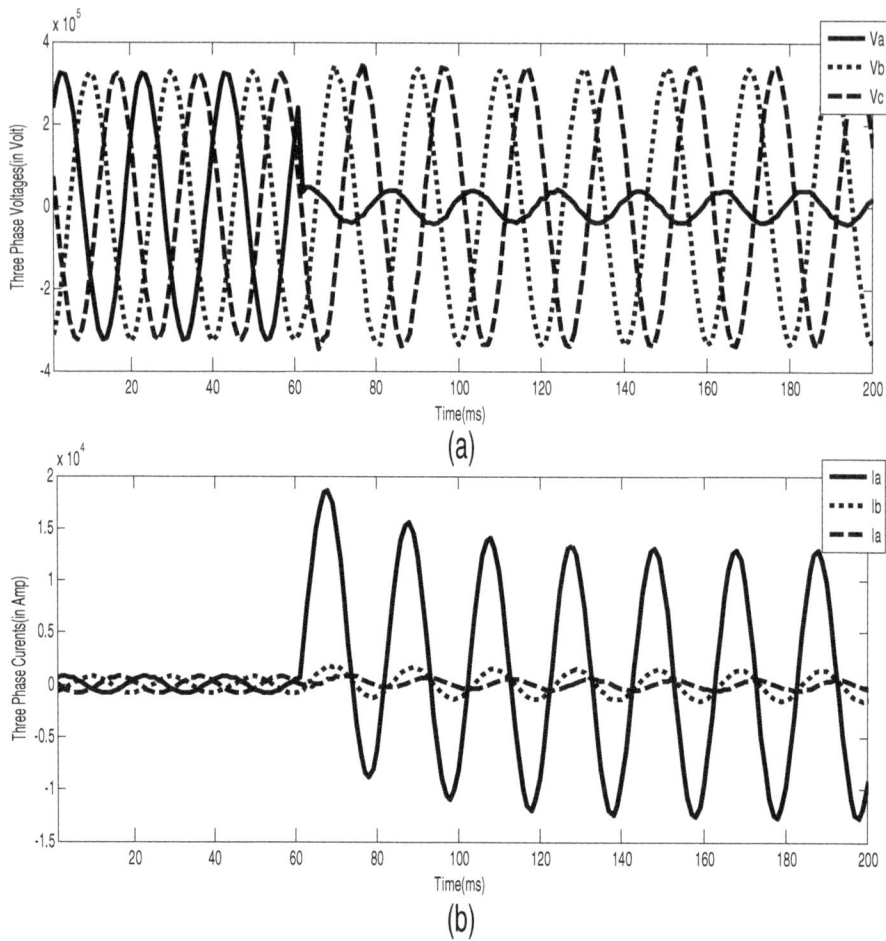

Fig. 3 a, b Three phase instantaneous voltages and currents during an AG fault at 5 km in Section 2 at 61 ms with Rf = 0.001 Ω respectively

$$X = \begin{bmatrix} V_{pf} \\ I_{pf} \end{bmatrix}$$

$$= \begin{bmatrix} V_{af(t)}, & V_{af(t+1)}, & \text{............................} & V_{agf(t+9)} \\ V_{bf(t)}, & V_{bf(t+1)}, & \text{............................} & V_{bf(t+9)} \\ V_{cf(t)}, & V_{cf(t+1)}, & \text{............................} & V_{cf(t+9)} \\ I_{af(t)}, & I_{a1f(t+1)}, & \text{............................} & I_{a1f(t+9)} \\ I_{bf(t)}, & I_{b1f(t+1)}, & \text{............................} & I_{b1f(t+9)} \\ I_{cf(t)}, & I_{c1f(t+1)}, & \text{............................} & I_{cf(t+9)} \end{bmatrix}$$

(2)

There are 2 outputs corresponding to the two sections S1 and S2 in the fault section identification module of ANN. Thus, the faulty section can be identified as shown in Eq. (3). Output of each section is '0' if no fault or '1' if there is fault.

$$Y1 = [S1, \ S2] \qquad (3)$$

In addition, the fault type as well as the faulty phase selection was determined by the fault classification

module of ANN. Fault classification module has two networks one for fault phase identification and one for ground identification. Thus, the 3- phases and neutral were considered as outputs given by the network for determining the faulty phase (A, B, C and G) presented in the fault loop. Depending on the fault kind occurring in the system, outputs must be '0' or '1'. Hence the fault classification outputs of network were shown in Eqs. (4) and (5)

$$Y2 = [A, \ B, \ C] \qquad (4)$$

$$Y3 = [G] \qquad (5)$$

Simulation of different fault cases to form training and test data set

Various fault parameters were varied for generating different fault cases. The 3-phase currents and voltages measured at single end of the single circuit line were used to compute DFT coefficients of the signal in

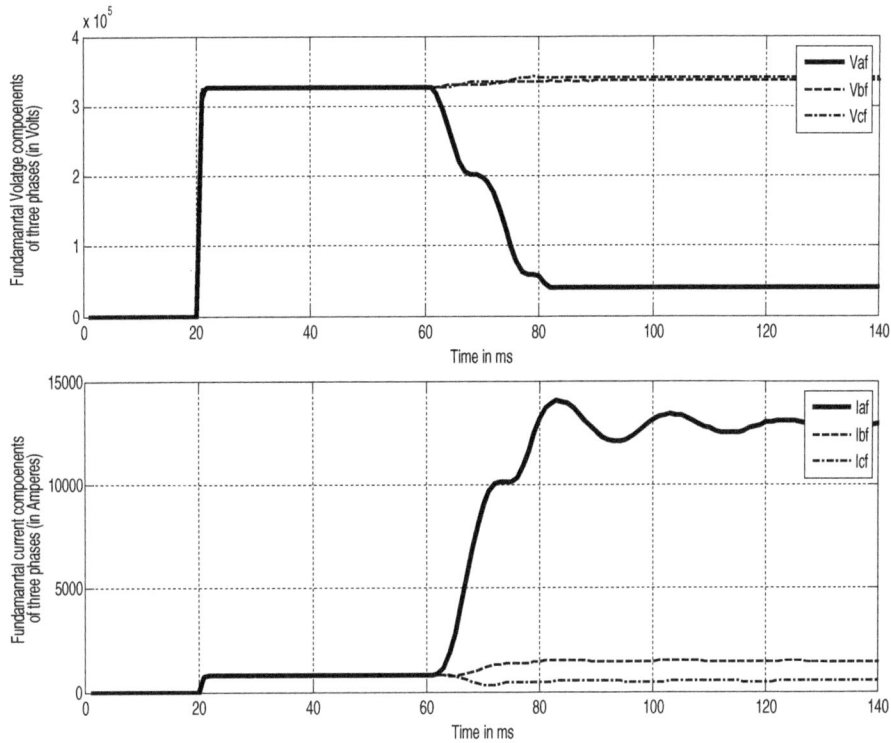

Fig. 4 Fundamental components of 3-phase voltage and current during an AG fault at 5 km in Section 2 at 61 ms with Rf = 0.001 Ω

MATLAB. DFT coefficients were applied as input to feed forward NN with Levenberg- Marquardt (LM) training algorithm to discriminate the fault, classify and also to identify the zone of faults. A large variety of fault cases have been studied using different parameters like fault inception angle/time, location, resistance, ground resistance, and fault types (LG, LLG, LL and LLL). Different fault parameter variation employed for training and testing the ANN was presented in Table 1. Fault case studies carried out to train the neural network involves 10 (fault type) × 23 (3 fault locations for section 1 and 20 for section 2) × 1 (fault inception angle) × 2 (fault resistance) =460 cases. The total numbers of samples utilized to train the neural network are 5082 to identify faulty section and 4089 for fault classification & phase identification and 4085 for ground identification.

Total number of fault cases for testing the neural network is 1 (fault section-2) × 10 (fault type) × 20 (fault locations) × 1 (fault inception angle) × (10 fault resistance for 6 type of LG and LLG fault + 2 fault resistance for 4 types of LL and LLL fault) = 1360 cases. Seven different fault locations in step of 10 such as (1, 11, 21....191; 2,

Table 1 Fault cases in training and testing purpose

Parameters	Training Data	Testing Data
Fault sections	Section-1 (S1) and Section-2 (S2)	section-1 (S1) and section-2 (S2)
Fault type	LG (AG, BG, CG), LLG (ABG, BCG, ACG), LL (AB, BC, CA), LLL (ABC)	LG (AG, BG, CG), LLG (ABG, BCG, ACG), LL (AB, BC, CA), LLL (ABC)
Fault location: FL in (km)	Section 1 (5,10 and 15 = 3 locations) and Section 2 (5,15,25,...195 = 20 locations) Total locations = 23	Different fault location between 1-17 km for section-1 and 1-199 km for section-2.
Fault inception angle: Φi in (°)	0	0, 45, 90, 135, 180, 225, 270, 315
Fault Resistance: Rf in (Ω)	0, 10 for phase faults and 0, 100 ohms for ground faults	0,10......100 = 10 fault resistance for 6 type of LG and LLG fault + 2 fault resistance for 4 types of LL and LLL fault
	Total fault cases = 460 cases	Total fault cases = 328 (section-1) + 9520 (section-2) = 9848

12, 22....192; 3, 13, 33...193; 4, 14, 24...194; 6, 16, 26... 196; 7, 17, 27...197; 8, 18, 28...198) were taken into consideration. So, total numbers of cases are $1360 \times 7 = 9520$ and thereafter 26 samples of each fault cases has been extracted which comes out to be $9520 \times 26 = 247520$ samples and 70 no fault samples has been added to form total 247590 samples/testing data set which has been used for testing the network.

Architecture and training of ANN with appropriate training data sets

After selecting the inputs and outputs for ANN, the number of layers and the number of neurons per layer were determined. Neuron numbers in hidden layer was decided by random investigation of 5, 10,...., 20 neurons. Then the transfer function was decided from the commonly used functions for instance, logsig, tansig, purelin, satlin etc. Different variations of fault scenarios were considered throughout the training process to make the ANN to learn accurately the fundamental problem and can respond accordingly. All the 10 types of shunt faults including LG, LLG, LL, and LLL have been simulated in the two sections of line (between 0-100% of line length) at 328 different fault locations with varied fault resistance (0,20,40,60,80 and 100 Ω) and fault inception angles (0 to 360°). The overall fault cases utilized for training and testing are 1380 and 9848 respectively. Several networks with a number of neurons in their hidden layer were trained with Levenberg-Marquardt (LM) algorithm. From earlier studies, it was observed that the ANNs trained with the LM algorithm bestow improved outcomes as compared to the outcomes of the ANNs trained with the Back propagation (BP) algorithm. The LM, a nonlinear least square algorithm utilized for the learning purpose of multilayer neurons. Therefore, the LM training algorithm is used for this purpose. Total fault samples taken for section identification are 5082. It has been observed that the neural network was having 6 neurons in input, 20 neurons in 1st hidden layer, 20 neurons in 2nd hidden layer and 1 neuron in output layer all comprising of "tansig" activation function for ANN based fault detector (6-20-20-2) can able to minimize the mean square error (mse) to an ultimate value of 0.00000989. Total fault samples taken for fault phase identification are 4089 and corresponding

to the input samples target is assigned. ANN network is trained using 2 hidden layers with 20 neurons in each layer as like as fault section identification training network. In the learning process of the network, the mse reduces in 94 cycles to 0.000000968. Total fault samples taken for ground identification are 4085. ANN network is trained using 2 hidden layers with 12 neurons in 1st layer and 10 neurons in 2nd layer. Here hyperbolic tangent sigmoid transfer function (tansig) and Levenberg-Marquardt algorithm were used. This network with 2 hidden layers is able to minimize the mse to an ultimate value of 0.000000979. The learning approach converges fast and through learning the mse reduces in 56 cycles to 0.000000979.

Results and discussions

There is necessity for testing of the ANN based fault section estimator and fault classifier/faulty phase identifier. This is required for identification of faults and variations in network parameters for which the network has not been trained earlier. Cases included in the validation data set are like faults nearer to protection zone boundary and high resistance faults. Testing of ANN was performed utilizing different types of faults like LG, LLG, LL and LLL in the two sections with different fault locations (L_f = 0-199 km in section-2 and 1-17 in section-1), fault inception angles (Φ_i = 0-360°) and fault resistance (R_f = 0-100 Ω). Table 2 shows the results based upon performance analysis of the whole scheme with respect to the percentage of correct answers and the detection time for various types of fault test cases. It is evident from Table 2 that percentage of accuracy and correct answers were high and the detection time was less than quarter cycle for most of the test cases in both the cases of fault detection/section estimation and fault classification modules. It was also less than half cycle for few cases only.

Evaluations of the proposed relay operation/fault detection time and reach setting were performed at faults near boundary in the 4.1 and 4.2 respectively. Proposed scheme was also verified using various faulted conditions like faults close to boundaries with high fault resistance, changeable faults inception angles. These test results were presented in following sub-sections.

Table 2 Detection time for fault section estimation and classification networks

ANN Network	Test Cases	Detection Time (ms)	Accuracy (correct answers)
Section estimation network	82%	Less than quarter cycle	100% (277137 cases)
	18%	Less than half cycle	
Fault classification network	80%	Less than quarter cycle	99.6% (247590 cases)
	20%	Less than half cycle	

Performance analysis

Evaluation of relay performance was done for ANN based fault detector, section estimator and classifier utilizing different fault cases based upon varied resistances, inception angles, types and locations. Performance of the scheme will be analyzed in terms of fault detection time and fault detection accuracy. Different fault resistances such as 0Ω, 20Ω, 40Ω, 60Ω and 80Ω were taken into consideration for identifying the fault, the faulty section and classification of fault. During a relay design, the key consideration was that its operation time should be less than a cycle time. In other schemes of conventional digital distance relaying, operating time is near about one cycle. In the present study we have considered ABC fault at 160 km away from bus-2 in section-2 with $R_f = 0\ \Omega$, $\Phi_i = 0°$ at 60 ms for evaluation of proposed relay operating time. Figure 5 depicts waveform related to output of ANN based fault section estimator for S1 and S2.

Outputs were low (zero) up to 60 ms in both the sections S1 and S2 which show that there is no occurrence of fault. The output of Output S2 of ANN based section estimator, after the occurrence ABC fault at 60 ms goes high (1) at 63 ms time in comparison with other output S1, which remained low (0) and unaffected. Thus the fault is in section-2. The operating time of relay can be calculated as follows:

Fault Inception time   =   60ms

Fault detection time   =   63ms

Hence time of operation   =   (63-60)ms
=   3ms (less than quarter cycle time)

Figure 6 depicts the output of ANN based fault classifier and faulty phase identifier during ABG fault at 180 km in Section-2 with Rf = 0Ω, φi = 0° at 60 ms. Fault locations of forward section are from 181 km to 198 km with 0° inception angles where fault classification takes less than a cycle of time as presented in Table 3. In Fig. 7, the output of ANN based phase identifier during BG fault at 183 km in Section-2 with $R_f = 20\Omega$, $\phi i = 0°$ at 60 ms represented.

It is worthwhile to mention here that, one full cycle recursive DFT has been used for estimating fundamental components of 3-phase currents and voltages in time domain. The increase and reduction in fundamental components of faulty phase currents and voltages respectively after fault inception at 60 ms can be detected by ANN. As a result of recursive DFT which computes the fundamental components values in time domain a continuous manner, the output of proposed ANN in corresponding faulty section rises to high output (1) at 63 ms as shown in Fig. 5.

Speed of the algorithm to detect the fault is less than quarter -half time, represents the response time of the algorithm given the test fault case input patterns to detect the fault i.e. transition of the output from no fault state (zero) to faulted state (one) after the initiation of the fault. Or in other word, quarter-half cycle samples of the voltages and currents are required to detect the fault. This speed does not account for the execution time of the simulation of the ANN algorithm using MATLAB/Simulink on a PC or workstation. Execution time of the simulation of the ANN algorithm will be affected by the computer tool i.e. either PC or workstation. Thus analysis in terms of the

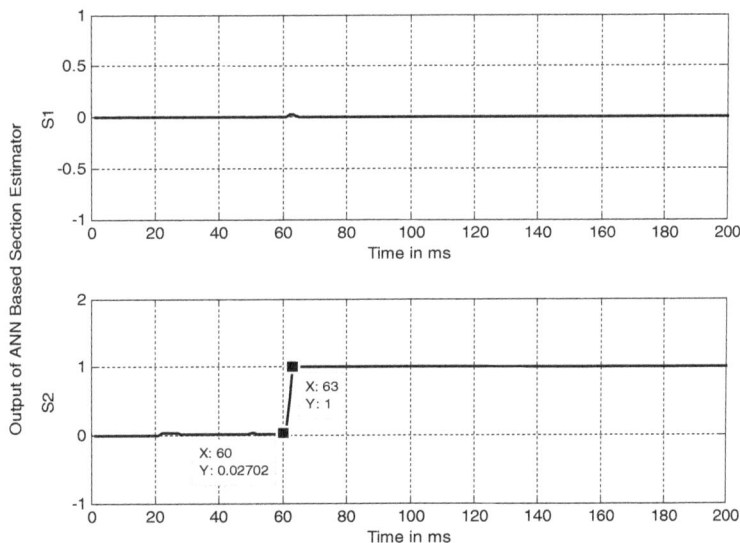

Fig. 5 Output of ANN based section estimator during ABC fault at 160 km in Section-2 with Rf = 0Ω, φi = 00 at 60 ms time

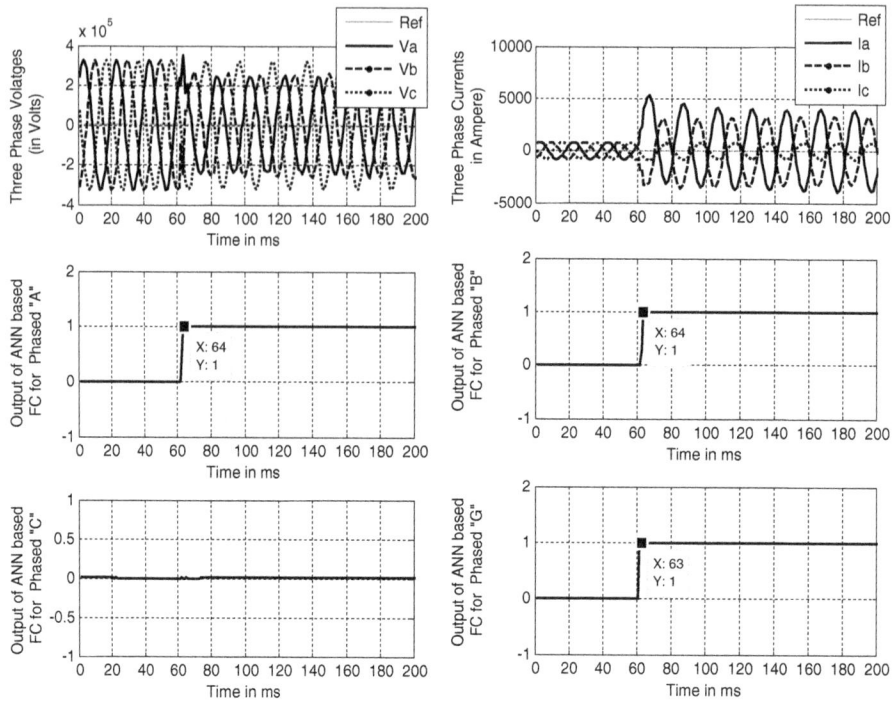

Fig. 6 Output of ANN based Fault Classifier during ABG fault at 180 km in Section-2 with Rf = 0Ω, φi = 00 at 60 ms

Table 3 Response for LL, LLL, LG, LLG faults with φi = 0⁰ at 60 ms and variable fault resistance

Fault type	Fault resistance	Fault location	Fault phase identification time				Fault section identification time
			A	B	C	G	
ABC	0	181	4	4	8	-	3
AC	10	182	4	-	6	-	3
BG	20	183	-	5	-	2	4
ABG	40	184	4	8	-	2	3
ACG	60	185	5	-	8	2	3
BCG	80	186	-	10	9	2	7
BC	0	187	-	4	10	-	4
BC	10	188	-	5	10	-	3
AB	10	189	4	5	-	-	3
CG	20	190	-	-	7	4	8
ABG	40	191	4	9	-	5	3
ACG	60	192	5	-	8	4	3
BCG	80	193	-	10	10	6	7
ABC	0	194	4	4	8	-	3
AC	10	195	4	-	6	-	4
AG	20	196	4	-	-	10	3
ABG	40	197	4	10	-	5	3
ACG	60	198	6	-	9	4	4

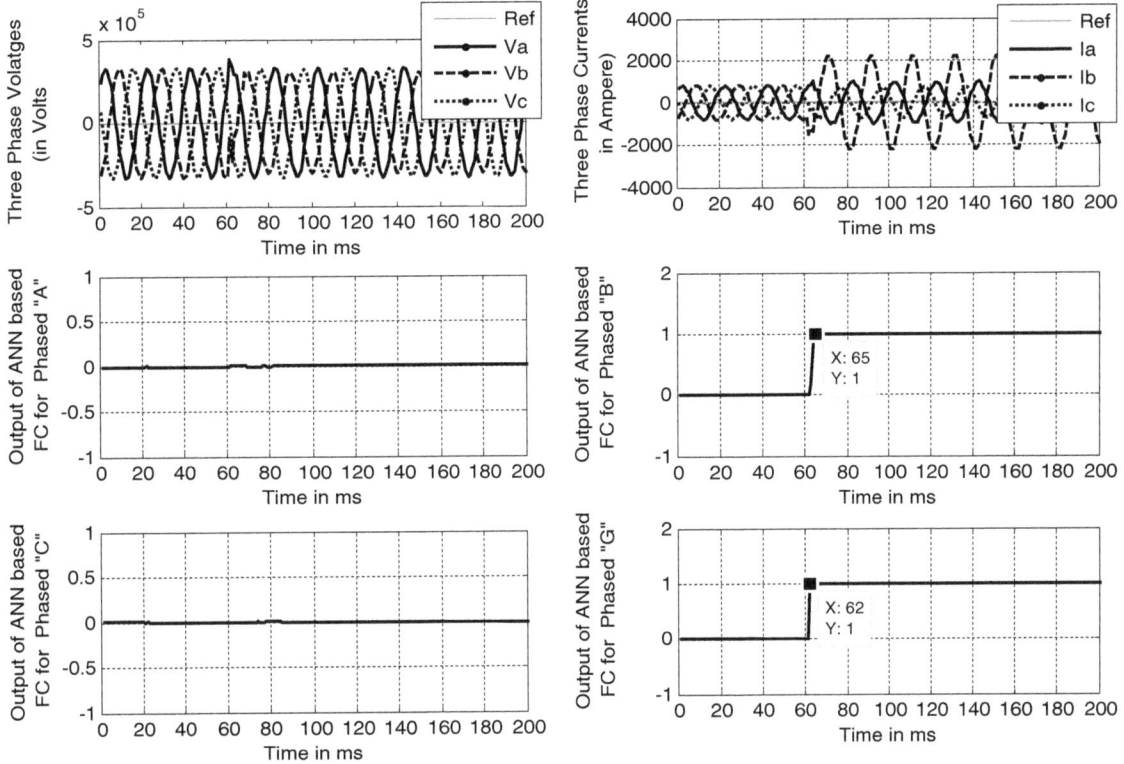

Fig. 7 Output of ANN based Fault Classifier during BG fault at 183 km in Section-2 with Rf = 20Ω, φi = 00 at 60 ms

execution time of the proposed algorithm has been carried out in a PC of with 2GB RAM and Intel(R) Core(TM), i7-3770 CPU, 3.4 GHz, 32bit processor and HP Z230 workstation with 4GB RAM and Intel(R) Xeon (R), CPU E3-1240 V3, 3.4 GHz, 64 bit processor. The average execution time for the simulation of the ANN algorithm implemented using MATLAB/Simulink on a PC and workstation is found to be 20.741 ms and 16.9 ms respectively.

Performance during far end boundary faults with high fault resistance

In a conventional manner, the digital distance relay's first zone reach setting is typically set as 80% of line length.

Table 4 Test results of far end boundary faults with high resistance (100Ω)

Fault type	Fault location	Fault section identification time		Fault phase identification time			
		S1	S2	A	B	C	G
AG	180	-	3	5	-	-	2
BG	182	-	9	-	10	-	2
CG	184	-	10	-	-	9	2
ABG	186	-	3	6	10	-	2
BCG	188	-	8	-	10	10	2
CAG	190	-	4	7	-	10	4
AG	192	-	4	7	-	-	5
BG	194	-	10	-	10	-	4
CG	196	-	11	-	-	10	5
ABG	**198**	**-**	**4**	12	12	-	6

Table 5 Test results for varying fault location with high fault resistance

Fault type	Fault resistance	Fault location	Fault section identification time	
			S1	S2
AG	100	-1	3	-
		-16	3	-
		1	-	2
		101	-	3
		198	-	4
ABG	100	-4	3	-
		-13	3	-
		33	-	3
		133	-	3
		170	-	3
ACG	100	-5	3	-
		-12	3	-
		44	-	2
		144	-	3
		160	-	3

Table 6 Performance in case of varying fault resistances

Fault type	Fault location	Fault resistance	Fault Phase identification time			Detection time of forward section
			A	C	G	
AG	40 km	0	3	-	2	2
		10	3	-	3	2
		20	3	-	2	2
		40	3	-	2	2
		60	3	-	2	2
		80	3	-	2	2
		100	3	-	2	2
ACG	120 km	0	3	6	3	3
		10	3	6	3	3
		20	4	7	3	3
		40	3	7	3	3
		60	4	7	3	3
		80	4	7	3	3
		100	4	7	3	3

This resulted in incapability for instant detection of faults close to remote end bus; however they were identified after some delay based on zone-2 timings. Various types of faults with $R_f = 100\,\Omega$ have been simulated with changeable fault locations in step of 1 km between 180-198 km in section-2 for studying relay performance for faults adjacent to remote end bus with high fault resistance. Table 4 summarises the responses of the protection scheme for far end faults with high fault resistance, and Table 5 depicts the test results for varying fault location with high fault resistance. In most of the cases, the

relay operation time is within half cycle time (10 ms), except in one case where it is 11 ms. The farthest end fault case of ABG fault at 198 km from the relay location at bus-2 with $R_f = 100\,\Omega$ at 60 ms was represented graphically in Fig. 8. Hence, it is summarised that the reach of the relay is approximately 99.5% in the first zone. The proposed scheme could detect the forward faults and its zone within half cycle time in all the cases.

Impact of fault resistance

The current study based on fault classification and phase selection using ANN module was tested for

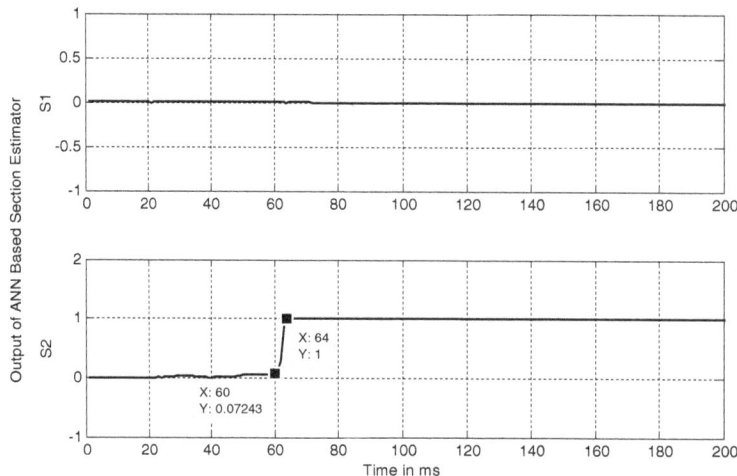

Fig. 8 Output of ANN based section estimator for ABG fault at 198 km from bus-2 in Section-2 with $R_f = 100\Omega$, $\phi i = 0^0$ at 60 ms time

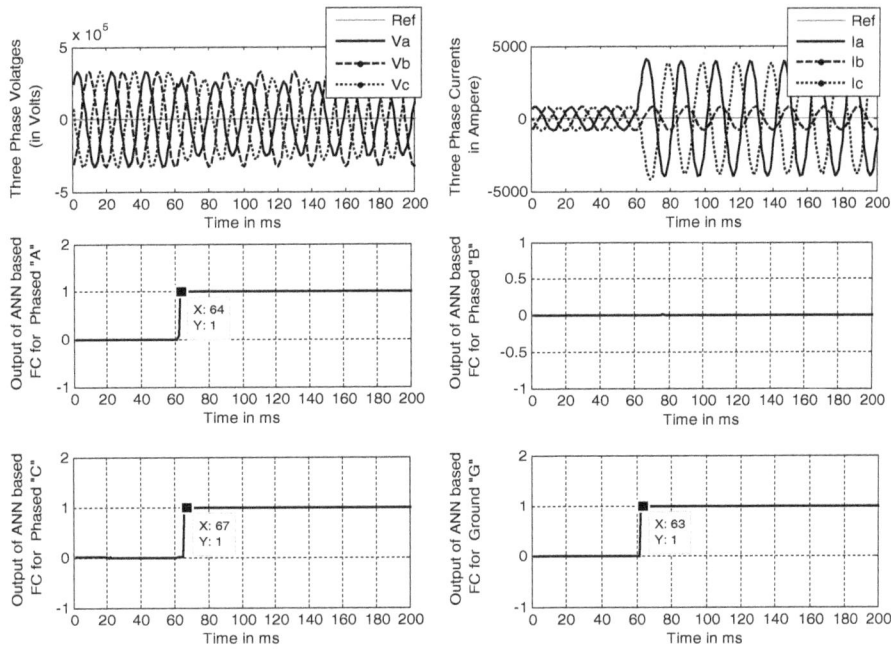

Fig. 9 Output of ANN based Fault Classifier during ACG fault at 120 km in Section-2 with $R_f = 20\Omega$, $\phi i = 0^0$ at 60 ms

different fault resistances. Different fault resistances such as 0Ω, 20Ω, 40Ω, 60Ω and 80Ω were taken into consideration for identifying the faulty phase and the type of fault and few test results for various fault resistances were demonstrated in Table 6 and Fig. 9 shows the Output of ANN based fault classifier during ACG fault at 120 km in Section-2 with $R_f = 20\Omega$,

$\phi i = 0^0$ at 60 ms. First two plots of Fig. 9 shows the instantaneous current and voltage signals. Output of ANN based fault phase identifier become high in A, C phases and G after 64 ms, 67 ms and 63 ms time respectively. Thus the fault is classified as LLG after 7 ms from the fault inception time and the faulty phases are A, C and G.

Table 7 Performance in case of different fault inception angle

Fault type	Fault location	Fault inception angle	Detection time of forward section	Fault phase identification time		
				A	B	G
AG	150	0	3	3	-	2
		45	2.5	3.5	-	3.5
		90	10	8	-	2
		135	4.5	5.5	-	3.5
		180	3	3	-	2
		225	2.5	3.5	-	3.5
		270	10	8	-	2
		315	4.5	5.5	-	3.5
ABG	150	0	3	3	4	3
		45	2.5	3.5	9.5	1.5
		90	5	8	5	2
		135	3.5	5.5	3.5	4.5
		180	3	3	4	3
		225	2.5	3.5	9.5	1.5
		270	5	8	5	2
		315	3.5	5.5	3.5	4.5

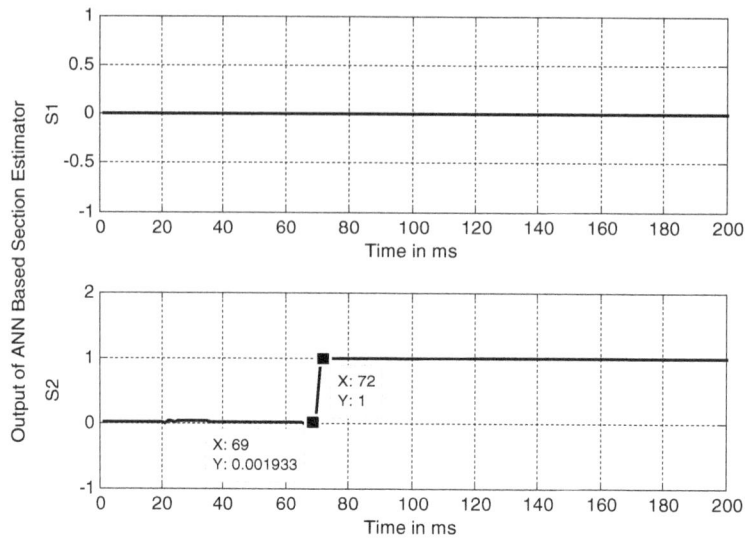

Fig. 10 Output of ANN based section estimator during AG fault at 150 km in Section-2 with $R_f = 0\Omega$, $\phi i = 135^0$ at 60 ms time

Impact of fault inception angle

Fault may occur at any instant of time in transmission system. So it is important that the designed scheme should work for the entire fault inception angle variation. For testing the effectiveness of method, fault inception angles between 0°- 315° in step of 45° were selected by keeping the fault resistance fixed at 0Ω. Some of the test results were shown in Table 7 for several inception angles of fault cases that substantiate the aptness of this current method for varied fault inception angles. Figure 10 shows the test results of the proposed scheme during AG fault at 150 km in Section-2 with $R_f = 0\Omega$, $\phi i = 135^0$ at 60 ms time.

Response of reverse fault in section-1

The performance evaluation of the proposed relay for faults in the reverse direction from the relay has been carried out which was located at bus-2, that is, in S1 and outcomes were presented in Table 8 and Fig. 11. As the fault locations are reverse faults from bus-2, so represented as negative. The relay requires maximum 4 ms for detecting the reverse fault. The proposed method is capable of detecting the reverse fault within quarter cycle of time.

Performance during variation in impedance of the source

Impedance of the source determines the strength of the source, if the impedance of the source is low, it is a strong source and if it is high it is a weak source. The impedance of the source connected to either side of the transmission lines is varied to check the performance of the proposed scheme under variation in impedance of the source. The impedance of the source is varied and different types of fault at different location are tested and few test results are depicted in Table 9. From the results, it can be seen that, the fault detection time is increases in some cases up to half cycle time however it correctly identifies the fault type and its section. Thus it is confirmed that the proposed relaying scheme is not affected by variation in the source impedance and it can detect the faulty phase and its section within half cycle time.

Test results of real time fault events

In this section, two case studies were interpreted by considering real time fault event data. We have collected data from 400 kV sub-station of PGCIL, Raipur, Chhattisgarh of two fault events occurred on

Table 8 Test results for reverse faults in section-1

Fault location (km)	Fault type	Fault detection time (ms)	Relay operation time (ms)
-16	BCG	63	3
-15	AG	64	4
-14	AG	62	2
-13	ABG	63	3
-12	ACG	63	3
-11	BCG	63	3

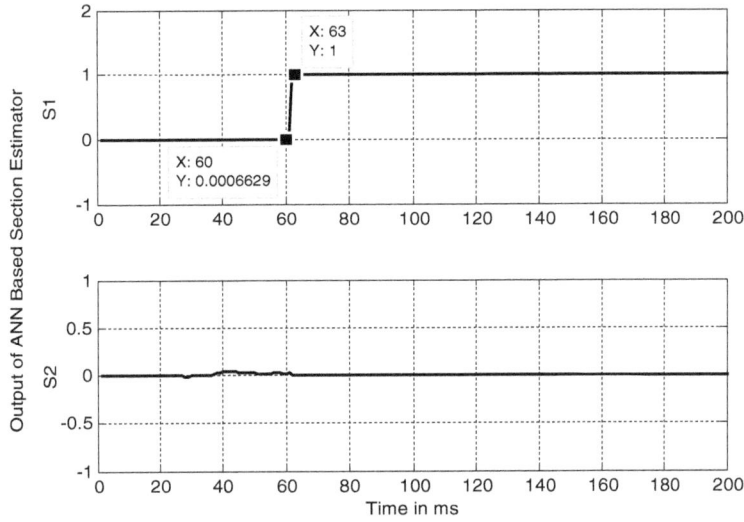

Fig. 11 Output of ANN based section estimator during BCG fault at -16 km in Section-1 with $R_f = 0\Omega$, $\phi i = 0^0$, at 60 ms time

22nd July 2014 and 24th April 2015 in 220 km line between Raipur-Korba. These data were preprocessed, and tested by applying the proposed ANN based fault detection, section estimation and fault classification schemes. The real time fault events case studies are discussed in detail in the following sub-sections.

Case Study1: Real time fault event occurred on 22nd July 2014

After pre-processing the fault data, the fundamental components of 3- phase currents and voltages are depicted in Fig. 12. Before the inception of fault, all the outputs of the fault detector and classifier are low (zero) and after some time of occurrence the fault, the outputs of ANN based classifier in the phase "A" and ground

"G" goes high at 294 ms and 293 ms time rspectively and other outputs remains low (zero) and unaffected as exemplified in Fig. 13. The fault type is classified as LG fault (AG).

The operating time of relay can be calculated as follows:

Fault inception time $=$ 291ms

Fault detection time (Maximum) of phase
A $-$ 294ms

Hence the time required by ANN based relay
$=$ (294-291) ms $=$ 3ms ($<$ Quarter cycle time)

Table 9 Response for variation in source impedance

Fault type	Fault inception angle(deg)	Source Impedance (in Ω)	Fault location (in km)	Fault phase identification time(ms)				Fault section identification time
				A	B	C	G	
AG	0	3.9 + j39.33	10	2			2	2
BG	45	4.1 + j41.20	30	-	7	-	5	7
CG	90	4.3 + j43.07	50	-	-	3	2	3
ABG	135	4.9 + j44.89	70	4	4	-	2	4
ACG	180	3.9 + j35.53	90	1	-	1	4	4
BCG	225	3.7 + j33.66	110	-	7	4	5	7
BC	270	2.8 + j31.86	130	-	4	5	-	5
CA	315	2.7 + j29.98	150	8		8	-	8
AB	360	3.7 + j41.24	170	3	5	-	-	5
ABC	0	5.3 + j42.97	190	4	6	9		9

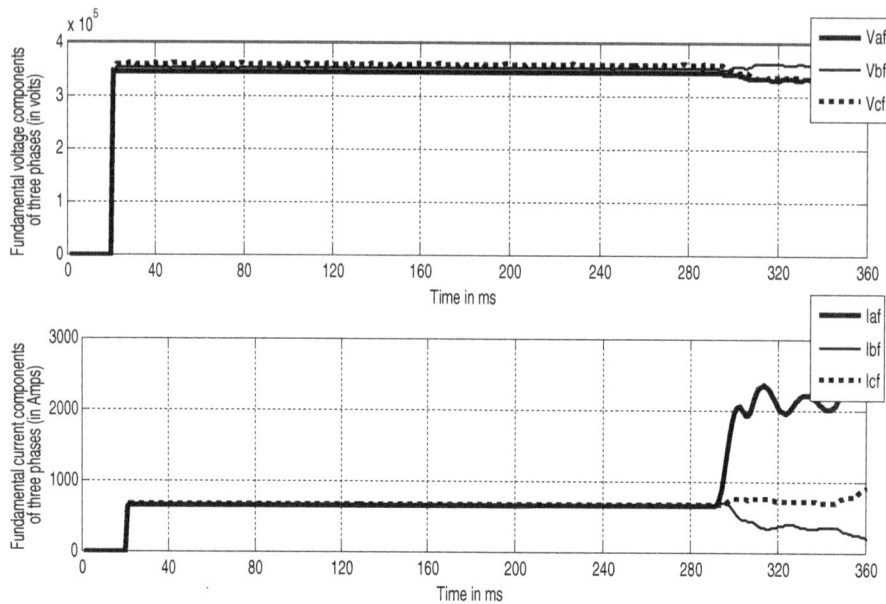

Fig. 12 Fundamental current and voltage waveforms in case of real time fault event occurred on 22nd July 2014

Case study2: real time fault event occurred on 24th April 2014

The Fig. 14 shows the fundamental components of 3-phase currents and voltages during a real time fault event on 24th April 2014. In this cases also, after the fault occurrence, based on the response time of the proposed scheme as shown in Fig. 15, the output of ANN based fault classifier in phase "A" goes high at 294 ms time and ground "G" becomes high at 293 ms time and other outputs remains low and unaffected. Thus this real time fault event is also classified as single line to ground fault

in "A" phase. The relay operating time can be computed as follows:

Fault inception time $=$ 287ms

Fault detection time (Maximum) of phase
A $=$ 294ms

Hence the time required by ANN based relay
$=$ (294-287) ms $=$ 7ms ($<$ Half cycle time)

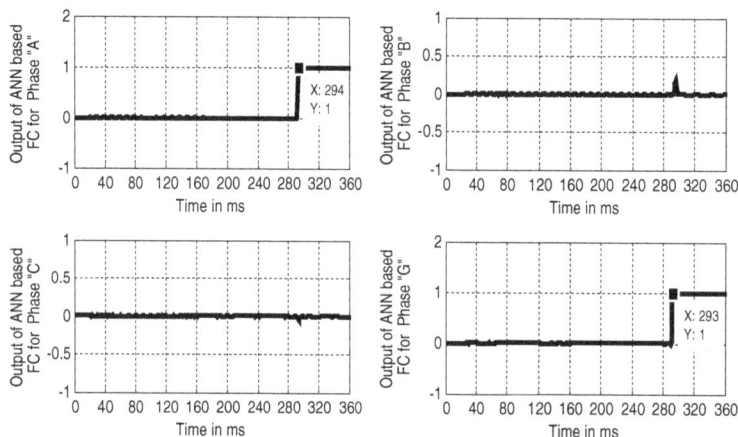

Fig. 13 Output of ANN based Fault Classifier during AG fault in case of real time fault event occurred on 22nd July 2014

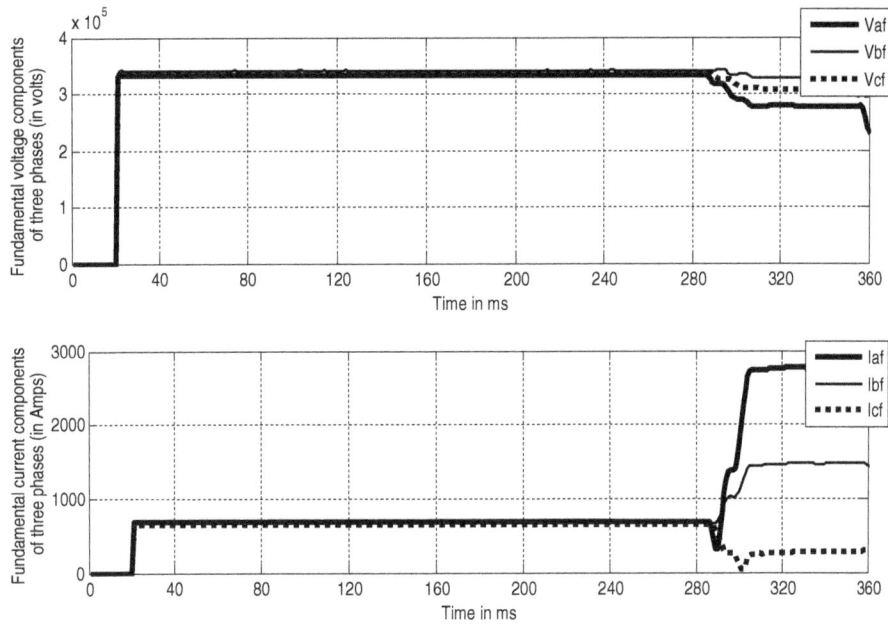

Fig. 14 Fundamental current and voltage waveforms in case of real time fault event occurred on 24th April 2015

Conclusion

This paper proposes an ANN based fault detection, section identification (direction discrimination), fault classification and faulty phase selection schemes which consider the fundamental components of current and voltage signals of 3-phase as input. Proposed ANN based method was tested with huge number of fault cases by varying different fault parameters. Test results shows that proposed relaying schemes can provide primary as well as back-up protection to the forward and reverse line sections respectively. Reach setting of the relay is 99.5% and

fault detection time is half cycle in most of the cases. As speed of the relay is an important criterion in directional relaying, proposed ANN based method will be efficient to use. The results based on extensive study indicate that the proposed scheme can reliably protect the transmission line against different fault situations and thus, is a potential candidate for effective protection measure. Moreover, the proposed scheme correctly identifies the faulty section and its direction when tested with real time fault events. Thus it can be implemented for protection of real power system networks as well.

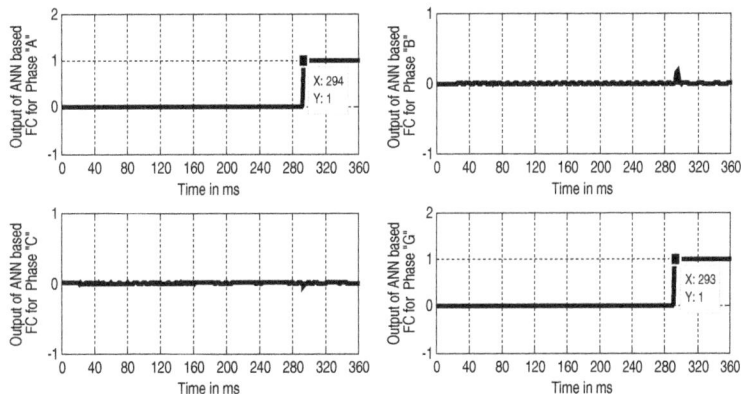

Fig. 15 Output of ANN based Fault Classifier during AG fault in case of real time fault event occurred on 24th April 2015

Acknowledgment

The authors acknowledge the financial support of Chhattisgarh Council of Science & Technology (CGCOST), Raipur for funding the project No. 8062/CGCOST/MRP/13, dtd. 27.12.2013. We also thank to the Department of Electrical Engineering, National institute of Technology, Raipur for providing the research facilities to conduct this project.

Authors' contributions

The first authors is principal investigator of the sponsored research project from Chhattisgarh Council of Science & Technology (CGCOST), Raipur for funding the project No. 8062/CGCOST/MRP/13, dtd. 27.12.2013. The second author is the project fellow appointed for carrying out the simulation of different fault events in the C.G. power system network. First author has designed the ANN model which involves selection of ANN architecture, training of the ANN and thereafter testing with the simulated as well as the real time fault event collected from state power utility. Finally the paper has been written by all the authors collectively. All authors read and approved the final manuscript.

Competing interests

The authors declare that they have no competing interests.

References

1. Yadav, A., & Dash, Y. (2014). An overview of transmission line protection by artificial neural network: fault detection, fault classification, fault location, and fault direction discrimination. *Advances in Artificial Neural Systems*. Article ID 230382, vol. 2014, pp. 20.
2. Phadke, A. G., & Thorp, J. S. (1988). *Computer Relaying for Power Systems*. 2nd edn, New York: Wiley, *5*, 137–186.
3. Ziegler. (2006). Numerical distance protection principles and applications. *SIEMENS*. 2nd edn, Wiley, *3*, 130–190.
4. Dalstein, T., & Kulicke, B. (1995). Neural network approach to fault classification for high speed protective relaying. *IEEE Transactions on Power Delivery, 10*, 1002–1011.
5. Bo, Z. Q., Aggarwal, R. K., Johns, A. T., Li, H. Y., & Song, Y. H. (1997). A new approach to phase selection using fault generated high frequency noise and neural networks. *IEEE Transactions on Power Delivery, 12*, 106–115.
6. Dong, X., Kong, W., & Cui, T. (2009). Fault classification and faulted-phase selection based on the initial current traveling wave. *IEEE Transactions on Power Delivery, 24*, 552–559.
7. Youssef, O. A. S. (2002). New algorithm to phase selection based on wavelet transforms. *IEEE Transactions on Power Delivery, 17*, 908–914.
8. Pradhan, A. K., Routray, A., Pati, S., & Pradhan, D. K. (2004). Wavelet fuzzy combined approach for fault classification of a series-compensated transmission line. *IEEE Transactions on Power Delivery, 19*, 1612–1618.
9. Kezunovic, M., Rikalo, I., & Sobajic, D. J. (1995). High-speed fault detection and classification with neural nets. *Electric Power Systems Research, 34*, 109–116.
10. Chowdhury, F. N., & Aravena, J. L. (1998). A modular methodology for fast fault detection and classification in power systems. *IEEE Trans. on Control Systems Technology, 6*, 623–634.
11. Agarwal, R. K., Xuan, Q. Y., Dunn, R. W., Johns, A. T., & Bennett, A. (1999). A novel fault classification technique of double circuit lines based on a combined unsupervised/supervised neural network. *IEEE Transactions on Power Delivery, 14*, 1250–1256.
12. Lin, W. M., Yang, C. D., Lin, J. H., & Tsay, M. T. (2001). A fault classification method by RBF neural network with OLS learning procedure. *IEEE Transactions on Power Delivery, 16*, 473–477.
13. Wang, H., & Keerthipala, W. W. L. (1998). Fuzzy-Neuro approach to fault classification for transmission line protection. *IEEE Transactions on Power Delivery, 13*, 1093–1104.
14. Dash, P. K., Pradhan, A. K., & Panda, G. (2000). A novel fuzzy neural network based distance relaying scheme. *IEEE Transactions on Power Delivery, 15*, 902–907.
15. Vasilic, S., & Kezunovic, M. (2005). Fuzzy ART neural network algorithm for classifying the power system faults. *IEEE Transactions on Power Delivery, 20*, 1306–1314.
16. Kamel, T. S., Hassan, M. A. M., & El-Morshedy, A. (2009). Advanced distance protection scheme for long Transmission lines in Electric Power systems using multiple classified ANFIS networks. In *Proc. 5th International Conference on Soft Computing, Computing with Words and Perceptions in System Analysis, Decision and Control* (pp. 1–5).
17. Silva, K. M., Souza, B. A., & Brito, N. S. D. (2006). Fault detection and classification in transmission lines based on wavelet transform and ANN. *IEEE Transactions on Power Delivery, 21*, 2058–2063.
18. Mahanty, R. N., & Dutta Gupta, P. B. (2006). Comparison of fault classification methods based on wavelet analysis and ANN. *Electric Power Components and Systems, 34*, 47–60.
19. Jamil, M., Kalam, A., Ansari, A. Q., & Rizwan, M. (2014). Generalized neural network and wavelet transform based approach for fault location estimation of a transmission line. *Applied Soft Computing, 19*, 322–332.
20. Jain, A., Thoke, A. S., & Patel, R. N. (2008). Fault classification of double circuit transmission line using artificial neural network. *International Journal of Electrical Systems Science and Engineering, WASET, USA, 1*, 230–235.
21. He, Z., Lin, S., Deng, Y., Li, X., & Qian, Q. (2014). A rough membership neural network approach for fault classification in transmission lines. *International Journal of Electrical Power and Energy Systems, 61*, 429–439.
22. Coury, D. V., Oleskovicz, M., & Aggarwal, R. K. (2002). An ANN routine for fault detection, classification and location in transmission lines. *Electrical Power Components and Systems, 30*, 1137–1149.
23. Gracia, J., Mazon, A. J., & Zamora, I. (2005). Best ANN structures for fault location in single-and double-circuit transmission lines. *IEEE Transactions on Power Delivery, 20*, 2389–2395.
24. Jiang, J. A., Chuang, C. L., Wang, Y. C., Hung, C. H., Wang, J. Y., Lee, C. H., & Hsiao, Y. T. (2011). A hybrid framework for fault detection, classification, and location- Part II: implementation and test results. *IEEE Transactions on Power Delivery, 26*, 1999–2008.
25. Yadav, A. (2012). Comparison of single and modular ANN based fault detector and classifier for double circuit transmission lines. *International Journal of Engineering, Science and Technology, 4*, 122–136.
26. Coury, D. V., & Jorge, D. C. (1998). Artificial neural network approach to distance protection of transmission lines. *IEEE Transaction on Power Delivery, 13*, 102–108.
27. Khaparde, S. A., Warke, N., & Agarwal, S. H. (1996). An adaptive approach in distance protection using an artificial neural network. *Electric Power Systems Research, 37*, 39–46.
28. Mazon, A. J., Zamora, I., Minambres, J. F., Zorrozua, M. A., Barandiaran, J. J., & Sagastabeitia, K. (2000). A new approach to fault location in two-terminal transmission lines using artificial neural networks. *Electric Power Systems Research Journal, 56*, 261–266.
29. Yadav, A., & Swetapadma, A. (2015). A single ended directional fault section identifier and fault locator for double circuit transmission lines using combined wavelet and ANN approach. *International Journal of Electrical Power and Energy Systems, 69*, 27–33.
30. Jongepier, A. G., & van der Sluis, L. (1994). Adaptive distance protection of a double-circuit line. *IEEE Transaction on Power Delivery, 9*, 1289–1297.
31. Martins, L. S., Martins, J. F., Piers, V. F., & Alegria, C. M. (2005). A neural space vector fault location for parallel double-circuit distribution lines. *Electrical Power and Energy Systems Journal, 27*, 225–231.
32. Skok, S., Marusic, A., Tesnjak, S., & Pevik, L. (2002). Double-circuit line adaptive protection based on kohonen neural network considering different operation and switching modes. *Large Engineering Systems Conference on Power Engineering, 2*, 153–157.
33. Bhalja, B. R., & Maheswari, R. P. (2007). High resistance faults on two terminal parallel transmission line: analysis, simulation studies, and an adaptive distance relaying scheme. *IEEE Transaction on Power Delivery, 22*, 801–812.
34. Oonsivilai, A., & Saichoomdee, S. (2009). Distance transmission line protection based on radial basis function neural network. *World Academy of Science, Engineering and Technology, 3*, 75–78.
35. Jiang, J. A., Yang, J. Z., Lin, Y. H., Liu, C. W., & Ma, J. C. (2000). An adaptive PMU based fault detection/location technique for transmission lines- Part I: Theory and algorithms. *IEEE Transaction on Power Delivery, 15*, 486–493.
36. Jiang, J. A., Lin, Y. H., Yang, J. Z., Too, T. M., & Liu, C. W. (2000). An adaptive PMU based fault detection/location technique for transmission lines- Part II: PMU implementation and performance evaluation. *IEEE Transaction on Power Delivery, 15*, 1136–1146.

37. Sidhu, T. S., Singh, H., & Sachdev, M. S. (1997). An artificial neural network for directional comparison relaying of transmission lines. In *Proc. 6th International Conference on Developments in Power System Protection* (pp. 282–285).

38. Gang, W., Jiali, H., Yao, L., Xiaodan, Y., & Zhongpu, L. (1997). Neural network application in directional comparison carrier protection of EHV transmission lines. In *Proc. 4th International Conference on Advances in Power System Control, Operation and Management* (Vol. 1, pp. 89–94).

39. Santos, R. C. D., & Senger, E. C. (2011). Transmission lines distance protection using artificial neural networks. *International Journal of Electrical Power and Energy Systems, 33,* 721–730.

3

Research on fault diagnosis for MMC-HVDC Systems

Zhiqing Yao[1*], Qun Zhang[2], Peng Chen[2] and Qian Zhao[2]

Abstract

Introduction: With the development of flexible HVDC technology, the fault diagnosis of MMC-HVDC becomes a new research direction.
Based on the fault diagnosis theory, this paper proposes a robust fault diagnosis method to study the fault diagnosis problem of MMC-HVDC systems.

Methods: By optimizing the gain matrix in the fault observer, fault detection with good sensitivity and robustness to disturbance is achieved. In the MMC-HVDC system, because of the inherently uncertain system and the presence of various random disturbances, the study of robust fault diagnosis method is particularly important.

Results: Simulation studies during various AC faults have been carried out based on a 61-level MMC-HVDC mathematical model. The results validate the feasibility and effectiveness of the proposed fault diagnosis method.

Conclusions: So this fault diagnosis method can be further applied to the actual project, to quickly achieve system fault diagnosis and accurately complete fault identification.

Keywords: Fault diagnosis, MMC-HVDC, Robust, State observer, Residual state equation

Introduction

In recent years, due to the increased size and complexity of control systems, fault diagnosis becomes particularly important, especially in power systems [1]. For instance, if line faults are not quickly detected and isolated, they can cause system failure or even lead to disastrous consequences [2]. Currently, there are many fault diagnosis methods, such as the diagnostic filter method, parameter identification method, expert system, artificial neural network etc. [3–5]. Although these methods can effectively detect failures in the system, they cannot precisely estimate these fault signals.

This paper proposes a fault diagnosis method based on full dimension state observer for flexible HVDC system. State estimation is carried out using the mathematical model of a three-phase 61-level modular multilevel converter (MMC) based HVDC (MMC-HVDC) system and the full dimension state observer. Residual state equation is established and is combined with the system output and system phasor. By optimizing the gain matrix in fault observer, the robustness of the fault detection is

investigated. In the MMC-HVDC system, because the system is inherently uncertain and there are various random disturbances, the study of robust fault diagnosis method is particularly important.

Methods
Model of MMC-HVDC
A three-phase MMC topology is shown in Fig. 1 [6–9]. According to KCL, the three-phase current can be expressed as

$$\begin{cases} i_a = i_{pa} + i_{na} \\ i_b = i_{pb} + i_{nb} \\ i_c = i_{pc} + i_{nc} \end{cases} \tag{1}$$

For three single-phase units, applying KVL to the upper and lower arms yields:

$$\begin{cases} u_a - \left(\dfrac{U_{dc}}{2} - u_{pa} \right) = 2L \dfrac{di_{pa}}{dt} + 2Ri_{pa} \\ u_b - \left(\dfrac{U_{dc}}{2} - u_{pb} \right) = 2L \dfrac{di_{pb}}{dt} + 2Ri_{pb} \\ u_c - \left(\dfrac{U_{dc}}{2} - u_{pc} \right) = 2L \dfrac{di_{pc}}{dt} + 2Ri_{pc} \end{cases} \tag{2}$$

* Correspondence: zhiqingy@dlwg.net
[1]Xuchang KETOP Electrical Research Institute, Xuchang, China
Full list of author information is available at the end of the article

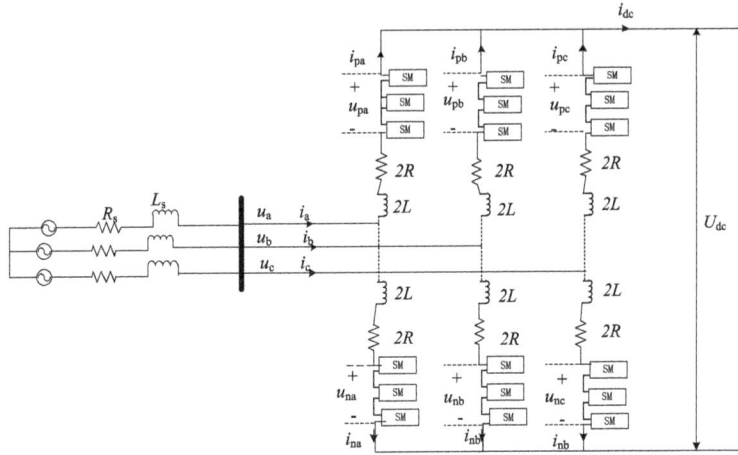

Fig. 1 Three phase MMC topology

$$\begin{cases} u_a - \left(u_{na} - \dfrac{U_{dc}}{2}\right) = 2L\dfrac{di_{na}}{dt} + 2Ri_{na} \\[2mm] u_b - \left(u_{nb} - \dfrac{U_{dc}}{2}\right) = 2L\dfrac{di_{nb}}{dt} + 2Ri_{nb} \\[2mm] u_c - \left(u_{nc} - \dfrac{U_{dc}}{2}\right) = 2L\dfrac{di_{nc}}{dt} + 2Ri_{nc} \end{cases} \quad (3)$$

In the above equations, 2L and 2R denote the equivalent arm inductance and resistance, respectively. Adding Eqs. (2) and (3) leads

$$\begin{cases} u_a - (u_{na} - u_{pa})/2 = L\dfrac{di_a}{dt} + Ri_a \\[2mm] u_a - (u_{nb} - u_{pb})/2 = L\dfrac{di_b}{dt} + Ri_b \\[2mm] u_c - (u_{nc} - u_{pc})/2 = L\dfrac{di_c}{dt} + Ri_c \end{cases} \quad (4)$$

According to Eq. (4), the time domain mathematic model of a MMC in abc coordinate is given by [10, 11].

$$\begin{cases} \dfrac{di_a(t)}{dt} = -\dfrac{R}{L}i_a(t) + \dfrac{1}{L}\left[u_a(t) - (u_{na}(t) - u_{pa}(t))/2\right] \\[2mm] \dfrac{di_b(t)}{dt} = -\dfrac{R}{L}i_b(t) + \dfrac{1}{L}\left[u_b(t) - (u_{nb}(t) - u_{pb}(t))/2\right] \\[2mm] \dfrac{di_c(t)}{dt} = -\dfrac{R}{L}i_c(t) + \dfrac{1}{L}\left[u_c(t) - (u_{nc}(t) - u_{pc}(t))/2\right] \end{cases}$$
$$(5)$$

Written Eq. (5) in phasor form yields

$$\frac{d}{dt}\begin{bmatrix} i_a \\ i_b \\ i_c \end{bmatrix} = \begin{bmatrix} -\dfrac{R}{L} & 0 & 0 \\[1mm] 0 & -\dfrac{R}{L} & 0 \\[1mm] 0 & 0 & -\dfrac{R}{L} \end{bmatrix}\begin{bmatrix} i_a \\ i_b \\ i_c \end{bmatrix}$$
$$+ \begin{bmatrix} -\dfrac{1}{2L} & \dfrac{1}{2L} & 0 & 0 & 0 & 0 \\[1mm] 0 & 0 & -\dfrac{1}{2L} & \dfrac{1}{2L} & 0 & 0 \\[1mm] 0 & 0 & 0 & 0 & -\dfrac{1}{2L} & \dfrac{1}{2L} \end{bmatrix}\begin{bmatrix} u_{na} \\ u_{pa} \\ u_{nb} \\ u_{pb} \\ u_{nc} \\ u_{pc} \end{bmatrix} \quad (6)$$
$$+ \begin{bmatrix} \dfrac{1}{L} & 0 & 0 \\[1mm] 0 & \dfrac{1}{L} & 0 \\[1mm] 0 & 0 & \dfrac{1}{L} \end{bmatrix}\begin{bmatrix} u_a \\ u_b \\ u_c \end{bmatrix}$$

Considering the uncertainty, external disturbances and system faults, the Eq. (6) can be written as follows

$$\begin{cases} \dot{\mathbf{x}}(t) = \mathbf{A}\mathbf{x}(t) + \mathbf{B}\mathbf{u}(t) + \mathbf{h}(t) + \boldsymbol{\Delta}(t) + \mathbf{B}_f\mathbf{f}(t) \\ \mathbf{y}(t) = \mathbf{C}\mathbf{x}(t) + \mathbf{D}\mathbf{M}(t) \end{cases} \quad (7)$$

where $\mathbf{x}(t) = \begin{bmatrix} i_a \\ i_b \\ i_c \end{bmatrix}$, $\mathbf{u}(t) = \begin{bmatrix} u_{na} \\ u_{pa} \\ u_{nb} \\ u_{pb} \\ u_{nc} \\ u_{pc} \end{bmatrix}$, $\mathbf{A} = \begin{bmatrix} -\dfrac{R}{L} & 0 & 0 \\[1mm] 0 & -\dfrac{R}{L} & 0 \\[1mm] 0 & 0 & -\dfrac{R}{L} \end{bmatrix}$,

$\mathbf{B} = \begin{bmatrix} -\dfrac{1}{2L} & \dfrac{1}{2L} & 0 & 0 & 0 & 0 \\[1mm] 0 & 0 & -\dfrac{1}{2L} & \dfrac{1}{2L} & 0 & 0 \\[1mm] 0 & 0 & 0 & 0 & -\dfrac{1}{2L} & \dfrac{1}{2L} \end{bmatrix}$, $\mathbf{h}(t) = \begin{bmatrix} \dfrac{1}{L} & 0 & 0 \\[1mm] 0 & \dfrac{1}{L} & 0 \\[1mm] 0 & 0 & \dfrac{1}{L} \end{bmatrix}\begin{bmatrix} u_a \\ u_b \\ u_c \end{bmatrix}$.

In Eq. (7), $\boldsymbol{\Delta}(t)$ represents uncertainty and external disturbances generated during the actual modeling process. $\mathbf{f}(t)$ represents the system faults which are required to detect and identify. $\mathbf{M}(t)$ is measurement noise introduced by the measurement system. $\mathbf{B}_f$, $\mathbf{C}$ and $\mathbf{D}$ are known matrices with appropriate dimensions. $\mathbf{y}(t)$ is the system output. If $\mathbf{\Omega}(t)$ is used to represent $\mathbf{\Delta}(t) + \mathbf{h}(t)$, Eq. (7) can be written as follows.

$$\begin{cases} \dot{\mathbf{x}}(t) = \mathbf{A}\mathbf{x}(t) + \mathbf{B}\mathbf{u}(t) + \mathbf{\Omega}(t) + \mathbf{B}_f\mathbf{f}(t) \\ \mathbf{y}(t) = \mathbf{C}\mathbf{x}(t) + \mathbf{D}\mathbf{M}(t) \end{cases} \tag{8}$$

For further analysis, the following assumptions are made:

1) Uncertainties in the system are norm-bounded, that is $\|\mathbf{\Delta}(t)\| \le V_\Delta$.
2) System matrix pair $(\mathbf{A}, \mathbf{C})$ is observable.
3) Uncertainties in the system meet the condition $\|\mathbf{\Omega}(t)\| \le \gamma\|\mathbf{x}(t)\| \le V_\Omega$.
4) The $\mathbf{M}(t)$ in the system is norm-bounded, that is $\|\mathbf{M}(t)\| \le V_M$.
5) The initial state of the system is zero.

Design of full dimension state observer

Fault diagnosis is carried out by full dimension state observer based on the state equation of flexible HVDC system [12–14]. It can be seen from Eq. (8) that $\mathbf{\Omega}(t)$ contains the system uncertainties and external disturbance, and the grid voltage u_a, u_b, u_c, which can be regarded as unknown nonlinear perturbation term.

A state observer is established according to Eq. (8):

$$\left\{ \dot{\hat{\mathbf{x}}}(t) = \mathbf{A}\hat{\mathbf{x}}(t) + \mathbf{B}\mathbf{u}(t) + \mathbf{H}(\mathbf{y}(t)-\hat{\mathbf{y}}(t))\hat{\mathbf{y}}(t) = \mathbf{C}\hat{\mathbf{x}}(t) \right. \tag{9}$$

In Eq. (9), H is a gain matrix to be designed. According to Eqs. (8) and (9), system residual state equation is given by:

$$\begin{cases} \dot{\mathbf{x}}(t)- \dot{\hat{\mathbf{x}}}(t) = (\mathbf{A}-\mathbf{HC})(\mathbf{x}(t)-\hat{\mathbf{x}}(t)) + \mathbf{\Omega}(t) + \mathbf{B}_f\mathbf{f}(t) \\ \quad -\mathbf{HD}\mathbf{M}(t)\mathbf{y}(t)-\hat{\mathbf{y}}(t) = \mathbf{C}(\mathbf{x}(t)-\hat{\mathbf{x}}(t)) + \mathbf{D}\mathbf{M}(t) \end{cases} \tag{10}$$

If $\mathbf{e}(t) = \mathbf{x}(t)-\hat{\mathbf{x}}(t)$ and $\mathbf{r}(t) = \mathbf{y}(t) - \hat{\mathbf{y}}(t)$, Eq. (10) can be written as follows:

$$\begin{cases} \dot{\mathbf{e}}(t) = (\mathbf{A}-\mathbf{HC})\mathbf{e}(t) + \mathbf{\Omega}(t) + \mathbf{B}_f\mathbf{f}(t)-\mathbf{HD}\mathbf{M}(t) \\ \mathbf{r}(t) = \mathbf{C}\mathbf{e}(t) + \mathbf{D}\mathbf{M}(t) \end{cases} \tag{11}$$

In the design of the full dimension state observer, the impact of the fault and external disturbance or uncertainties on the residual should be considered. The transfer function from system uncertainties and external disturbance to residual $\mathbf{r}(t)$ is presented by $\mathbf{T}_{rd}$ whereas $\mathbf{T}_{rf}$ presents the transfer function from system fault to residual $\mathbf{r}(t)$. If the value of $\|\mathbf{T}_{rd}\|$ is small enough and

the value of $\|\mathbf{T}_{rf}\|$ is big enough in the design, the designed observer will be robust. Therefore, the following performance index is used to design the fault diagnosis observer:

$$J = \frac{\|\mathbf{T}_{rf}\|}{\|\mathbf{T}_{rd}\|} \tag{12}$$

In practical applications, the above performance index can be equivalent to the following equation:

$$\|r(t)\|_\infty \le \beta\|d(t)\|_\infty, \|r(t)\|_- \ge \eta\|f(t)\|_- \tag{13}$$

Equation (13) can be solved by using multi-objective optimization methods and thus the robust control theory will be used to solve the performance optimization problem proposed by Eq. (13).

Theorem 1 The system represented by Eq. (8) satisfies the assumptions (1)-(5). Optimization index β is given and satisfies the condition $\mathbf{D}^T\mathbf{D} - \beta^2\mathbf{I} < 0$. In case of no fault in the system, if there are a positive definite symmetric matrix R and a constant factor $\varepsilon_1 > 0$ meeting the following linear matrix inequality:

$$\begin{bmatrix} \mathbf{R}(\mathbf{A}-\mathbf{HC}) + (\mathbf{A}-\mathbf{HC})^T\mathbf{R} + \varepsilon_1^{-1}\mathbf{R}^T\mathbf{R} + \mathbf{C}^T\mathbf{C} & 0 & \mathbf{C}^T\mathbf{D}-\mathbf{RHD} \\ 0 & \varepsilon_1\gamma^2\mathbf{I} & 0 \\ \mathbf{D}^T\mathbf{C}-(\mathbf{RHD})^T & 0 & \mathbf{D}^T\mathbf{D}-\beta^2\mathbf{I} \end{bmatrix} \tag{14}$$

the residual system (11) is asymptotically stable and meets $\|r(t)\|_\infty \le \beta\|d(t)\|_\infty$.

Theorem 2 The system represented by Eq. (8) satisfies the assumptions (1)-(5). Optimization index η is given and satisfies the condition $\mathbf{D}^T\mathbf{D} - \eta^2\mathbf{I} < 0$. In case of no fault in the system, if there are a positive definite symmetric matrix T and a constant factor $\eta_1 > 0$ meeting the following linear matrix inequality:

$$\begin{bmatrix} \mathbf{R}(\mathbf{A}-\mathbf{HC}) + (\mathbf{A}-\mathbf{HC})^T\mathbf{R} + \eta_1^{-1}\mathbf{R}^T\mathbf{R} + \mathbf{C}^T\mathbf{C} & 0 & -\mathbf{C}^T\mathbf{D} + \mathbf{RHD} \\ 0 & \eta_1\gamma^2\mathbf{I} & 0 \\ -\mathbf{D}^T\mathbf{C} + (\mathbf{RHD})^T & 0 & \eta^2\mathbf{I}-\mathbf{D}^T\mathbf{D} \end{bmatrix} \tag{15}$$

The residual system (11) is asymptotically stable and meets $\|r(t)\|_- \ge \eta\|f(t)\|_-$.

Theorem 3 The system represented by Eq. (8) satisfies the assumptions (1)-(5). Optimization index $\beta > 0$, $\eta > 0$ are given. If there are positive definite symmetric matrix R, T,H, constant factor $\varepsilon_1 < 0$ and $\eta_1 > 0$ meeting the following linear matrix inequality:

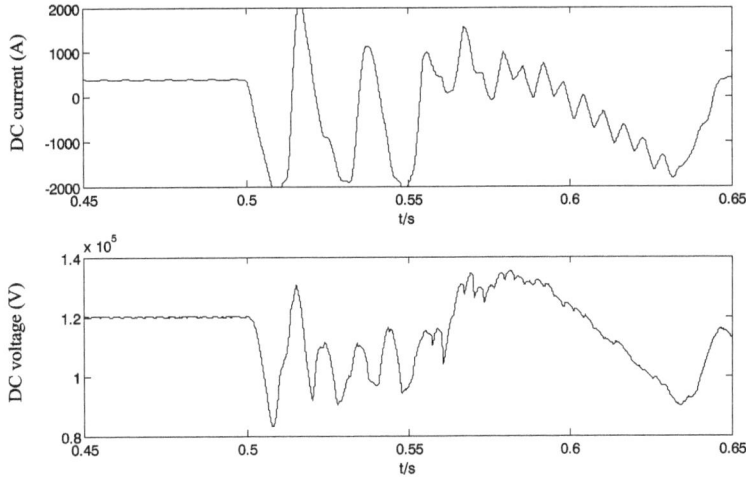

Fig. 2 Phase-A to ground fault

$$\begin{bmatrix} \mathbf{R}(\mathbf{A}-\mathbf{HC}) + (\mathbf{A}-\mathbf{HC})^T\mathbf{R} + \varepsilon_1^{-1}\mathbf{R}^T\mathbf{R} + \mathbf{C}^T\mathbf{C} & 0 & \mathbf{C}^T\mathbf{D}-\mathbf{RHD} \\ 0 & \varepsilon_1\gamma^2\mathbf{I} & 0 \\ \mathbf{D}^T\mathbf{C}-(\mathbf{RHD})^T & 0 & \mathbf{D}^T\mathbf{D}-\beta^2\mathbf{I} \end{bmatrix} \quad (16)$$

$$\begin{bmatrix} \mathbf{R}(\mathbf{A}-\mathbf{HC}) + (\mathbf{A}-\mathbf{HC})^T\mathbf{R} + \eta_1^{-1}\mathbf{R}^T\mathbf{R} + \mathbf{C}^T\mathbf{C} & 0 & -\mathbf{C}^T\mathbf{D}+\mathbf{RHD} \\ 0 & \eta_1\gamma^2\mathbf{I} & 0 \\ -\mathbf{D}^T\mathbf{C}+(\mathbf{RHD})^T & 0 & \eta^2\mathbf{I}-\mathbf{D}^T\mathbf{D} \end{bmatrix} \quad (17)$$

The residual system (11) is asymptotically stable and meets $\|r(t)\|_\infty \le \beta\|d(t)\|_\infty$ and $\|r(t)\|_- \ge \eta\|f(t)\|_-$.

The gain matrix H of the full dimension state observer can be obtained from Theorem 3. The design process of gain matrix H takes into account the impact of system disturbances and uncertainties on the residual, and the sensitivity of the residual for faults.

Determination of fault detection threshold

When disturbances or uncertainties exist in the system, a fault detection threshold method is usually used to determine whether the system has a fault [15–17]. This section provides fault detection thresholds calculated according to the definition of norm.

Theorem 4 The system represented by Eq. (8) satisfies the assumptions (1)-(5). If the residual system (11) meets $\|\mathbf{r}(t)\| > (a\mathbf{V}_\Omega + b\mathbf{V}_M)(t-t_0) + c\mathbf{V}_M$, then system fault is detected, where $a = \sup_{t\in[t_0,t]}\|\mathbf{C\Psi}(t,s)\|$, $b = \sup_{t\in[t_0,t]}\|\mathbf{C\Psi}(t,s)\mathbf{HD}\|$, $c = \sup_{t\in[t_0,t]}\|\mathbf{D}\|$.

Proof According to the residual state Eq. (11) the following two equations can be derived:

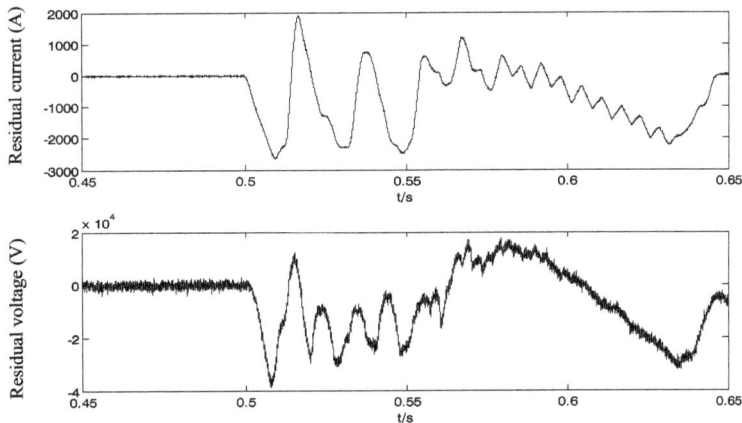

Fig. 3 Residual error curve during phase-A to ground fault

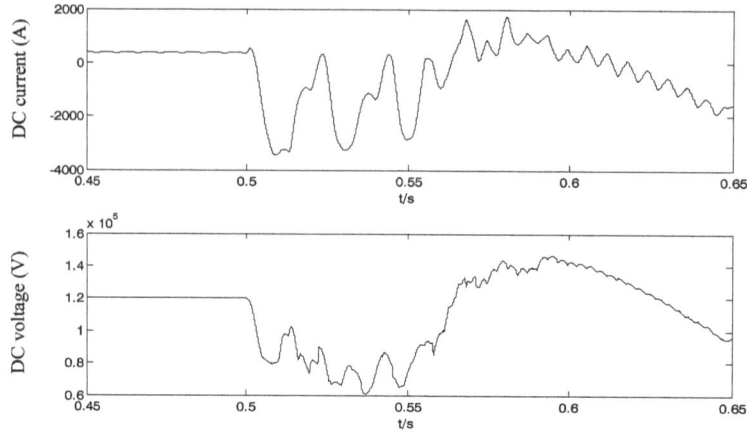

Fig. 4 Phase-AB to ground fault

$$\mathbf{e}(t) = \mathbf{\Psi}(t,t_0)\mathbf{e}(t_0) + \int_{t_0}^{t} \mathbf{\Psi}(t,s)\big[\mathbf{\Omega}(t) + \mathbf{B}_f\mathbf{f}(t) - \mathbf{HDM}(t)\big]ds$$

(18)

$$\mathbf{r}(t) = \mathbf{C}\left\{\mathbf{\Psi}(t,t_0)\mathbf{e}(t_0) + \int_{t_0}^{t} \mathbf{\Psi}(t,s)\big[\mathbf{\Omega}(t) + \mathbf{B}_f\mathbf{f}(t) - \mathbf{HDM}(t)\big]ds\right\}$$
$$+ \mathbf{DM}(t)$$

(19)

According to assumption (5), $\mathbf{e}(t_0) = 0$. Thus

$$\mathbf{r}(t) = \int_{t_0}^{t} \mathbf{C}\mathbf{\Psi}(t,s)\big[\mathbf{\Omega}(t) + \mathbf{B}_f\mathbf{f}(t) - \mathbf{HDM}(t)\big]ds + \mathbf{DM}(t)$$

(20)

Equation (21) can be obtained by seeking norm in both sides as

$$|\mathbf{r}(t)| \le \int_{t_0}^{t} \|\mathbf{C}\mathbf{\Psi}(t,s)\mathbf{\Omega}(t)\|ds + \int_{t_0}^{t} \|\mathbf{C}\mathbf{\Psi}(t,s)\mathbf{B}_f\mathbf{f}(t)\|ds$$
$$+ \int_{t_0}^{t} \|\mathbf{C}\mathbf{\Psi}(t,s)\mathbf{HDM}(t)\|ds + \|\mathbf{DM}(t)\|$$

(21)

where $a = \sup_{t\in[t_0,t]} \|\mathbf{C}\mathbf{\Psi}(t,s)\|$, $b = \sup_{t\in[t_0,t]} \|\mathbf{C}\mathbf{\Psi}(t,s)\mathbf{HD}\|$, $c = \sup_{t\in[t_0,t]} \|\mathbf{D}\|$

According to assumptions (1)-(4), Eq. (22) is obtained as

$$\|\mathbf{r}(t)\| \le \int_{t_0}^{t} a\|\mathbf{\Omega}(t)\|ds + \int_{t_0}^{t} b\|\mathbf{M}(t)\|ds + \|\mathbf{DM}(t)\|$$

(22)

Equation (22) can be simplified as

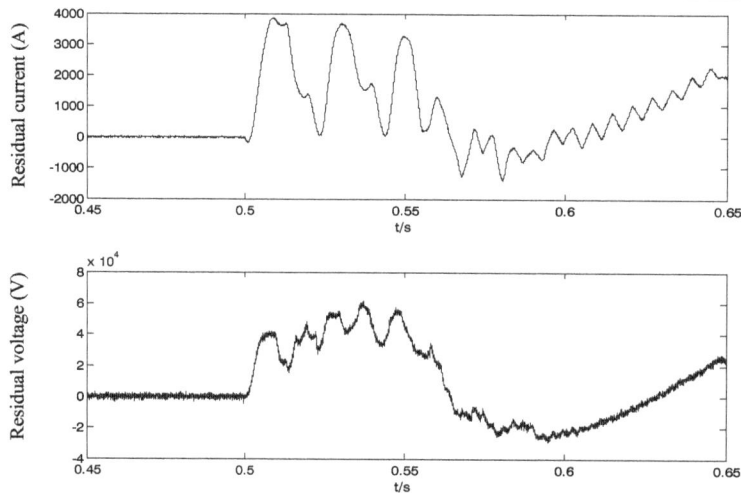

Fig. 5 Residual error curve during phase-AB to ground fault

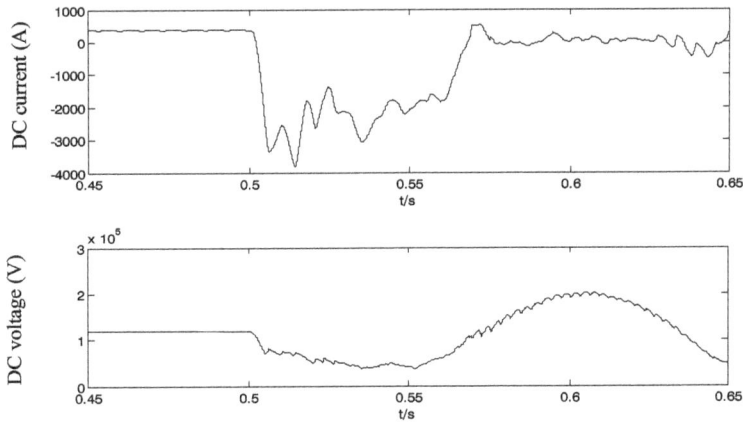

Fig. 6 Three phase to ground fault

$$
\begin{aligned}
\|\mathbf{r}(t)\| &\le \int_{t_0}^{t} a\mathbf{V}_\Omega ds + \int_{t_0}^{t} b\mathbf{V}_M ds + c\mathbf{V}_M \\
&= a\mathbf{V}_\Omega(t{-}t_0) + b\mathbf{V}_M(t{-}t_0) + c\mathbf{V}_M \\
&= (a\mathbf{V}_\Omega + b\mathbf{V}_M)(t{-}t_0) + c\mathbf{V}_M
\end{aligned}
\tag{23}
$$

Proof is done.

Results

Simulation results

According to the MMC-HVDC system state Eq. (7), common types of fault, e.g. phase A to ground fault, phase-AB to ground fault, three phase to ground fault are simulated based on the 61-level MMC-HVDC system using Matlab/Simulink and RT-LAB.

The rated DC voltage of the system is 120 kV and the rated active power is 50 MW. The rectifier controls active power control and the inverter controls the DC voltage. AC fault is simulated during 0.5 to 0.55 s. The simulation output voltage and current waveforms are shown in Figs. 2, 3, 4, 5, 6 and 7.

Phase-A to ground fault

Figure 2 shows the DC bus voltage and current waveform when the phase-A to ground fault occurs in the rectifier side. The fault is introduced at 0.5 s when the system is running at steady state. The residual characteristics of the DC bus voltage and current can be calculated by the MMC-HVDC system residual state Eq. (13), as shown in Fig. 3. The baseline reference points of the HVDC fault observer output voltage and current are zero, which can be seen by comparing Figs. 2 and 3. When the system operates normally, taking into account the existing uncertainties and various random disturbances, residual curve should fluctuate within a limited range. From Fig. 3 it can be seen that, after a system fault, the system output curves (the DC bus voltage and current) significantly deviate from zero and their normal

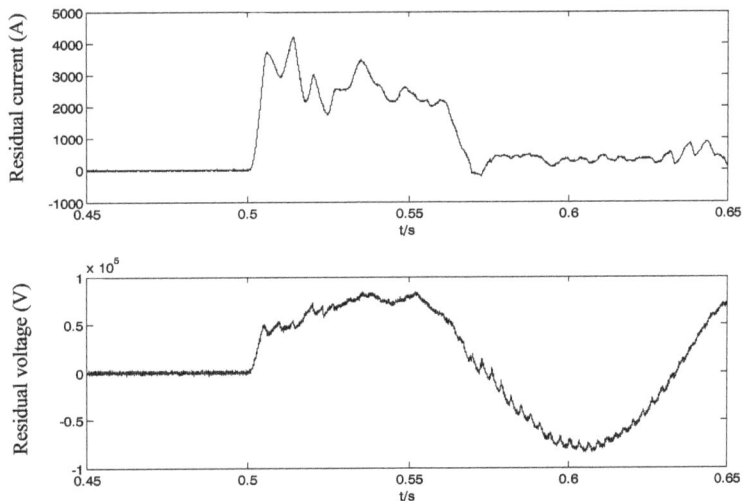

Fig. 7 Residual error curve during three phase to ground fault

bands. The system's fault diagnostic threshold can be calculated according to Theorem 4. If the residual waveform fluctuates within the calculated threshold, it is judged that the system is operating normally. If the residual waveform exceeds the calculated threshold, a fault is detected and immediate protection action should be taken. Furthermore, the fault identification and fault estimation can also be studied according to the residual waveform. As seen from the simulation, when $t > 0.52$ s, it detects a fault in the system.

Phase-AB to ground fault

The waveforms during phase-AB to ground fault are shown in Fig. 4 where the fault is introduced at 0.5 s. Due to the existence of various disturbances and uncertainties in the system, measurement noise is inevitably introduced when measuring system output. The residual voltage and residual current waveforms are shown in Fig. 5. According to Theorem 4, the fault detection threshold is Jth = 0.136 * 104. As seen from the simulation, when $t > 0.52$ s, the fault is detected.

Three phase to ground fault

The waveforms during three phase to ground fault is shown in Fig. 6. The fault is again introduced at 0.5 s. The residual voltage and residual current waveforms calculated according to residual system state equation are shown in Fig. 7. According to Theorem 4, the fault detection is Jth = 2.62*104. As seen from the simulation, when $t > 0.52$ s, the system fault is detected.

Conclusions

In this paper, a robust fault diagnosis method is proposed to detect system fault in MMC-HVDC power transmission systems. The advantage of the proposed method is that the state observer considers the system uncertainties and fault sensitivity. Therefore, the proposed method can quickly achieve system fault diagnosis and accurately complete fault identification. The simulation results have proved the effect of robust fault diagnosis.

Authors' contributions
ZQY analyzed and proposed the robust fault diagnosis method. QZ, PC and QZ completed the simulations on RT-LAB and the written of this paper. ALL the authors read and approved the final manuscript.

Competing interests
The authors declare that they have no competing interests.

Author details
[1]Xuchang KETOP Electrical Research Institute, Xuchang, China. [2]XJ Electric Co., Ltd, Xuchang, China.

References
1. Zhang, ST, & Hao, JJ. (2012). Fault diagnosis based on multiple-models particle filter. *Control Engineering of China, 19*(5), 864–871.
2. Song, ZM, Guo, YH, Xun, TS, et al. (2012). Reaserch on HVDC transmission system fault analysis based on wavelet transform. *Power System Protection and Control, 40*(3), 100–104.
3. Bian, L, & Bian, CY. (2014). Review on intelligence fault diagnosis in power networks. *Power System Protection and Control, 42*(3), 146–158.
4. Li, HW, Yang, DS, Sun, YL, et al. (2013). Study review and prospect of intelligent fault diagnosis technique. *Computer Engineering and Design, 34*(2), 632–637.
5. Wang, JL, Xia, L, Wu, ZG, et al. (2010). State of arts of fault diagnosis of power systems. *Power System Protection and Control, 38*(18), 210–216.
6. Dai, GF, Zhao, D, Lin, PF, et al. (2015). Study of control strategy for active power filter based on modular multilevel converter. *Power System Protection and Control, 43*(8), 74–80.
7. Gnanarathna, UN, Gole, AM, & Jayasinghe, RP. (2011). Efficient modeling of modular multilevel HVDC converters (MMC) on electromagnetic transient simulation programs. *IEEE Transactions on Power Delivery, 26*(1), 316–324.
8. Peralta, J, Saad, H, Dennetiere, S, et al. (2012). Detailed and averaged models for a 401-level MMC-HVDC system. *IEEE Transactions on Power Delivery, 27*(3), 1501–1508.
9. Simon, P.T. Simplified dynamic model of a voltage sourced converter with modular multilevel converter design. IEEE/PES Power Systems Conference and Exposition. Seattle, USA, 2009: 1–6
10. Simon, P.T. Modeling the trans bay cable project as voltage-sourced converter with modular multilevel converter design. IEEE Power and Energy Society General Meeting. Michigan, USA, 2011: 1–8
11. Guan, M, & Xu, Z. (2012). Modeling and control of modular multilevel converter based HVDC system under unbalanced grid conditions. *IEEE Transactions on Power Electronics, 27*(12), 4858–4867.
12. Zhu, QL, Peng, CH, Li, JF, et al. (2012). Fuzzy DTC of PMSM based on full order observer. *Power Electronics, 46*(1), 87–89.
13. Yang, JQ, & Zhu, FL. (2014). Linear-matrix-inequality observer design of nonlinear systems with unknown input and measurement noise reconstruction. *Control Theroy & Applications, 31*(4), 538–544.
14. Gao, H, & Cai, XS. (2013). Functional obsever design for a class of nonlinear systems. *Control Theory & Applications, 30*(9), 1207–1210.
15. Zhao, Y, Dong, S, & Li, TY. (2010). A new adaptive threshold algorithm to partial discharge processing based on HHT-MDL criterion. *Power System Protection and Control, 38*(5), 45–50.
16. Zhu, XH, Li, YH, Li, N, et al. (2013). Novel observer-based on robust fault detection method for nonlinear uncertain systems. *Control Theory & Applications, 30*(5), 644–648.
17. Qin, LG, He, X, & Zhou, DH. (2015). A fault estimation method based on robust residual generators. *Journal of Shanghai Jiaotong University, 49*(6), 768–744.

4

New development in relay protection for smart grid

Baohui Zhang[1*], Zhiguo Hao[1] and Zhiqian Bo[2]

Abstract

This series of papers report on relay protection strategies that satisfy the demands of a strong smart grid. These strategies include ultra-high-speed transient-based fault discrimination, new co-ordination principles of main and back-up protection to suit the diversification of the power network, optimal co-ordination between relay protection and auto-reclosure to enhance robustness of the power network. There are also new development in protection early warning and tripping functions of protection based on wide area information.

In this paper the principles, algorithms and techniques of single-ended, transient-based and ultra-high-speed protection for EHV transmission lines, buses, DC transmission lines and faulty line selection for non-solid earthed networks are presented. Tests show that the methods presented can determine fault characteristics with ultra-high-speed (5 ms) and that the new principles of fault discrimination can satisfy the demand of EHV systems within a smart grid.

Keywords: Smart grid, Fault transient component, Ultra-high-speed protection

Introduction

Relay protection is the key to the safe operation of a power system. The functions of relay protection have been developed along with enhancements to electrical power systems and the implementation techniques developed with the related areas of science and technology. Ensuring the function of a relay to satisfy the requirements of the development of the smart grid and perform the protection task with high reliability, involves a series of key technical issues. These issues, for example, include among others, the principles, criteria and algorithms for discriminate internal and external faults. In order to remove the faulty component with minimum area of interruption, it requires the setup of multiple protection relays and their technical coordination. In order to meet the above requirements, protection devices based on various hardware platforms, different techniques and the on site operational management of these devices are adopted. Each stage is instrumental in ensuring that the relay operates correctly. This paper mainly outlines the research and development undertaken for the protection

system based on the latest digital protection technology and these developments will greatly influence the type of relays produced in the future.

China is building a strong smart grid and is constructing a 750/1000 kV transmission network to provide super power transmission capability. To utilize this kind of transmitting capacity considering the transient stability limits, it relies on the performance of the relay protection. Therefore, the research and development of ultra fast protection with response speed within 5 ms has great practical significance. With the development of high speed Digital Signal Processor (DSP) embedded system techniques and the application of optical sensor, the latest protection hardware posses the ability to record and compute the detailed fault transients. The mathematical tool Wavelet transform used to analyze the characteristics of the non-periodic sudden changes of a signal provides a powerful tool for the analysis and the computation of fault transient data. The fault transient components consist of information including the fault location and type. The above mentioned device and algorithms can be used to develop ultra high speed protection relays to identify the fault location and fault type based on detecting and processing the high frequency components of such fault transients.

* Correspondence: bhzhang@mail.xjtu.edu.cn
[1]School of Electrical Engineering, Xi'an Jiaotong University, Xi'an 710049, China
Full list of author information is available at the end of the article

For over a century, many protection principles have been developed based on the lumped parameter model and utilizing the differences between the power frequency voltage and current during the normal operation and fault states. The vector and symmetrical component analysis methods have also been developed to constitute the theoretical basis for fault identification. In order to eliminate the non-power frequency components of the fault transient period, many filtering methods have been developed and have become an important component of the protection algorithms. The maturity of these theories and algorithms over the years enables many power system fault conditions to be correctly and selectively removed very quickly. However, to obtain complete information over a power frequency period (50Hz) requires at least 20 ms to collect. To remove a fault faster, an approximation technique is adopted that uses the information from ½ or ¼ of the period to represent the whole period to reduce the operational zone of the protection but still guarantee the selectivity.

In the 1960's, people started to develop protection using the fault generated transient wave front, which marked the start of using fault transients for protection. Due to the limitation of the available technical means and the complexity in identifying the reflections of the wave front from the fault point, the prototype relays developed at that time do not offer today's desired reliability and speed. However, site records indicate that the fault transients contain an abundant amount of information about the fault location and the type of fault in the fault record. Through the examination and analysis of these characteristics, new protection principles can be developed to solve the problems which can not

be resolved by power frequency based protection principles. This approach can also further increase the response speed of the protection relay to meet the requirements of the UHV/EHV power system, which is in line with the development trends of the technology and today's power systems.

Although the concept of the smart grid was only raised recently, researchers have already undertaken research into many of the aspects needed to meet the development requirements of the latest power grid technology. This paper will introduce the research work over the past 10 years conducted by the authors' team on: the new ultra high speed protection principle, device implementation and test results obtained by utilizing the transient characteristics of the AC and DC transmission systems. Due to the space limitation, this paper will concentrate on the research results, while the detailed principles, algorithms and development process of the techniques can be found in the references.

Transient based protection for ultra high voltage transmission lines

The protection diagram is shown in Fig. 1 Discriminating the internal and external faults is achieved by utilizing the differences in the magnitude and direction of the high frequency voltages and currents.

Main protection component and the principle employed [1–12]

(1) Starting Unit(SU): based on the sudden change in the characteristic (Lipschitz coefficient) of the initial traveling wave caused by a fault on the transmission line, it adopts the "wavelet modular change and

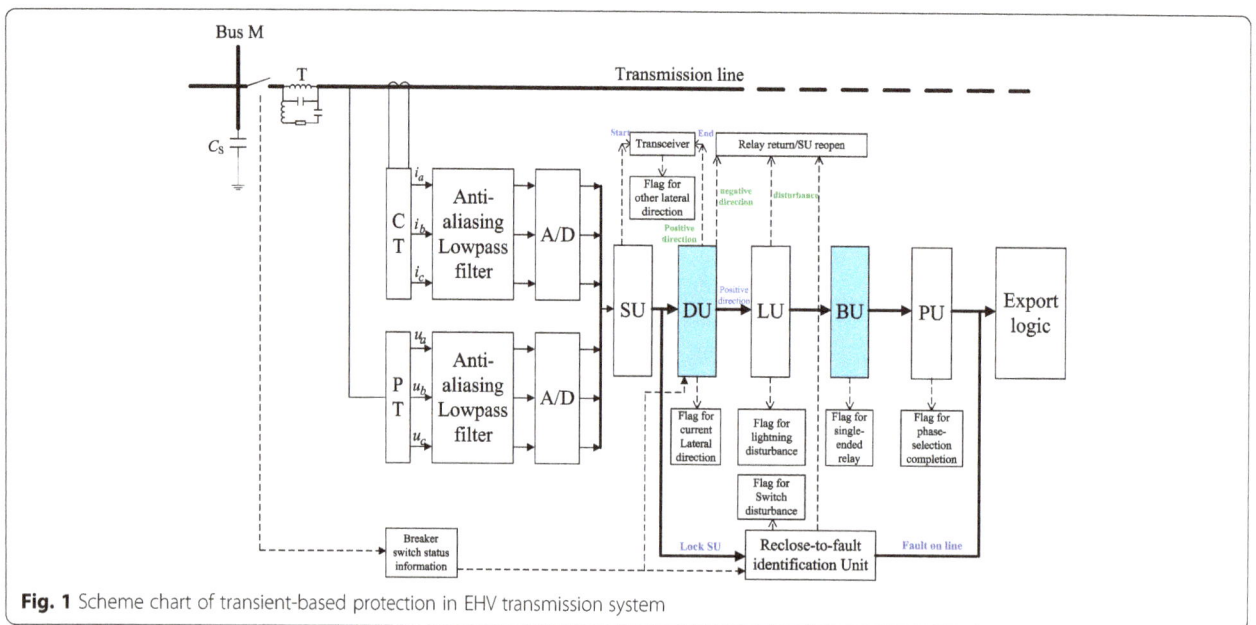

Fig. 1 Scheme chart of transient-based protection in EHV transmission system

Wavelet Transform Module Sum" method to detect the signal singularity and to construct the fault start and interference not start element.

(2) Direction Unit(DU): when the fault is from the forward direction, the ratio of the positive direction traveling wave to the reverse direction traveling wave is $\Delta uf / \Delta ub = kr$, where kr is the reflection coefficient and $0 < |kr| < 1$. For the reverse direction fault, $\Delta uf / \Delta ub \rightarrow \infty$. Based on this the principle and algorithm for identification of fault direction can be constructed.

(3) Boundary Unit: the protection boundary is composed of the line trap and the equivalent grounding capacitance of busbar. The magnitude of the high frequency signal for an internal fault is significantly higher than that of an external one (for an external fault, the high frequency signal has difficulty passing through the boundary). Utilizing the magnitude ratio of the high to low frequencies of the reverse voltage or current traveling waves, the internal and external fault can be distinguished.

(4) Lightning Strike Identification Unit(LU): the algorithm extracts and computes the spectral energies of the 0–5 kHz and 5–10 kHz signals. The ratio of the spectral energies in each of two frequency bands is used to distinguish between a normal fault and a lightning strike. A higher ratio indicates a fault, a lower ratio a lightning strike.

(5) Reclose-to-fault identification Unit: there are significant differences in the time intervals of the initial traveling wave and the reflected one, the signal polarity and the frequency domain characteristics of the current for a faulty line and a healthy line. These differences in such characteristics can be used to develop the principles and algorithms for this element.

(6) Phase selection Unit(PU): the phase selection unit can be categorized into two types:1) traveling wave phase selection method, which utilizes the magnitude and polarity of the traveling wave in the current waveform; 2) the transient energy based phase selection method, which measures the relative magnitudes of the transient energies of the three phase current and determines the faulted phase based on the coupling relationship between the faulted and the un-faulty phases.

Experimental device used for the transient based protection [3, 4, 13]

There are two devices. One of them is the prototype relay for transient based protection, which is able to simultaneously sample 8 channels of transient signal with 400 kHz sampling rate and 12 bit A/D. The other one is the analogue transient signal generator, which is able to rapidly convert the digital data obtained from on site fault recorders or EMTP simulations into analogue signals. Based on over 800 groups of simulation data obtained from EMTP, 4000 real-time tests were conducted. The entire protection algorithms include 34728 multiplier and 31785 additional operations. Tests show that this prototype relay is able to make correct decisions within 4 ms for all cases.

Discussions on relevant issues of transient based protection

(1) Instrument transformer [14]:

Tests were conducted for electromagnetic types of instrument transformers. The high frequency signals are transferred through the windings turn to turn transformer action and the turn to earth capacitances of the primary and secondary windings. Tests show that the voltage transformer is able to transfer high frequency signals under 700 kHz and the current transformer under 400 kHz. The cut-off frequency is related to the method of construction of the instrument transformer regardless whether the transformer has an iron core or not.

(2) The influence of the switching operation and the high frequency power line carrier signal [15].

Extensive simulation tests show that there is no maloperation for switching operations on both the protected and neighboring lines. The energy in the power line carrier signal is significantly lower than that in the generated fault signal in the high frequency band of interest. When both signals exist in the system, the sensitivity of the fault discrimination for the boundary element will be slightly lower, but a correct decision can still be made for the external fault.

Studies on the whole hvdc transmission line ultra high speed protection based on single end measurement [16–18]

The connection diagram of the HVDC system is shown in Fig. 2. Within the dashed line is the HVDC transmission line. The DC filtering group and the smoothing reactors are installed at the two ends of the lines, which provide a natural boundary for the high frequency transients. For the internal and external fault with respect to the location at the dashed line where the protection relays are installed, there are distinguishing differences in the high frequency voltage and current signals. The single end ultra high speed protection for DC line in this situation is more reliable than that for AC lines and is not influenced by the fault instance.

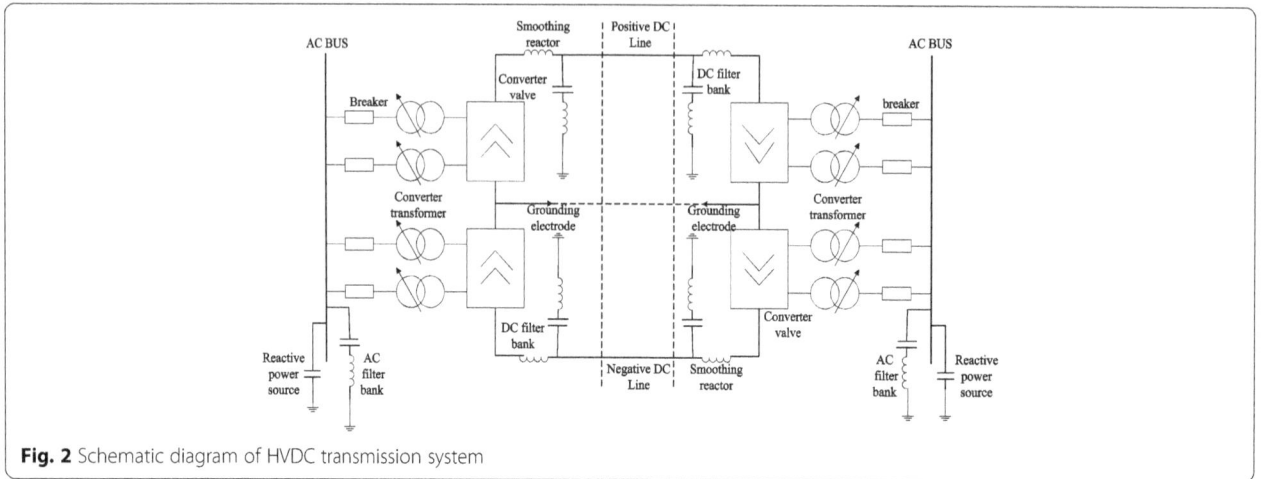

Fig. 2 Schematic diagram of HVDC transmission system

At present DC line protection is mainly provided by manufacturers e.g. ABB and Siemens. The discrimination criterion of the single end protection algorithm is based on the super-imposed voltage, the rate of change of the voltage traveling wave, and the gradient of the change in the current. On 27th August 2007, the voltage rate of change of the traveling wave protection did not operate for a high resistance fault (fault resistance 295Ω) on the GeNan HVDC transmission line. Similarly, during the operations of the TianGuang HVDC transmission line project in recent years, the voltage rate of change of the traveling wave protection has often failed to operate for high impedance faults. Since the rate of change protection is affected by fault location and fault resistance, its operation threshold is set to ensure no mal-operation for external fault. As a result of this it can not guarantee the reliable operation for internal fault.

The principle diagram of the single ended ultra high speed protection principle, utilizing the characteristics of the high frequency signal for a DC transmission line is shown in Fig. 3. From Fig. 3 it follows that:

(1) The principle and algorithm of the start element: First it detects the change of current and adopts the improved gradient algorithm. Using the sum of the current 3 samples and subtracting this sum from the last 3 samples, protection will start when the change of the current gradient exceeds a certain value. The calculation is simple but provides high sensitivity for high resistance fault.

(2) Principle and algorithm of boundary element: The characteristic of the boundary frequency for a DC transmission line is shown in Fig. 4. This figure shows that the magnitude of the high frequency signal for the detected internal faults is 50db higher than that for external faults. This difference can be used to distinguish between internal and external faults. Using the db3 wavelet transform to extract

the fault components Δu and Δi of the voltage and current, the high frequency forward voltage Δu_f and reverse voltage Δub of the traveling wave are computed. In this way the energies of the high frequency fault components can be used to establish the protection criteria to distinguish between internal and external faults.

(3) Fault polarity detection: The two pole HVDC transmission line is a double-circuit construction. Due to the coupling between the two lines, a fault on one pole will induce relatively high electromagnetic transients of voltage and current in the healthy line. Therefore, it is difficult to identify the faulty line based on the measurement of the magnitude of the high frequency signal at one end. When comparing the signals on the two lines with a low frequency range,

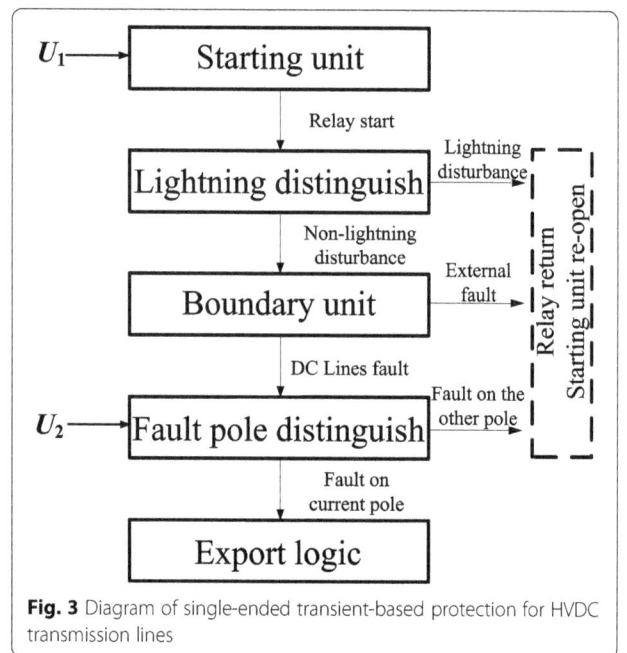

Fig. 3 Diagram of single-ended transient-based protection for HVDC transmission lines

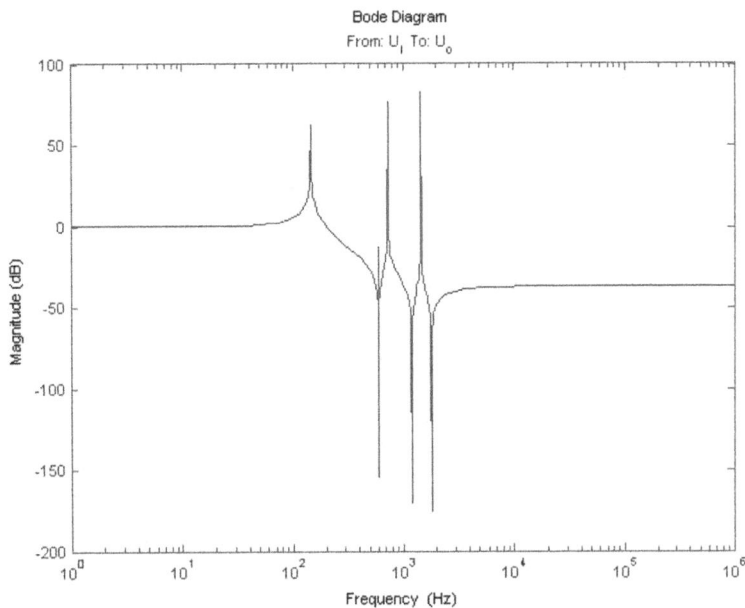

Fig. 4 Amplitude-frequency characteristic of the DC line boundary ($10^0 \sim 10^6$Hz)

the signal of the faulty phase is higher than that of the healthy ones. This criterion can be used to distinguish whether a fault is on its protected line.

Simulation tests of the prototype relay: A typical 1043 km long, ±500 kV HVDC transmission system is simulated using the EMTP software. The sampling frequency is 100 kHz and the db3 wavelet transform is used. The entire protection algorithm requires 5760 multiplier and 4800 summation calculations and can be computed within 300 μs using C32 series of DSP. By taking 1 ms as the data window, the protection can be easily computed within 2 ms. Simulation tests were conducted for over one thousand fault cases in the test system. The protection remained stable for all cases. For a lightning strike on the DC lines and 500Ω fault resistance, the protection is able to provide high sensitivity with an operation threshold 10 times higher than the setting threshold. However, the protection algorithm based on the rate change of voltage du/dt can not distinguish between internal and external faults when the fault resistance is over 300Ω.

Ultra high speed busbar protection [19, 20]
The PSCAD/EMTDC software was used to assess the performance of the transient based busbar protection. The equivalent busbar model, where all the busbar stray capacitances are represented by one capacitor, can not truly represent the transient process for busbar internal fault. For this situation the correct models, which take into consideration of the stray capacitance in various locations for the transformer, arrester, CVT and line trap

should be used. The typical 3/2 connection diagram of the simulation is shown in Fig. 5.

The sampling frequency is 400Hz/s and the 4 order B-spline derivative function is used as the foundation dyadic wavelet transform. The performances of the three algorithms were examined to compare the polarity of the traveling wave current, the differential traveling wave and the traveling wave energy for the comparison of the busbar protection. Two out of three modes should meet the criteria when 3 modes are used. Simulation results show that all 3 principles for busbar protection can distinguish between internal and external faults within 2 ms and the traveling wave differential protection possesses the highest sensitivity and reliability. The algorithm is also less influenced by any branch traveling wave current due to the transformer, which can result in difficulties when determining the polarity and direction of the traveling wave current. The reliability, therefore, of the algorithm requires further research. The main advantages of the algorithm are the fast speed, and the fact that it is not influenced by CT saturation and fault path resistance.

Faulted line selection device for non-effective earth systems [21–23]
For many years, stable state quantities have been used for line selection for non-effective earthed systems. However, the magnitudes of the stable state quantities are small, and can become even smaller when taking into consideration of the compensation effect of the Petersen coil, the fault path resistance and any CT errors. As a result, the success rate for this type of faulty line selection device is very low, especially when large fault

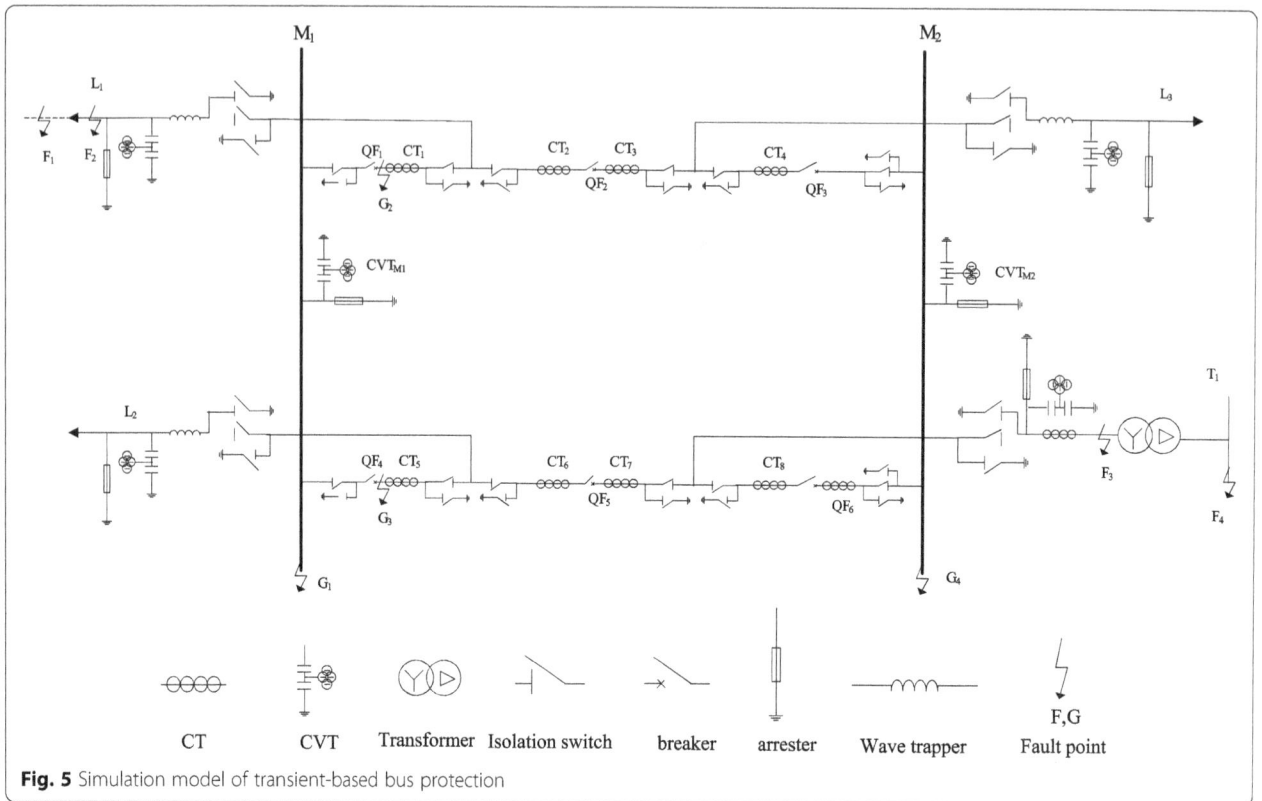

Fig. 5 Simulation model of transient-based bus protection

resistance and arcing fault are involved. Although there are many installations on-site, power engineers have generally lost faith in this type of device due to its low accuracy.

The magnitude of the transient capacitance current can reach up to 10 times or even higher of the stable state capacitance current without the effect of a Petersen coil. Figure 6 compares the direction and magnitude of the zero sequence transient currents for every feeder.

The current in each feeder follows into the busbar and the one with the highest magnitude is from the faulty feeder. However, the above mentioned characteristics are only true for a particular frequency band, and this band varies and is also influenced by network parameters and fault modes etc. Therefore, adaptively selecting the frequency band is the key for accurate and reliable faulty line selection.

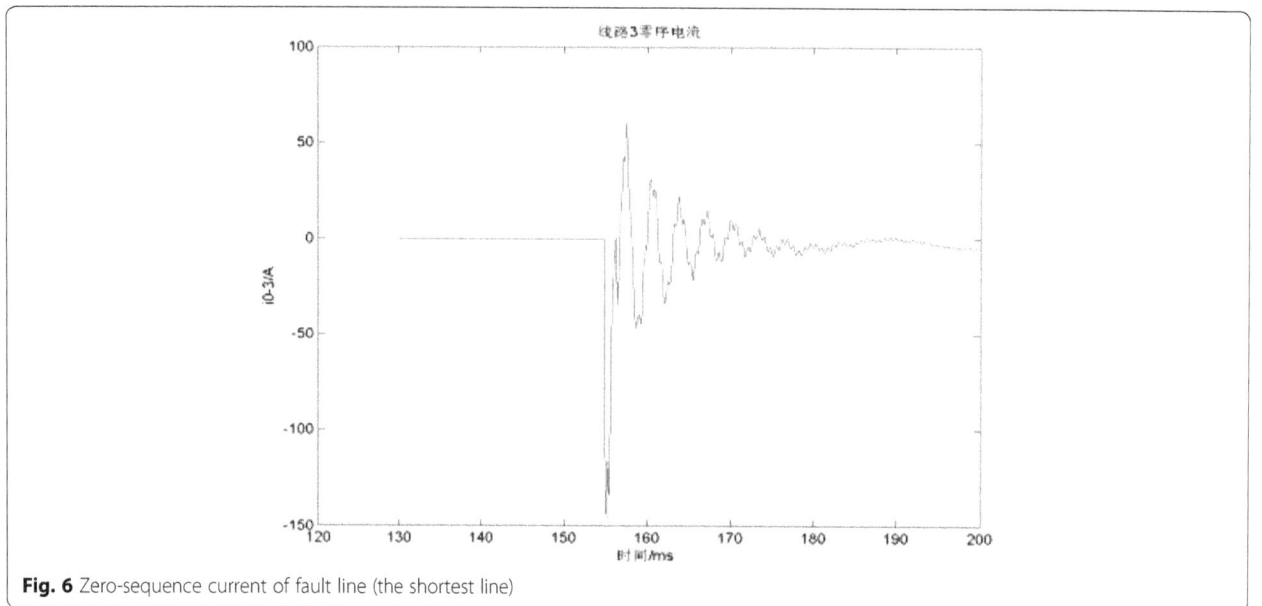

Fig. 6 Zero-sequence current of fault line (the shortest line)

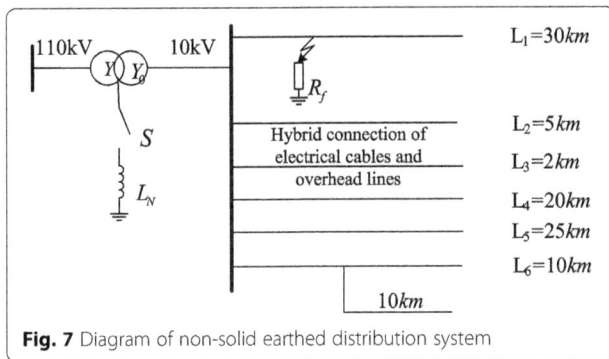

Fig. 7 Diagram of non-solid earthed distribution system

For the mixture overhead line and underground cable system shown in Fig. 7, RTDS simulation was conduced for over 10 thousand cases for unearthed, neutral connection and earth connection through a Petersen coil. This was done for under and over compensation and the accuracy was found to be over 97 %. There were no wrong line selection record for a number of devices installed on-site either.

Conclusion

The construction of the strong smart grid gives rise to demanding requirement for the protection relay in respect to both speed of operation and reliability. With the development of high speed DSP and the application of optical sensor, the utilization of fault transient information for fault identification is no longer a difficult task. The implementation of protection principles based on transient information has already become possible.

Due to the inconsistent nature of the wave impedance for various power apparatuses and the reflection and refraction characteristics of their interconnection points, the fault transients contain abundant information about the fault location and type. By correct analysis and the full utilization of such information it is possible to construct ultra high speed and highly sensitive AC, DC and busbar main protection. In addition faulty line selection for neutral non-effective earth systems using transient characteristics offers high sensitivity and reliability that is not influenced by the ways of neutral compensation.

Authors' contributions
BHZ investigated the framework of relative relay protection for smart grid, and drafted the manuscript. ZGH summarized the history and recent development of smart grid relay protection. ZQB participated in typesetting and revision of the manuscript. All authors read and approved the final manuscript.

Competing interests
The authors declare that they have no competing interests.

Author details
¹School of Electrical Engineering, Xi'an Jiaotong University, Xi'an 710049, China. ²XUJI Group Corporation, Xuchang 461000, China.

References

1. Ha, H.X. (2002). The study of boundary protection for EHV transmission lines. Thesis: Xi'an Jiaotong University.
2. Duan, J.D. (2005). Study of ultra-high-speed transient- based protection for extra-high-voltage transmission system. Thesis: Xi'an Jiaotong University.
3. Zhang, S. X. (2004). *Research on principle simulation platform of transmission line single-ended transient based protection*. Xi'an: Xi'an Jiaotong University.
4. Zhou Yi. (2006). Research on implemention technology of utra-high-speed transient-based protection for transmission line. Thesis: Xi'an Jiaotong University.
5. Duan, J. D., Zhang, B. H., & Zhou, Y. (2005). Study of ultra-high-speed transient-based directional relay. *Proceedings of the CSEE, 25*(4), 7–12.
6. Duan, J. D., Zhang, B. H., Li, P., et al. (2007). Single-ended transient-based protection for EHV transmission lines: principle and algorithm. *Proceedings of the CSEE, 27*(3), 45–51.
7. Duan, J. D., & Zhang, B. H. (2004). Study of starting algorithm using traveling-waves. *Proceedings of the CSEE, 24*(9), 30–36.
8. Duan, J. D., Ren, J. F., Zhang, B. H., et al. (2006). Study of transient approach of discriminating lightning disturbance in ultra-high-speed protection. *Proceedings of the CSEE, 26*(23), 7–13.
9. Duan, J. D., Zhang, B. H., Ren, J. F., et al. (2007). Single-ended transient-based protection for EHV transmission lines: basic theory. *Proceedings of the CSEE, 27*(1), 38–43.
10. Duan, J. D., Luo, S. B., Zhang, B. H., et al. (2007). Study of discriminating approach for switching into fault line in ultra-high-speed protection. *Proceedings of the CSEE, 27*(10), 78–84.
11. Duan, J. D., Zhang, B. H., & Zhou, Y. (2005). Study of fault-type identification using current traveling waves in extra-high-voltage transmission lines. *Proceedings of the CSEE, 25*(7), 58–63.
12. Duan, J. D., Zhang, B. H., Zhou, Y., et al. (2006). Transient-based faulty phase selection in EHV transmission lines. *Proceedings of the CSEE, 26*(3), 1–6.
13. Zhang, S. X., Zhang, B. H., & Duan, J. D. (2004). Research on development platform for transient-based protection. *Relay, 32*(15), 34–38.
14. Yin, X.C. (2006) Testing and simulation on secondary electromagnetic transformer for high frequency character. Thesis: Xi'an Jiaotong University.
15. Xue, J., Zhao, J. Q., Zhang, B. H., et al. (2007). Research on anti-interfering capability of ultra-high-speed transient- based Protection for transmission line. *High Voltage Apparatus, 43*(5), 374–377.
16. Cao, R.F. (2006). Study of Non-unit transient-based protection for HVDC transmission line. Thesis: Xi'an Jiaotong University.
17. Cao, R. F., You, M., Zhang, B. H., et al. (2009). *Ultra-high-speed transient-based protection for HVDC lines*. Beijing: 2009 International Conference on UHV Transmission.
18. You, M., Zhang, B. H., Cao, R. F., et al. (2009). *Study of non-unit transient-based protection for HVDC transmission lines*. Wuhan: Asia-Pacific Power and Energy Engineering Conference.
19. Chen, J. (2004). Simulation of protection principle of the busbar protection, by used of fault transient. Thesis: Xi'an Jiaotong University.
20. Duan, J. D., Zhang, B. H., & Chen, J. (2004). Study on the current traveling- wave differential bus protection. *Automation of Electric Power Systems, 28*(9), 43–48.
21. Zhao, H. M., Zhang, B. H., Duan, J. D., et al. (2006). A new scheme of faulty line selection with adaptively capturing the feature band for power distribution networks. *Proceedings of the CSEE, 26*(2), 41–46.
22. Zhang, B. H., Zhao, H. M., Zhang, W. H., et al. (2008). Faulty line selection by comparing the amplitudes of transient zero sequence current in the special frequency band for power distribution networks. *Power System Protection and Control, 36*(13), 5–11.
23. Xu, J. D., Zhang, B. H., You, M., et al. (2009). Fault line selection device for non-solid earthed network based on transient zero-sequence current features. *Electric Power Automation Equipment, 29*(4), 101–105.

5

Integrated protection based on multi-frequency domain information for UHV half-wavelength AC transmission line

Qingping Wang[*], Zhiqian Bo, Xiaowei Ma, Ming Zhang, Yingke Zhao, Yi Zhu and Lin Wang

Abstract

Half-wavelength AC Transmission (HWACT) can improve capability of AC transmission significantly. According to the basic principle of HWACT, the electromagnetic transient model of HWACT is built to analyze the fault transient process of transmission line. Based on fault characteristics of HWACT, the adaptability of traditional protections for transmission lines is analyzed briefly, such as current differential protection, distance protection and over current protection. In order to solve the problems of conventional protection caused by HWACT, a novel integrated protection based on multi-frequency domain information is proposed in this paper, which uses both the power frequency information and transient information. The integrated protection based on multi-frequency domain information takes advantages of power frequency and transient protections, which can not only improve the performance of traditional protection of AC transmission line but also realize fast fault judgment by transient travelling wave protection.

Keywords: Half-wavelength AC transmission, Relaying protection, Electromagnetic transient simulation, Travelling wave

Introduction

In order to support UHV long-distance transmission, the Half-wavelength AC Transmission (HWACT) is presented to improve capability of AC transmission, whose electric transmission distance is close to a half of the power frequency wavelength. It is a three-phase AC transmission technology used for extremely long distance, such as three thousand kilometers at 50Hz or two kilometers and five hundred kilometers at 60Hz. In a vast country, HWACT technology is very attractive because it's a way of long distance, large capacity transmission. In recent years, many countries have launched the positive research on it, such as Brazil and Korea. Under the development strategy of Global Energy Internet of China, long distance, large capacity transmission is an inevitable way, but the relay protection technology is one of problems troubling the engineering application of HWACT technology all the time [1].

Voltage and current distribution of half-wavelength line is very different from the characteristic of short line [2–6]. The voltage and current of two terminals don't follow the Kirchhoff's law based on the lumped parameters no longer, the measured impedance are no longer monotonic with distance [7–10]. The traditional protection principle cannot be used on half-wavelength line. In a word, the traditional relay protection principle cannot be directly applied on half-wavelength line. HWACT technology is proposed by the Soviet scholars in the 1940s [11], but research on relay protection of half-wavelength power system, is very less at home and abroad. A line current differential protection principle based on Bergeron model for half-wavelength AC transmission line is proposed in [7] and [10], based on the measured values and calculated values with Bergeron model on one side of the line, but it is impressionable by the line parameter. Paper [8] analyzes the applicability of current differential protection and distance protection, and proposes compensation algorithms of differential current calculation and interphase impedance calculation, but don't involve ground impedance calculation.

* Correspondence: yqwqpxixi@163.com
Corporate R&D Center, XJ group, State Grid Corporation of China, Beijing, China

Overall, the research on relay protection technology of half-wavelength transmission is in the initial stage. It need much deeper research on a large number of technical problems about protection principle and algorithm, the protection configuration and setting, and so on [12–16].

According to the basic principle of HWACT, the electromagnetic transient model of HWACT is built to support fault transient process analysis. Based on fault characteristics of HWACT, the adaptability of traditional protections for transmission lines is analyzed briefly, such as current differential protection, distance protection and over current protection. A novel integrated protection based on multi-frequency domain information is proposed in this paper, which uses both the power frequency information and transient information. The advantages and implementation of integrated protection for UHV half-wavelength AC transmission line are described in this paper at last.

Methods
Principle of half-wavelength AC transmission
Structure & principle of half-wavelength AC transmission
The transmission distance of half-wavelength AC transmission is close to a half of the power frequency wavelength, which is 3000 km at 50Hz. The half-wavelength power system researched is show in Fig. 1. The length of line MN is 3000 km, and the left side is connected to a power station group, the right side is connected to a big power grid. The direction of power flow is from M to N at normal operation.

When the line is equal to or longer than 250 km, we must consider the influence of the distributed parameters, to get more precise model of the line. The distributed parameter model of HVAC transmission line is shown in Fig. 2.

Set line series impedance of per length unit is equal to z, shunt admittance is equal to y, then z = r + jωL, y = g + jωC. The voltage and current of any point on the transmission line with x far from the end is as follows:

$$V(x) = \cosh\gamma x \ V_R + Z_c \sinh\gamma x I_R \quad (1)$$

$$I(x) = \frac{1}{Z_c} \sinh\gamma x \ V_R + \cosh\gamma x I_R \quad (2)$$

Z_c is the characteristic impedance, be equal to $\sqrt{\frac{z}{y}}$, γ is the propagation constant, be equal to $\alpha + j\beta = \sqrt{zy} = \sqrt{(r + j\omega L)(g + j\omega C)}$, and the real part α is the

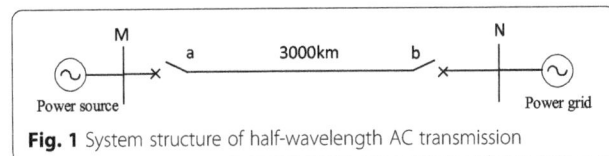

Fig. 1 System structure of half-wavelength AC transmission

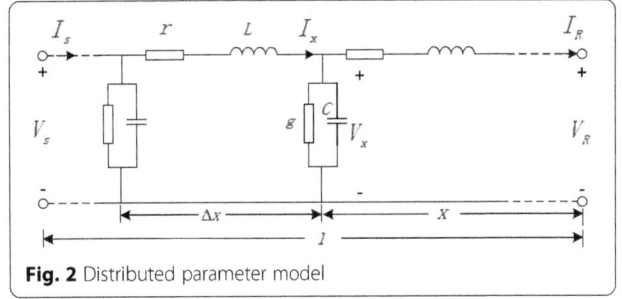

Fig. 2 Distributed parameter model

attenuation constant, the imaginary part β is the phase constant. Under lossless situation, the relation of voltage and current at the head and end is:

$$V_s = \cosh\beta l \ V_R + Z_c \sinh\beta l I_R \quad (3)$$

$$I_S = \frac{1}{Z_c} \sinh\beta l \ V_R + \cosh\beta l \ I_R \quad (4)$$

Here $Z_c = \sqrt{\frac{L}{C}}$, $\beta = \omega\sqrt{LC}$. The transmitting power is:

$$P_l = \frac{V_s V_r}{Z_c \sin(\beta l)} \sin\delta = \frac{P_n}{\sin(\beta l)} \sin\delta \quad (5)$$

V_s and V_r are the voltages of the head and end, P_n is natural power, δ is the angle V_s ahead of V_r. It thus appears that, when βl is equal to 180°, length of line is half-wavelength line, and P_l can reach infinity in ideal conditions. Because $\beta = \omega\sqrt{LC} = \omega/(3 \times 10^8)$ l is equal to 3000 km at 50Hz.

Characteristics of half-wavelength AC transmission
As a kind of super long distance transmission scheme, half-wavelength AC transmission technology has unique economic and technical advantages.

1. *Half-wavelength transmission line needn't reactive power compensator. Regardless of how power flow changes, the voltages of sending end and receiving end are always consistent.*
2. *Transmission capacity of half-wavelength transmission line is bigger, can reach infinity in ideal conditions.*
3. *Half-wavelength transmission needn't add switching station on the line, because it needn't reactive compensation and has high stability margin.*
4. *The economy of half-wavelength transmission is better. According to preliminary estimates by Brazil, 1000 kV HWACT's transmission cost of per unit length and per unit power is 29.8% of 500 kV EHV's [6].*

Although half-wavelength AC transmission technology has so many advantages, it still has some technical

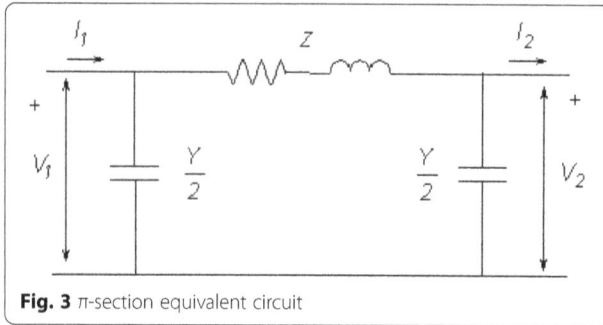

Fig. 3 π-section equivalent circuit

difficulties in the engineering application, described as follows:

1. *Overvoltage problem. It includes steady overvoltage and fault overvoltage.*
2. *Secondary arc current problem. When single-phase earth fault occurrd, secondary arc current of half-wavelength transmission line is bigger than the conventional line.*
3. *Security and stability problems. When half-wavelength line suffers from a large disturbance, it will appear power angle stability problem and dynamic stability problem.*
4. *Protection problem. The fault electrical characteristics of voltage and current are very different from the short line's, due to the long line and large distributed capacitance, it will have a large effect on tranditional protection principle and setting principle.*

Electromagnetic transient model of half-wavelength AC transmission

In order to analyze the fault transient process, the electromagnetic transient model of half-wavelength AC transmission line is built in this paper. The normal electromagnetic transient models of transmission line include π-sections model, Bergeron model and frequency dependent model.

π-section model

A π-section model will give the correct fundamental impedance, but cannot accurately represent other frequencies unless many sections are use, shown in Fig. 3. It is suitable for very short lines where the travelling wave models cannot be used, due to time step constraints. It is definitely unsuitable to the half-wavelength AC transmission line which is the extra long distance transmission system about 3000 km.

Bergeron model

The Bergeron model represents the system L and C in a distributed manner, shown in Fig. 4. In fact, it is roughly equivalent to using an infinite number of series-connected p-sections except that the total system resistance R is lumped. As with π-sections, the Bergeron model accurately represents system parameters at the fundamental frequency.

Because the Bergeron model is not frequency-dependent (calculates at a single frequency), it is suitable for studies where frequencies other than the fundamental are of little or no concern. The Bergeron model is accurate enough to research the protection based on fundamental frequency, but cannot be used to analyze transient travelling waves.

Frequency dependent model

The frequency dependent model is a type of distributed traveling wave model. The system resistance R is distributed across the system length instead of lumped at the end points. The Frequency Dependent Phase model is interfaced to the electric network by means of a Norton equivalent circuit, shown in Fig. 5.

The history current source injections I_{hisk} and I_{hism} are updated each time step, given the node voltages V_k and V_m. The steps by which this is accomplished by the Frequency Dependent Phase Model time-domain interface routine is as given in below:

$$I_k(n) = G \cdot V_k(n) - I_{hisk}(n) \tag{6}$$

$$I_{kr}(n) = I_k(n) - I_{ki}(n) \tag{7}$$

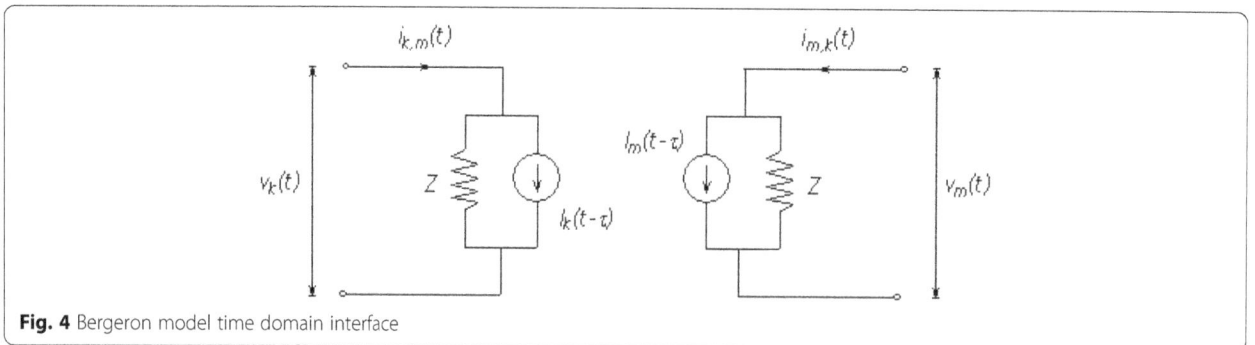

Fig. 4 Bergeron model time domain interface

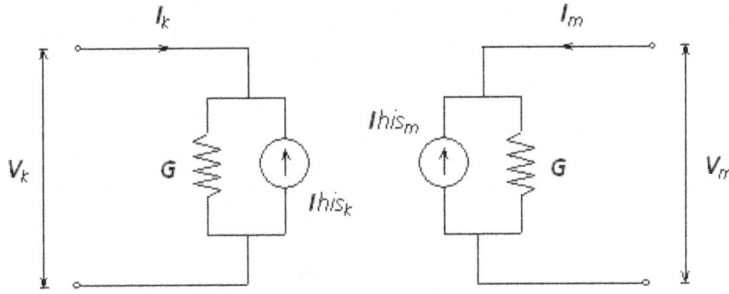

Fig. 5 Frequency dependent model time domain interface

$$I_{ki}(n+1) = H * I_{mr}(n-\tau) \tag{8}$$

$$I_{hisk}(n+1) = Y_c' * V_k(n) - 2 \cdot I_{ki}(n+1) \tag{9}$$

The frequency dependent models are solved at a number of frequency points, which considers the frequency dependence of internal transformation matrices. Unlike the Bergeron model, these models also represent the total system resistance R as a distributed parameter (along with a distributed system L and C), providing a much more accurate representation of attenuation. It is the ideal electromagnetic transient model for half-wavelength AC transmission lines, which can be used to research transient protection based on travelling wave.

Fault transient analysis for half-wavelength AC transmission

In order to analyze the adaptability of traditional protection for the half-wavelength transmission line, the fault characteristics at different fault points are researched in this paper. The three-phase short circuit characteristics are shown as an example. It can be seen from the figures that, when faults happen along the line, the short circuit characteristics of the half-wave length AC transmission line is totally different from normal transmission line.

Voltage & current characteristics of short circuit
When the fault happens on arbitrary position along the Half-Wave transmission line, the amplitude of voltages and current at different fault point is shown in Figs. 6 and 7.

As shown in Fig. 6, the voltages increase when the distance between the beginning and the fault position increases until the distance reaches 2900 km. Then the measured voltage decreases quickly when the fault happens in the last 100 km.

The current decreases firstly with distance increasing shown in Fig. 7, until the distance reaches 1500 km. Then the current increases until the distance reaches 2900 km, and then decreases as quickly as the voltage.

By analyzing the short circuit characteristics, the short circuit current is similar to load current when fault occurs at the middle of the line. It might be the dead zone of the current protection which installed on the beginning of the line.

Impedance characteristics of short circuit
The measuring impedance is quite different to normal AC transmission line, shown in Fig. 8. If the fault occurs within 1500 km from the beginning, the impedance amplitude increases until the middle of transmission line. Once the distance is more than 1500 km, the impedance amplitude will decrease quickly.

The locus in the complex plane of measured impedance is shown in Fig. 9. With the increase of the distance between the beginning and the fault position, the impedance will switch from the first quadrant to the fourth quadrant at 1500 km and its amplitude reaches the maximum.

When the fault occurs at the middle of half-wavelength, the distance protection will mis-operate. And the distance protection might mal-operate when the external fault occurs at adjacent lines or buses.

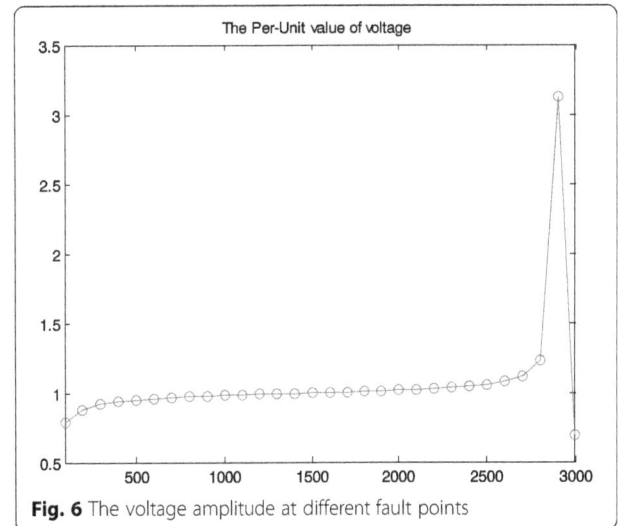

Fig. 6 The voltage amplitude at different fault points

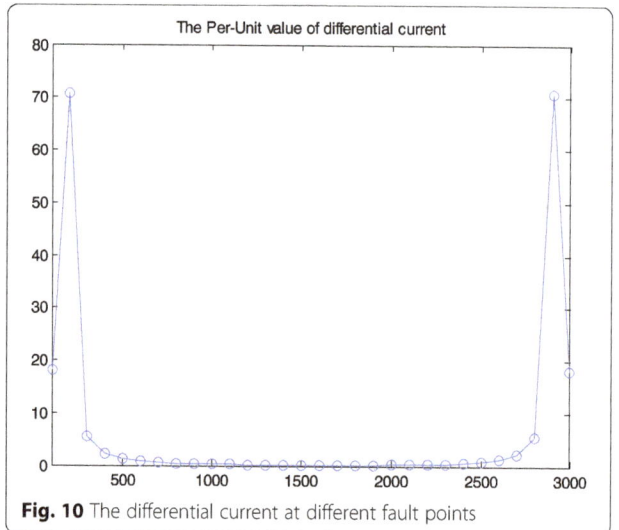

Fig. 7 The curent amplitude at different fault points

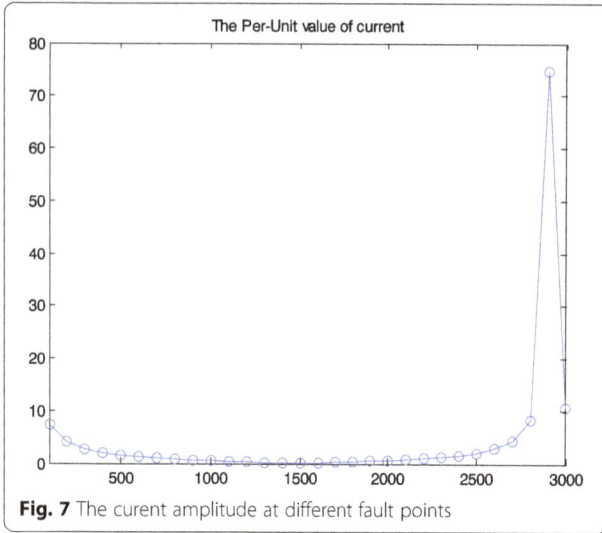

Fig. 9 The impedance locus at different fault points

Differential current characteristics of short circuit

The differential current at different fault points is also different to normal AC transmission, shown in Fig. 10. When fault occurs at both ends of the line, the differential current is very high. But the fault occurs in the range of 500 km to 2500 km, the amplitude of the differential current is closed to zer0. Therefore, the range of 500 km to 2500 km might be the dead zone of the conventional current differential protection.

Results

Principle of integrated protection based on multi-frequency domain information

Adaptability analysis of conventional protection

According to characteristic of half-wavelength AC transmission, the conventional protection principle line is not suitable to the half-wavelength AC transmission line anymore. The reasons are shown as follow,

1) *Mis-operation caused by fault at the middle of line*
    According to the analysis of the short circuit characteristics above, the conventional protection principles, including current differential protection, distance protection, over current protection and under voltage protection, have a large range of dead zone at the middle of the half-wavelength transmission line. It might cause mis-operation of the conventional protection when the fault point is located from 500 km to 2500 km.

2) *Mal-operation caused by adjacent external fault*
    On the other hand, the short circuit characteristics of both the end of line and the beginning of adjacent line are similar to the fault characteristics at the initial point of the half-wavelength transmission line. It is difficult to distinguish internal fault and external fault only based on existing protection principle. The conventional protection might mal-

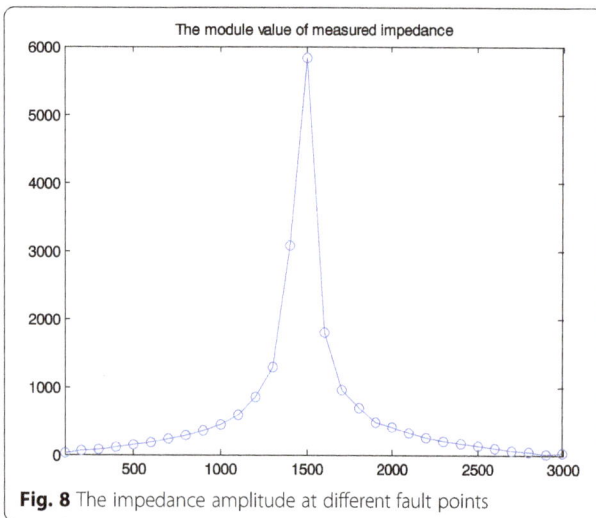

Fig. 8 The impedance amplitude at different fault points

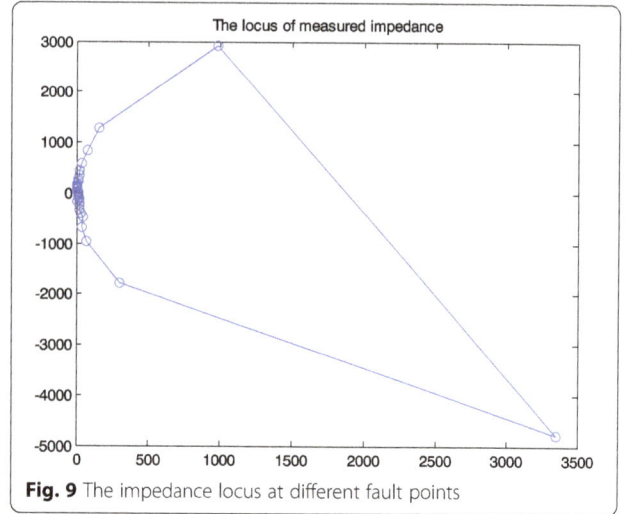

Fig. 10 The differential current at different fault points

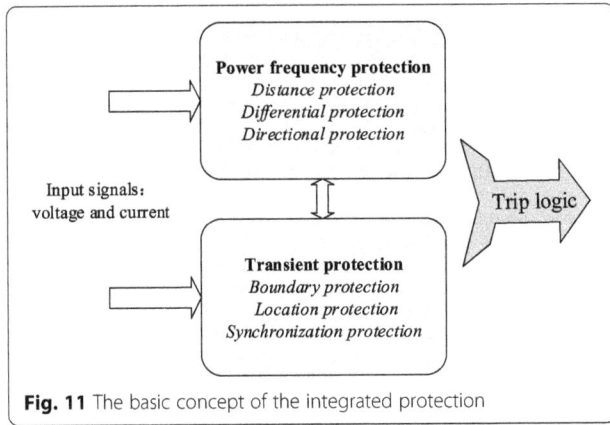

Fig. 11 The basic concept of the integrated protection

operate in condition of external fault at the beginning of adjacent line.

3) *Influence of fault transient process*
In addition, it can be found during the electromagnetic transient simulated calculation that the duration of transient process will be up to 60 ms or even 100 ms because of distributed capacitor of extra long-distance transmission line, after the fault occurs. Therefore, in order to avoid mal-operation caused by transient process, the protection must be delayed accordingly, which cannot meet requirement of fast fault isolation for UHV transmission.

4) *Time delay caused by long-distance communication*
The protections based on two terminals information, such as current differential protection and pilot protection, the time delay of the communication will be at least 10 ms, because the length of the half-wavelength transmission line is about 3000 km. Therefore, the fast non-communication protection is necessary to be researched to improve the operation speed of line protection for the half-wavelength transmission.

The integrated protection principle based on both power frequency and transient information

On the adaptability analysis of existing protection, we have the following conclusion: the conventional protection based on the power frequency information is not completely adaptive to the half-wavelength AC transmission line. In order to improve the performance of the conventional protection to meet new requirements for half-wavelength transmission line, the integrated protection based on both power frequency and transient information is presented in this paper.

According to the electromagnetic theory, the energy transmits in the form of travelling wave. In process of the fault transient, there will be traveling wave of voltage and current on the transmission line. The transient information of the traveling wave can be used to implement the high-speed transient protection. The transient protection is not influenced by the special fault characteristics of the half-wavelength AC transmission line, which cause mal-operation or mis-operation of the conventional protection. It is not only faster to trip then the conventional protection, but also more sensitive without being affected by distributed capacitor. However, the power frequency protection principles are still valid when the faults occurred in the dead zone of the transient protection. Therefore, the integrated protection principle which combined the power frequency protection principles and the transient protection principles is proposed for the protection of the half-wave length AC transmission line, shown in Fig. 11.

The integrated protection implements calculation with the input signals, including voltages and currents of two terminals. The configurable principles based on power frequency include distance protection, differential protection and directional protection. The configurable principles based on transient information include boundary protection, location protection, signal synchronization protection. The configuration should follow the following principles.

1. *Multiple protection principles coordinate according to protected area and different fault types.*
2. *The operation speed can be improved by applying the transient protection principle based on the local information.*
3. *The design of trip logic should be designed based on the coordination of multiple protection principles.*

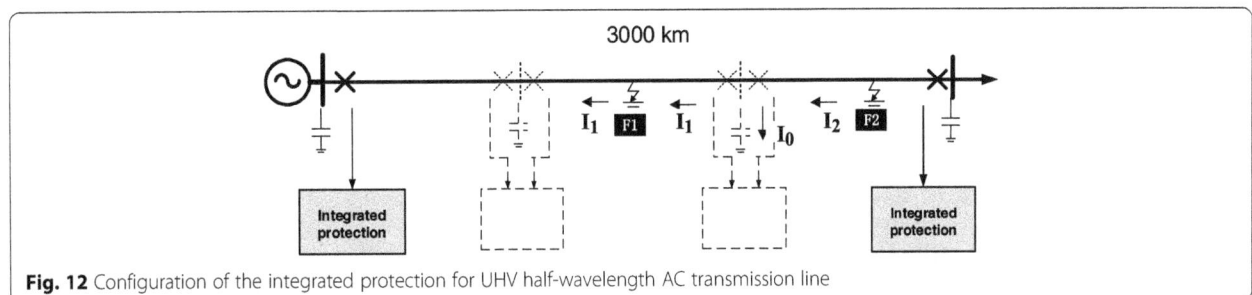

Fig. 12 Configuration of the integrated protection for UHV half-wavelength AC transmission line

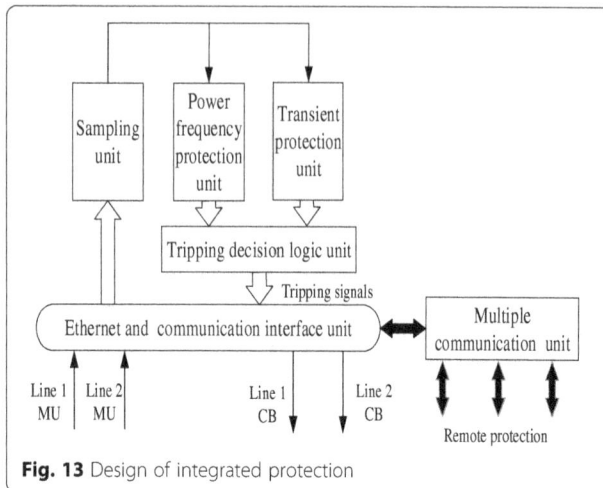
Fig. 13 Design of integrated protection

4. *The transient protection is designed to cover the faults at the middle and far-end of the line, including the high impedance fault.*
5. *The influence of the surge arrester to the transient protection should be taken in to account.*

Implementation of integrated protection for UHV half-wavelength AC transmission line

The integrated protections based on multi-frequency domain information are installed on both ends of line, shown in Fig. 12.

The design of integrated protection is shown in Fig. 13. The sampling unit of power frequency and transient protection adopts the integrated design. The input signals of sampling unit are from the process bay equipment, communicated by Ethernet and communication interface unit.

Power frequency protection unit and transient protection unit work respectively, and transfer their results into the tripping decision logic unit. The tripping decision logic unit coordinates the outputs of two protections and decides the final tripping signals, and trips the corresponding circuit breaker. In addition, the integrated protection can communicate with other remote protection by multiple communication unit.

Conclusion

A new integrated protection principle based on multi-frequency domain information is proposed to solve the protection problems of UHV half-wavelength AC transmission line. Fault transient characteristics of half-wavelength AC transmission is quite different from the characteristics of the normal transmission line. The integrated protection proposed in this paper can improve the protection performances of the half-wavelength AC transmission line. The essential benefits are shown as follow,

1) *To protect the full line of the half-wavelength AC transmission line for different fault types;*
2) *To improve reliability and sensitivity;*
3) *To implement high speed protection for full line;*
4) *To avoid the time delay caused by communication and transient process.*

Authors' contributions
QW carried out the system design, participated in the simulation analysis and drafted the manuscript. ZB participated in the design of the study and coordination. XM and MZ carried out the simulation and helped to draft the manuscript. YZ, YZ and LW participated in its design and coordination and helped to perform the statistical analysis. All authors read and approved the final manuscript.

Competing interests
The authors declare that they have no competing interests.

References
1. Zheng, J. C. (2009). Smart power devices and half-wave-length AC transmission technology[J]. *Power and Electrical Engineers, 3,* 12–15.
2. Qin, X. H., Zhang, Z. Q., Xu, Z. X., Zhang, D. X., & Zheng, J. C. (2011). Study on the steady state characteristic and transient stability of UHV AC half-wave-length transmission system based on quasi-steady model[J]. *Proceedings of the CSEE, 31*(31), 66–75.
3. Portela, C., & Tavares, M. C. (2002). *"Modeling simulation and optimization of transmission lines[C]." Applicability and limitations of some used procedures* IEEE (pp. 1–38). Recife: IEEE.
4. Dias, R., Santos, G., & Aredes, M. (2005). *"Analysis of a series tap for half-wavelength transmission lines using active filters[C]." IEEE 36th Power Electronics Specialists Conference* (pp. 1894–1900). Recife: IEEE.
5. Sokolov, N. I., & Sokolova, R. N. (1999). The feasibility of using half-wave power transmission lines at higher frequencies[J]. *Electrical Technology Russia, 1,* 66–84.
6. Wang, G., Li, Q. M., & Zhang, L. (2010). *"Research status and prospects of the half-wavelength transmission lines[C]." Power and Energy Engineering Conference* (pp. 1–5). Chengdu: PEEC.
7. Xiao, S. W., Cheng, Y. J., & Wang, Y. (2011). A Bergeron model based current differential protection principle for UHV half-wavelength AC transmission line[J]. *Power System Technology, 35*(9), 46–50.
8. Liu, J. H. (2013). Study on protection Principle for Half Wavelength AC Transmission Line Based on Distributed Parameters[D]. Beijing: North China Electric Power University.
9. Wang, Y. (2011). Research on longitudinal differential protection principle for half wavelength AC transmission line[D]. Beijing: North China Electric Power University.
10. Cheng, Y. J. (2012). Analysis of the fault and the relay protection for half wavelength AC transmission line[D]. Beijing: North China Electric Power University.
11. WOLF, A. A., & SHCHERBACHEV, O. V. (1940). On normal working conditions of compensated lines with half-wave characteristics (in Russian) [J]. *Elektrichestvo, 1,* 147–158.
12. Fang, Y. (2013). Research on half wave-length AC transmission line protection principle[J]. *Science & Technology Information, 5,* 390–392.
13. Li, B., He, J. L., Yang, et al. (2007). Improvement of distance protection algorithm of UHV long transmission line[J]. *Automation of Electric Power Systems, 31*(1), 43–46.
14. Li, B., He, J. L., Chang, W. H., et al. (2010). Bergeron model based distance protection for long transmission lines[J]. *Automation of Electric Power Systems, 34*(23), 52–55.
15. Duan, J. D., Zhang, B. H., Li, P., et al. (2007). Principle and algorithm of non-unit transient-based protection for EHV transmission lines[J]. *Proceedings of the CSEE, 27*(7), 45–51.
16. Zhang, W. J., He, B. T., & Shen, B. (2007). Traveling-wave differential protection on UHV transmission line with shunt reactor[J]. *Proceedings of the CSEE, 27*(10), 56–61.

6

Control principles of micro-source inverters used in microgrid

Wenming Guo[*] and Longhua Mu

Abstract

Since micro-sources are mostly interfaced to microgrid by power inverters, this paper gives an insight of the control methods of the micro-source inverters by reviewing some recent documents. Firstly, the basic principles of different inverter control methods are illustrated by analyzing the electrical circuits and control loops. Then, the main problems and some typical improved schemes of the ωU-droop grid-supporting inverter are presented. In results and discussion part, the comparison of different kinds of inverters is presented and some notable research points is discussed. It is concluded that the most promising control method should be the ωU-droop control, and it is meaningful to study the performance improvement methods under realistic operation conditions in the future work.

Keywords: Mirogrid, Micro-source inverter, Droop control, Control principle

Introduction

Recently, with the increased concern on environment and intensified global energy crisis, the traditional centralized power supply has shown many disadvantages.Meanwhile, the high-efficiency, less-polluting distributed generation (DG) has received increasing attentions [1, 2]. Microgrids [3–5], which comprise micro-sources, energy storage devices, loads, and control and protection system, are the most effective carrier of DGs. When a microgrid is connects to the utility grid, it behaves like a controlled load or generator, which removes the power quality and safety problems caused by DGs' direct connection. Microgrids can also operate in islanded mode, thus increase system reliability and availability of the power supply.

Proper control is a precondition for microgrids' stable and efficient operation. The detailed control requirements come from different aspects, such as voltage and frequency regulation, power flow optimization etc. Since these requirements are of different importance and time scale, a three-level microgrid control structure is proposed in [6]. As the foundation of microgrid control system, the primary control is aimed at maintaining the basic operation of the microgrid without communication, which has become a hot research topic recently. Since most microsources utilize inverters to convert electrical energy, the primary control is essentially the management of power

inverters. Micro-source inverters are required to work in a coordinated manner based only on local measurements and the control strategies decide the roles of each microsource. According to the principle of master–slave control, the micro-source inverters can be divided into grid-feeding, grid-forming, and PQ-droop grid-supporting inverters. From the perspective of peer control, the ωU-droop grid-supporting invertershelp to realize microgrids' plug and play function. Although being widely discussed in the technical literatures, it still lacks a sufficient practical control method andexisting control technologies need to be further studied and improved. This paper describes the control principles of several typical microsource inverters and compares their characteristics so as to provide a fundamental understanding of microgrids' primary control.

Method

Grid-feeding inverter

The control objective of grid-feeding (GFD) [11] inverter is to track the specified power references. Figure 1 illustrates the control block diagram of the most common current controlled GFD inverter. For dispatchable microsources, such as micro-turbine and fuel-cells, the inverter power references can be set directly according to practical requirements. For non-dispatchable micro-sources, such as photovoltaic cells, the active power reference is usually decided by the voltage controller of the inverter's DC bus.

* Correspondence: gwmsch@163.com
Department of Electrical Engineering, Tongji University, Shanghai, China

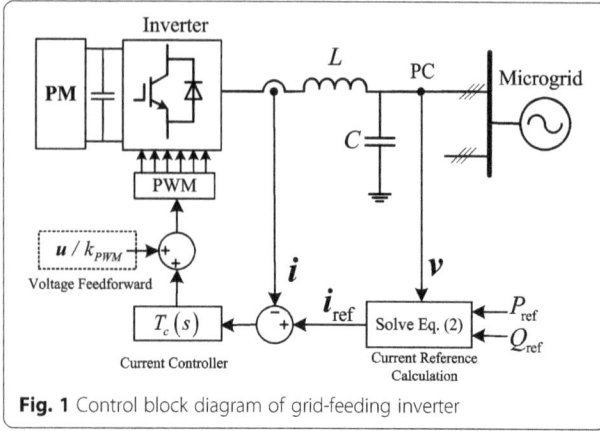

Fig. 1 Control block diagram of grid-feeding inverter

In addition, this type of sources can also export reactive power without affecting maximum power point tracking.

The GFD inverter's power referencetrackingis realized by adjusting the output currents. The control system calculates the output current references based on the relationships among the inverter's output power, output current and the voltage at the point of connection (PC). The three-phase voltages at the PC are represented by vector v, and the inverter's output currents are represented by vector i as

$$v = [v_a \ v_b \ v_c]^T$$

$$i = [i_a \ i_b \ i_c]^T$$

Neglecting the power consumed on the filter inductor, the output power of GFD inverter is calculated according to instantaneous power theory [7] as:

$$\begin{cases} P = v \cdot i \\ Q = -|v \times i| \end{cases} \tag{1}$$

If the current controller in Fig. 1 is properly designed, the output currents of the GFDinverter will follow their references. Thus the current reference vector, i_{ref}, can be obtained by solving the following equation:

$$\begin{cases} P_{ref} = v \cdot i_{ref} \\ Q_{ref} = -|v \times i_{ref}| \end{cases} \tag{2}$$

The output currents of the GFD inverter are the same as the currents flownthrough the filter inductor. In natural reference frame, there exists the following relation:

$$sLi = u_{inv} - u \tag{3}$$

The voltages at the PC, u, are measured using voltage transducers, the output voltages of the inverter, u_{inv}, can then be adjusted based on u (see the voltage feedforward in Fig. 1) to control the voltage drop on the filter inductor. This implies that the filter inductor's currents can be controlled indirectly. If the potential at the middle point of the inverter's DC bus equal zero and ignore

the delay of PWM process, the inverter is equivalent to a proportional element with gain k_{PWM}. Thus, the closed-loop transfer function of the inverter's current control can be written as:

$$G_c(s) = \frac{k_{pwm}T_c(s)}{sL + k_{pwm}T_c(s)} \tag{4}$$

$T_c(s)$ needs be designed in a way to ensure $G_c(s)$ have sufficient bandwidth. Meanwhile, the gain and phase shift of $G_c(s)$ around fundamental frequency should be close to 0 dB and 0 degree respectively. Therefore, the output current of the GFD inverter can track their references quickly and accurately.

For three-phase balanced operation cases, the control system of the GFD inverter is usually designed in dq reference frame, where the voltages and currents are DC signals. In this case, using PI controller can realize the output current tracking without steady-state error. In dq reference frame, the Park transformation will result in coupling between the d and q axis inductor currentcomponents, as shown in Eq. (5). Therefore, the control system must comprise dq decoupling modules. The detailed control block diagram in dq reference frame is illustrated in Fig. 2.

$$sLI_{dq} = U_{inv,dq} - U_{dq} - \begin{bmatrix} 0 & -\omega L \\ \omega L & 0 \end{bmatrix} I_{dq}$$

$$\tag{5}$$

For unbalanced operation cases, the GFD inverters need simultaneously controlthe positive and negative sequence currents [8, 9]. Under such condition, using PR controller [10] in $\alpha\beta$ reference frame might be a better choice as a single PR controller can regulate both the positive and negative sequence currents, and the control effect is similar to that of using two PI controllers in double positive/negative dq reference frames.

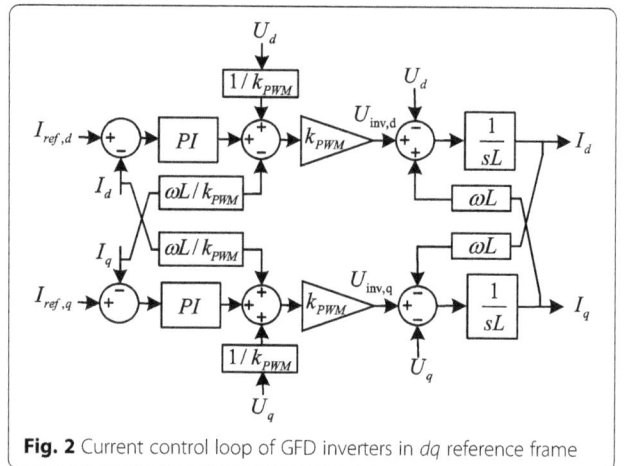

Fig. 2 Current control loop of GFD inverters in dq reference frame

Grid-forming inverter

The control objective of the grid-forming (GFM) [11] inverters is to maintain stable voltage and frequency in a microgrid. GFM inverters are characterized by their low output impedance, and therefore they need a highly accurate synchronization system to operate in parallel with other GFM inverters [11]. GFM inverters usually equips with energy storage on their DC sides, therefore they can respond to the change of load in a short time. The control block diagram of a GFM inverter is shown in Fig. 3, including an inner inductor current loop, which is identical to that of the GFD inverter, and an outer capacitor voltage loop. GFM inverters achieve their control objective by regulating the filter capacitor's voltage, u. In natural reference frame, there exists the following relation:

$$sCu = i - i_o \qquad (6)$$

where i and i_o are the inductor and grid currents, respectively.

According to the above analysis, the GFM inverters can also precisely control their inductor current by a properly designed inner current loop. The impact of the grid current on capacitor voltage is removed by current feedforward and thus, u is fully controlled by adjusting i.

As shown in Eq. (7), in the dq reference frame, the d and q axis component of the filter capacitor voltage are also coupled. Similarly, it is necessary to introduce dq decoupling modules in the voltage control loop as illustrated in Fig. 4.

$$sCU_{dq} = I_{dq} - I_{o,dq} - \begin{bmatrix} 0 & -\omega C \\ \omega C & 0 \end{bmatrix} U_{dq}$$

$$(7)$$

Grid-supporting inverter

There exists an approximate linear droop relation between the P-ω and Q-U of traditional synchronous generators. By emulating this output characteristics, grid-supporting (GS) [11] inverters, aimed at sharing load proportional to their

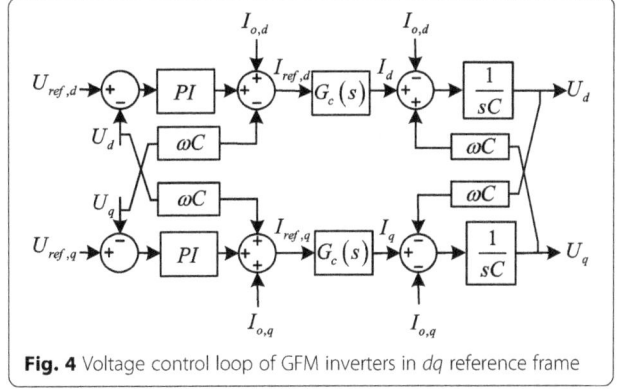

Fig. 4 Voltage control loop of GFM inverters in dq reference frame

power capacities, can deploy two different droop control structures, namely "PQ-droop" and "ωU-droop". The PQ-droop GS inverter adjusts its output power as a function of the variation of the microgrid's voltage and frequency. In this case, the inverter behaves like a power source and its control system is designed based on that of the GFD inverter, as shown in Fig. 5(a). On the contrary, the voltage and frequency at the PC of the ωU-droop GS inverter are adjusted according to the variations of its output power. The ωU-droop GS inverter behaves as a controlled voltage source and its control system is based on that of the GFM inverter, as shown in Fig. 5(b).

In Fig. 5, ω_0 and U_0 represent the no-load frequency and no-load voltage, k_P and k_Q represent the active and reactive power droop coefficients, respectively. In steady state, the frequency of the microgrid is a global quantity, and the voltages at different points of the microgrid are almost identical. If "ω_0" and "U_0" of each inverter are identical, then both the PQ-droop GS inverter and the ωU-droop GS inverter can share load variations as follows:

$$k_{p1} \Delta P_1 = k_{p2} \Delta P_2 = ... = k_{pn} \Delta P_n \qquad (8)$$

$$k_{Q1} \Delta Q_1 = k_{Q2} \Delta Q_2 = ... = k_{Qn} \Delta Q_n \qquad (9)$$

where k_{Pi} and ΔP_i (i = 1,2,...,n) represent the active power droop coefficient and output active power variation of the ith GS inverter, respectively. k_{Qi} and ΔQ_i represent the reactive power droop coefficient and output reactive power variation of the ith GS inverter, respectively. Although both types of GS inverters shown in Fig. 5 have a good load-sharing performance, the PQ-droop GS inverter cannot operate by itself. In contrast, the ωU-droop GS inverter is controlled as a voltage source, and thus can work independently regardless of the microgrid operation mode. The ωU-droop GS inverter has acquired extensive attentions for its excellent features though some problems still exist, including:

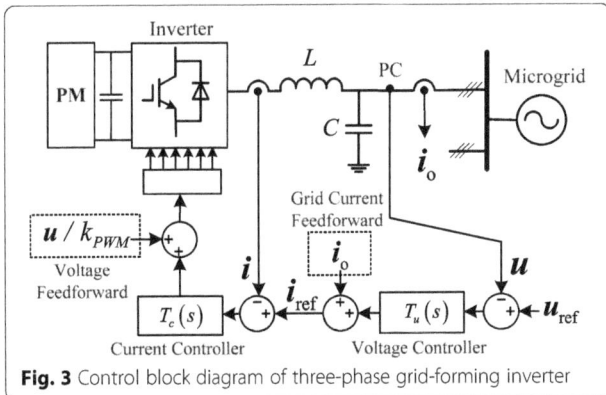

Fig. 3 Control block diagram of three-phase grid-forming inverter

(a) Control block diagram of PQ-droop grid-supporting inverter

(b) Control block diagram of ωU-droop grid-supporting inverter

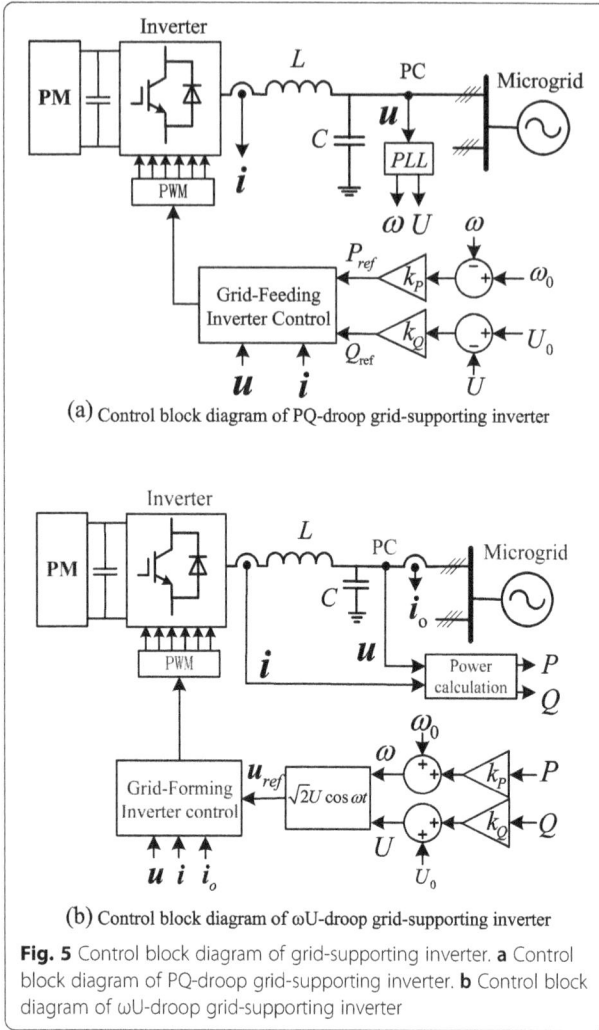

Fig. 5 Control block diagram of grid-supporting inverter. **a** Control block diagram of PQ-droop grid-supporting inverter. **b** Control block diagram of ωU-droop grid-supporting inverter

- the line impedance of a low-voltage microgrid has a large resistive component, thus P-ω and Q-U droop control is no longer suitable.
- the voltages at the PCs of each inverter are not completely equal, thus the GS inverters cannot share reactive power precisely.

Many researchers have proposed various improved methods to deal with the above problems and some typical schemes will be presented in the following sections.

A. Decoupling transformation method
   As depicted in Fig. 6, the voltage at the PC of theωU-droop GS inverter is denoted by U∠δ, and the voltage at the microgrid bus is denoted by E∠0. Z_L is the line impedance between the inverter's filter capacitor and the microgrid bus with an impedance angle of θ.

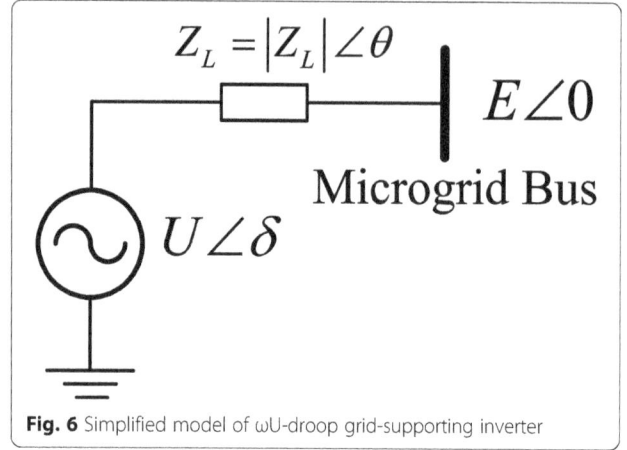

Fig. 6 Simplified model of ωU-droop grid-supporting inverter

Due to the small power angle δ, it is assumed that:

$$\sin\delta = \delta, \cos\delta = 1 \qquad (10)$$

Thus, the output power of the GS inverter can be expressed as:

$$\begin{cases} P = \dfrac{EU\cos\theta - E^2\cos\theta + EU\delta\sin\theta}{Z_L} \\ Q = \dfrac{EU\sin\theta - E^2\sin\theta - EU\delta\cos\theta}{Z_L} \end{cases} \qquad (11)$$

If both the resistive and reactive components of the line impedance cannot be ignored, the output active and reactive power of the inverter will be dependent on both δ and U. In this case, the P-ω(δ) and Q-U decoupling relation will no longer valid. To solve this problem, the virtual power P', Q' and the transformer matrix T_{PQ} are introduced in [12, 13]:

$$\begin{bmatrix} P' \\ Q' \end{bmatrix} = T_{PQ}\begin{bmatrix} P \\ Q \end{bmatrix} = \begin{bmatrix} \sin\theta & -\cos\theta \\ \cos\theta & \sin\theta \end{bmatrix}\begin{bmatrix} P \\ Q \end{bmatrix} \qquad (12)$$

According to Eq. (11) and (12), it can be derived that:

$$\begin{cases} P' = \dfrac{EU}{Z}\delta \\ Q' \dfrac{EU - E^2}{Z} \end{cases} \qquad (13)$$

The ωU-droop control based on the virtual power is given as:

$$\begin{cases} \omega = \omega_0 - k_P P' \\ U = U_0 - k_Q Q' \end{cases} \qquad (14)$$

Similarly, transforming ω(δ) and U with the matrix $T_{\omega U}$ [14] gives the virtual frequency (phase angle) and voltage as:

$$\begin{bmatrix} \omega'(\delta') \\ U' \end{bmatrix} = T_{\omega E}\begin{bmatrix} \omega(\delta) \\ U \end{bmatrix} = \begin{bmatrix} \sin\theta & \cos\theta \\ -\cos\theta & \sin\theta \end{bmatrix}\begin{bmatrix} \omega(\delta) \\ U \end{bmatrix} \qquad (15)$$

According to Eq. (11) and (15), it can be derived that:

$$\begin{cases} P = \dfrac{EU\delta' + (U - U^2 - E^2)E \quad \cos\theta}{Z} \\ Q = \dfrac{EUU' + (U - U^2 - E^2)E \quad \sin\theta}{Z} \end{cases} \quad (16)$$

Since E and U are constant when the droop control process reaches steady-state, the output active and reactive power of the GS inverter will be regulated by the virtual frequency and voltage respectively. Thus, a novel ωU-droop can be established:

$$\begin{cases} \omega' = \omega_0' - k_P P \\ U' = U_0' - k_Q Q \end{cases} \quad (17)$$

In Eq. (17), ω_0' and U_0' is the corresponding virtual no-load frequency and voltage. The droop control block diagram of the GS inverters applying two types of decoupling transformation method is shown in Fig. 7.

It is worth noting that the first decoupling method is designed to share the virtual power rather than the real power. So there exists a complicated relation between the variations of each inverter's output power and their droop coefficients when the load in the microgrid changed. The second decoupling method avoids this problem considering that all inverters have the same ω' and U', i.e. the R/X of each line in the microgrid must be identical. In addition, the variables directly controlled by Eq. (17) are ω' and U', and thus, it is necessary to carefully select the droop coefficients [14] to ensure that the real frequency and voltage are located in reasonable ranges.

B. Virtual impedance method

The coupling between the output active and reactive power of the conventional ωU-droop control can be mitigated by introducing virtual impedance [15], as illustrated in Fig. 8. The voltage at the inverter's PC is expressed as:

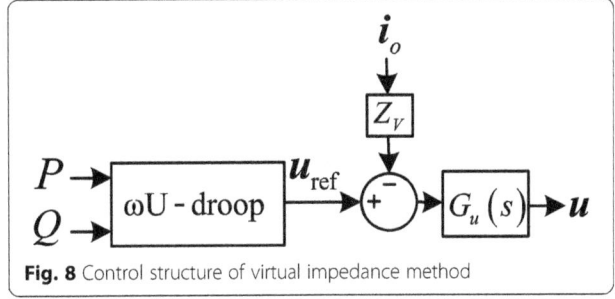

Fig. 8 Control structure of virtual impedance method

$$U = G_u(s) \ U_{ref} - G_u(s) \ Z_V I_o \quad (18)$$

where $G_u(s)$ is the voltage closed-loop transfer function of the ωU-droop GS inverter, and Z_V is virtual impedance.

The total impedance between the equivalent voltage source of the inverter and the microgrid bus can be written as:

$$Z = G_u(s) \ Z_V + Z_L \quad (19)$$

where Z_L is the line impedance.

If the magnitude of the virtual impedance is much larger than the line impedance, the total impedance will be largely decided by the virtual impedance. However, the large total impedance may cause the microgrid voltage to reduce substantially. In [16], a novel method was proposed to solve this problem by introducing a negative resistive component into the virtual impedance. As the virtual negative resistor counteracts the line resistor, the total impedance can be designed to be mainly inductive and of small magnitude. According to Eq. (11), if the total impedance is mainly inductive [17], the GS inverter should adopt P-ω and Q-U droop control. However, if the total impedance is mainly resistive [18], P-U and Q-ω droop control should be applied.

C. Reactive power sharing method based on communication

To improve the reactive power sharing accuracy, a common method is to revise the GS inverters' droop control parameters, including no-load voltage and droop coefficient. The following analysis takes the inductive line ($\cos\theta \approx 0, \sin\theta \approx 1$) as examples. According to Eq. (11), the relation between the output reactive power and the voltage of the GS inverter's PC is shown as:

$$U = E + \frac{Z_L}{E} Q \quad (20)$$

In the Q-U plane, the intersection of the operational curve described by Eq. (20) and the reactive power droop curve is the GS inverter's stable operating point [19].

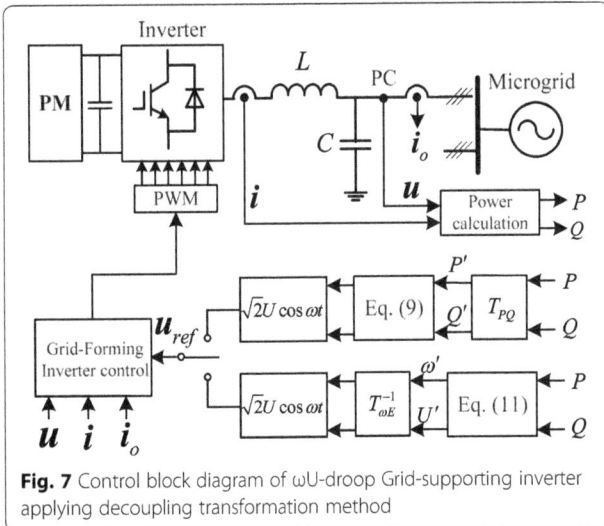

Fig. 7 Control block diagram of ωU-droop Grid-supporting inverter applying decoupling transformation method

As illustrated in Fig. 9, there are two inverters, namely 1# and 2#, with the same droop coefficient. The total impedance between these two inverters' equivalent voltage sources and the microgrid bus are Z_1 and Z_2, respectively. If Z_1 is not equal to Z_2, the inverters' operating points will be different. Increasing inverters' droop coefficient leads to new operating points. The voltage of the microgrid bus moves from E to E', and the inverters' output power changes move from Q_1 and Q_2 to Q_1' and Q_2', respectively. It can be seen that the reactive power sharing accuracy is improved with the increase of the inverter's droop coefficient. Decreasing the GS inverter's no-load voltage can also increase reactive power sharing accuracy, as shown in Fig. 10. To adjust each inverter's droop curve parameters in a coordinated manner [19, 20], it is necessary to employ a centralized control system.

Different with the method of adjusting droop parameter, reference [21] proposed an improved control structure by introducing integral module, as shown in Fig. 11.

In Fig. 11, U_0 is the inverter no-load voltage; E is the voltage of microgrid bus; k_Q is the reactive power droop coefficient; K_u is the integral gain. The transfer function of the inverter's output reactive power can be written as:

$$Q(s) = \frac{K_u[U_o - E(s)]\ E(s) - sE^2(s)}{sZ_L + K_u k_Q E\ (s)} \tag{21}$$

and its steady-state value can be calculated as:

$$\lim_{t \to \infty} Q(t) = \lim_{s \to 0}\ sQ(s)$$
$$= \lim_{s \to 0} \frac{(U_0 - E)\ K_u E - sE^2}{sZ_L + K_u k_Q E} = \frac{U_0 - E}{k_Q} \tag{22}$$

In this method, the output reactive power of each GS inverter is independent to the line impedance Z_L. By delivering the voltage information of the microgrid bus to each GS inverter, accurate reactive power sharing can be

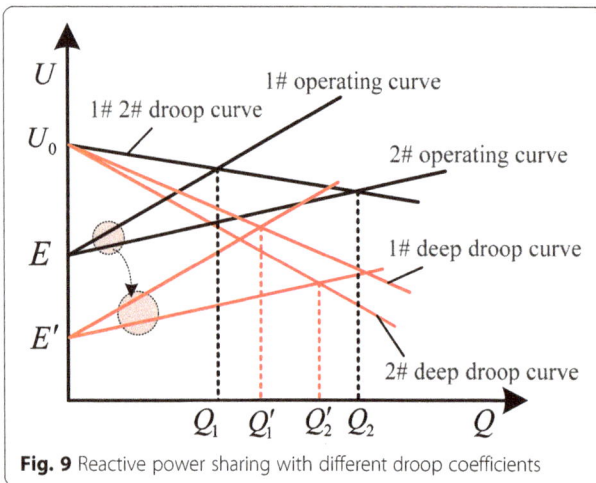

Fig. 10 Reactive power sharing with different no-load voltages

realized. This method doesn't require a central controller to participate, avoiding the usage of complicated algorithms. Besides, the additional parameter, K_u, can be used to adjust the dynamic response of reactive power control.

Results and Discussion

As can be seen from the above sections, the GFD inverter behaves as constant power source and it participates neither in voltage regulation nor in load variations sharing. The GFM inverter behaves as constant voltage source and it is responsible not only for maintaining the microgrid's voltage and frequency, but also for keeping power balance. Load sharing among the GFM inverters is a function of the impedances between the inverters and microgrid bus. The PQ-droop and ωU-droop GS inverters can be regarded as the upgraded version of the GFD and GFM inverters, and they behave as controlled power source and controlled voltage source, respectively. When the microgrid operation conditions change, they can adaptively adjust the output power or voltage to achieve a more flexible load sharing. Currently the most promising control method is the ωU-droop control, because it can make the system autonomy and achieve seamless mode switching. When the microgrid is operated in islanded mode, any

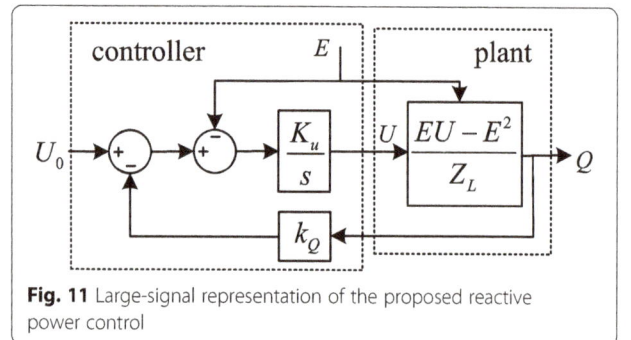

Fig. 9 Reactive power sharing with different droop coefficients

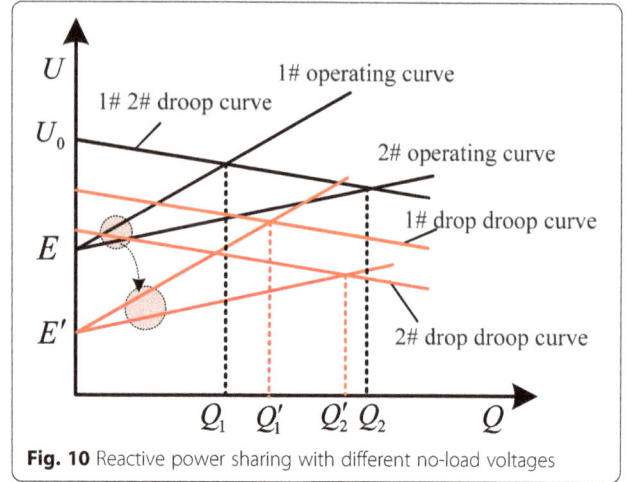

Fig. 11 Large-signal representation of the proposed reactive power control

addition or reduction of a single ωU-droop GS inverter do not influence the configuration of the original system. When the microgrid operated in grid-connected mode, the ωU-droop GS inverter can output the specified power by modifying its no-load voltage and frequency. However, this autonomous control method is not widely applied among numerous experimental microgrids, because there still exist many practical problems, such as the dynamic response speed, the impact of control parameters on system stability, the capability to deal with unbalanced and non-linear loads, and control strategies under fault conditions. In addition, it can be seen from the above analysis that the performance of the ωU-droop GS inverter operating with no communication is inferior. In order to enhance the accuracy of reactive load sharing, it is worthwhile to study the design of the control algorithms with reduced communication requirements.

Conclusions

This paper illustrates the control principles of micro-source inverters, including grid-feeding, grid-forming, and grid-supporting inverters. The PQ-droop and ωU-droop grid-supporting inverters can be regarded as the upgraded version of grid-feeding and grid-forming inverters with a more flexible load sharing capability. Since the conventional ωU-droop control exists some shortages, several improved methods of ωU-droop based grid-supporting inverters are presented. The comparison of various inverters is carried out and the valuable research points are also discussed.

Acknowledgments
This work was supported in part by Nation Natural Science Foundation of China (51407128) and the key technologies research project on distribution network reconfiguration of State Grid Hunan Electric Power Company (5216A1300JV).

Competing interests
The authors declare that they have no competing interests.

Authors' contributions
WG and LM conceived and designed the study. WG wrote the paper. All authors read and approved the final manuscript.

About the authors
W. M. Guo was born in 1989 in Hunan, China. He received his B.S. degrees in electrical engineering from Tongji University in 2011, where he is currently working towards a Ph.D. degree. His current research interests are microgrid protection and control.
L. H. Mu was born in 1963 in Jiangsu, China. He is currently a full professor in the Department of Electrical Engineering, Tongji University, Shanghai, China. His current research interests include protective relaying of power system, microgrid and power quality.

References
1. Kroposki, B., Pink, C., Deblasio, R., et al. (2010). Benefits of power electronic interfaces for distributed energy systems. *IEEE Transactions on EnergyConversion, 25*, 901–908.
2. Walling, R. A., Saint, R., Dugan, R. C., et al. (2008). Summary of distributed resources impact on power delivery systems. *IEEE Transactions on PowerDelivery, 23*, 1636–1643.
3. Lopes, J. A. P., Moreira, C. L., & Madureira, A. G. (2006). Defining control strategies for MicroGrids islanded operation. *IEEE Transactions on Power Systems, 21*, 916–924.
4. Nikkhajoei, H., & Lasseter, R. H. (2009). Distributed generation interface to the CERTS microgrid. *IEEE Transactions onPower Delivery, 24*, 1598–1608.
5. Olivares, D. E., Mehrizi-Sani, A., Etemadi, A. H., et al. (2014). Trends in microgrid control. *IEEE Transactions onSmart Grid, 5*, 1905–1919.
6. Ali, B., & Ali, D. (2012). Hierarchical structure of microgrids control system. *IEEE Transactions onSmart Grid, 3*, 1963–1976.
7. Fangzheng, P., & Jih-Sheng, L. (1996). Generalized instantaneous reactive power theory for three-phase power systems. *IEEE Transactions on Instrumentation and Measurement, 45*, 293–297.
8. Camacho, A., Castilla, M., Miret, J., et al. (2015). Active and reactive power strategies with peak current limitation for distributed generation inverters during unbalanced grid faults. *IEEE Transactions on Industrial Electronics, 62*, 1515–1525.
9. Miret, J., Camacho, A., Castilla, M., et al. (2013). Control scheme with voltage support capability for distributed generation inverters under voltage sags. *IEEE Transactions onPower Electronics, 28*, 5252–5262.
10. Zmood, D. N., Holmes, D. G., & Bode, G. H. (2001). Frequency-domain analysis of three-phase linear current regulators. *IEEE Transactions onIndustry Applications, 37*, 601–610.
11. Rocabert, J., Luna, A., Blaabjerg, F., et al. (2012). Control of power converters in AC microgrids. *IEEE Transactions onPower Electronics, 27*, 4734–4749.
12. De Brabandere, K., Bolsens, B., Van den Keybus, J., et al. (2007). A voltage and frequency droop control method for parallel inverters. *IEEE Transactions onPower Electronics, 22*, 1107–1115.
13. Zhou, X., Rong, F., Lu, Z., et al. (2012). A coordinate rotational transformation based virtual power V/f droop control method for low voltage microgrid. *Automation of Electric Power Systems, 36*, 47–51.
14. Li, Y., & Li, Y. W. (2011). Power management of inverter interfaced autonomous microgrid based on virtual frequency-voltage frame. *IEEE Transactions onSmart Grid, 2*, 30–40.
15. Guerrero, J. M., Vicuña, D., García, L., et al. (2005). Output impedance design of parallel-connected UPS inverters with wireless load-sharing control. *IEEE Transactions onIndustrial Electronics, 52*, 1126–1135.
16. Zhang, P., Shi, J., Ronggui, L. I., et al. (2014). A control strategy of virtual negative impedance for inverters in Low-voltage microgrid. *Proceedings of the CSEE, 34*, 1844–1852.
17. Chen, Y., Luo, A., Long, J., et al. (2013). Circulating current analysis and robust droop multiple loop control method for parallel inverters using resistive output impedance. *Proceedings of the CSEE, 33*, 18–29.
18. Matas, J., Castilla, M., et al. (2010). Virtual impedance loop for droop-controlled single-phase parallel inverters using a second-order general-integrator scheme. *IEEE Transactions onPower Electronics, 25*, 2993–3002.
19. Han, H., Liu, Y., Sun, Y., et al. (2014). An improved control strategy for reactive power sharing in microgrids. *Proceedings of the CSEE, 34*, 2639–2648.
20. Jin, P., Xin, A. I., & Wang, Y. (2012). Reactive power control strategy of microgird using potential function method. *Proceedings of the CSEE, 32*, 44–51.
21. Sao, C. K., & Lehn, P. W. (2005). Autonomous load sharing of voltage source converters. *IEEE Transactions on Power Delivery, 20*, 1009–1016.

7

Protection of large partitioned MTDC Networks Using DC-DC converters and circuit breakers

Md Habibur Rahman[1], Lie Xu[1]* and Liangzhong Yao[2]

Abstract

This paper proposes a DC fault protection strategy for large multi-terminal HVDC (MTDC) network where MMC based DC-DC converter is configured at strategic locations to allow the large MTDC network to be operated interconnected but partitioned into islanded DC network zones following faults. Each DC network zone is protected using either AC circuit breakers coordinated with DC switches or slow mechanical type DC circuit breakers to minimize the capital cost. In case of a DC fault event, DC-DC converters which have inherent DC fault isolation capability provide 'firewall' between the faulty and healthy zones such that the faulty DC network zone can be quickly isolated from the remaining of the MTDC network to allow the healthy DC network zones to remain operational. The validity of the proposed protection arrangement is confirmed using MATLAB/SIMULINK simulations.

Keywords: DC circuit breaker, DC-DC converter, DC fault, HVDC, Modular multilevel converter (MMC), Network partition

Introduction

HVDC is an economic solution of transmitting large amount of power over a long distance compare to the traditional HVAC transmission system due to less transmission losses and smaller cable size for given power level. Voltage source converter (VSC) technology is becoming the main focusing area of recent HVDC research due to its inherent flexible ability of independent active and reactive power control, AC voltage support, and black–start capabilities [1, 2].

There are different network topologies to develop a large MTDC network. At present radial and meshed type network configurations are the most common ones. Radial network configuration is similar to the traditional AC distribution system. Meshed network configuration is more reliable than radial due to redundant supply channels but incur higher cable cost. Any topology can be configured, once technical and economic benefits of the network operators have been dealt with [3].

The major challenges towards the protection of an MTDC network in the event of a fault at the DC side of

the network include fault detection, fault location and isolation [4–6]. Due to low impedance of the DC network there is steep rise in fault current and fast DC voltage collapse which can potentially damage the power electronics converter and disrupt power transmission across the whole DC network. Therefore, a robust and accurate protection system is required which can detect the fault and its location and isolate the faulty section in a selective manner allowing fast restoration of the system following a DC fault [6–9].

Various protection strategies have been proposed for MTDC networks [10–14]. In [10] a handshaking' protection method for VSC based MTDC network is proposed where DC switch and AC circuit breakers (ACCBs) are used to protect the entire system though the complete network has to be shutdown and de-energized to allow the DC switches to isolate the faulty branches. The proposed concept can pose major operational problems for a large MTDC network and connected AC systems as the large 'loss of infeed' may cause large excursion in AC frequency. Thus, fast and reliable protection is mandatory for fast fault clearance to avoid complete shutdown of the entire network so as to minimize the disturbances to the connected AC networks. Some

* Correspondence: lie.xu@strath.ac.uk
[1]University of Strathclyde, Glasgow, UK
Full list of author information is available at the end of the article

MMC topologies [15–17] can block or control DC fault current but resulting additional capital cost and power loss. In addition, such converter topologies cannot isolate fault from the MTDC network apart from protecting themselves from over-current so additional DC protection equipment is still required.

For MTDC network protection, different types of DC circuit breakers (DCCBs) such as slow mechanical DCCBs, solid state DCCBs and hybrid DCCBs, have been proposed [18, 19]. Some DCCBs, e.g., solid state and hybrid types are capable of operating within a few milliseconds but to avoid complete shutdown of the entire MDTC network they will have to be used at every cable branch leading to excessively high capital cost, larger footprint and high on state losses (for solid state DCCB only). In contrast, slow mechanical DCCBs incur low capital cost and low loses [20].

DC-DC converters allow DC sections with different DC voltage levels to be interconnected. They can also isolate DC faults rapidly and allow the healthy part of the network remains operational. Various studies have been conducted with different DC-DC converter topology [21, 22] for MTDC network though its high capital cost and power loss, and larger footprint mean its use has to be carefully considered.

The main contribution of this paper is on the use of DC network partition but interconnected using DC-DC converters at strategic locations. The partitioned DC network zones can be protected by means of ACCBs, DC Switches and slow DCCBs depending on their individual network configuration. The main purpose of the protection arrangement is to minimize the use of expensive DCCBs and DC-DC converters to reduce the capital cost of large MTDC networks. The rest of the paper is structured as follows: Section II describes the fault behaviour of half bridge MMC based converters. Large MTDC network partition with protection arrangement is outlined in section III and detailed system configuration is considered in section IV. In section V, two case studies in Matlab/Simulink environment is performed to demonstrate the validity of the proposed protection arrangement and finally, section VI draws the conclusions.

DC fault behavior

In a MTDC network, a single DC fault can cause serious consequences due to low impedance of the DC network leading to high fault current propagating throughout the entire network that could enforce the complete shutdown of the network for prolonged period [23]. As soon as a DC fault occurs, there is a step rise in DC current due to the discharge of the DC cable capacitor and AC side current starts feeding through the freewheeling diodes which could cause damage to the power electronics devices [24]. There are various reasons, which can

lead to a DC fault such as lightening strike in case of overhead lines, ship anchors for undersea cables, electrical stress, cable aging, physical damage, environmental stress etc.

To analyse the DC fault behaviour, an equivalent circuit of a half bridge MMC is shown in Fig. 1, and a number of stages can be considered which has been well documented in [25]. Unlike the conventional two-level VSCs, half bridge MMC experiences considerably lower DC fault current due to relatively small cable capacitance and absence of large DC link capacitors at converter terminals. A DC line-to-line fault is applied at 1.1 s and the MMC is blocked 1 ms after the fault initiation. Figure 2 illustrates the system response during the fault period. Long (Fig. 2(a)-(e)) and short (Fig. 2(f)-(j)) duration time-scale waveforms have been presented for ease of analysis. The DC link voltages depicted in Fig. 2(a) and (f) show immediate drop with oscillations after the fault initiation and reach to negative values due to the presence of the arm reactance. In the meantime, there is a step rise in the DC link current as shown in Fig. 2(b) and (g). Fig. 2(c), (h) and (d), (i) show the upper and lower arm currents respectively, and Fig. 2(e) and (j) show large AC fault currents feeding from the AC networks during the fault period. This AC fault current flows through the freewheeling diodes before being interrupted by the opening of the ACCB whose tripping time is set at 80 ms after the detection of AC over current.

MTDC network partition and protection
Network partition

Significant challenges need to overcome to protect a large scale MTDC network in the event of a single DC fault. The straight forward solution is to install fast acting DCCBs at every DC cable connection points though it will incur huge cost in system protection. Therefore, a number of facts need to be considered while configuring a large MTDC network protection such as infrequency of DC fault events, inconsistency of power generation from wind farms and investment in protection cost. To rationalize the cost and reliable protection, a large MTDC network can be partitioned into a number of small DC network zones. In case of a fault event in a particular zone, the fault can be isolated by clearance from AC side protection and DC switches [13, 26, 27].

The acceptable permanent 'loss of infeed' is 1.8 GW currently in UK according to the system operator in order to maintain network stability. The 'loss of infeed' could be due to a fault event or regular maintenance. Therefore, while partitioning a large MTDC network, each DC network zone should be configured in such a way that the total permanent 'loss of infeed' is kept below the maximum power loss criterion in the event of a fault. But this partitioning reduces the operational

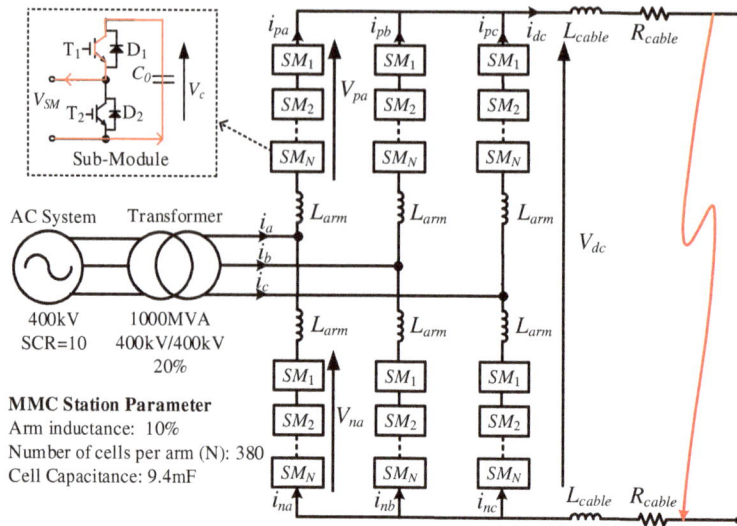

Fig. 1 Equivalent circuits of half bridge MMC during DC line-to-line fault

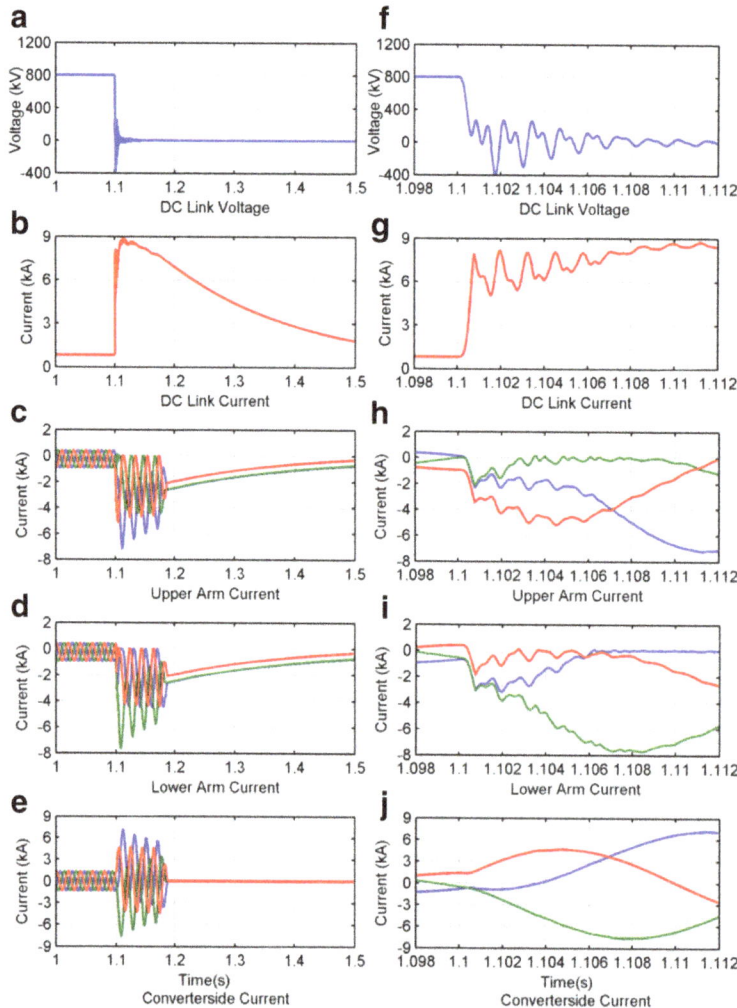

Fig. 2 Response from a half bridge MMC converter during DC line-to-line fault

flexibility of the MTDC network. An alternative option to protect a large MTDC network is to use DC-DC converters located at strategic locations, joining different DC network zones allowing the entire network to be operated interconnected at pre-fault condition but partitioned into islanded DC network zones following any fault events. An example DC network configuration is shown in Fig. 3 where MMC based DC-DC converters are used. In case of a fault event in one of the DC network zones at least two DC network zones can remain operational. Each network zone can be protected using different protection arrangement depending on their configurations.

Protection arrangement

There are different options to clear the DC side faults without causing a large loss of infeed in a large MTDC network. The main purpose of this work is to keep the healthy zone in the large MTDC system operational all times following a DC fault by means of using DC-DC converter at strategic location. Each DC network zone can be protected using any combination of slow mechanical DCCBs, ACCBs and DC switches depending on the zone configuration. The following steps are considered for the proposed system to clear a DC fault allowing the healthy zones to remain operational:

Step 1: Using local current measurement in the converter arms and DC sides to detect the fault.
Step 2: If converter arm currents reach above predetermined set value converter will be blocked. This applies to all the AC-DC and DC-DC converters and will isolate the faulty zone from the healthy ones.
Step 3: By opening the ACCBs and DC switches or the slow DCCBs in the faulty zone, the DC fault can be isolated.

Step 4: After isolating the faulty part within the faulty zone remaining part of the network can be restarted and can be reconnected to the healthy zones depending on fault location.

System configuration

Figure 4 shows the six-terminal MTDC system considered in the paper consisting of 6 half bridge MMCs connected to AC systems. The system contains two DC network zones (one radial and one meshed network respectively) which are interconnected by DC cables equipped with a DC-DC converter. No Fast acting DCCBs are used within any DC network zones so as to reduce the cost and power loss. Here DC network Zone 1 (±320 kV DC) which is a radial network is protected using ACCBs and DC switches. As for the DC network Zone 2 (±400 kV DC), a meshed network configuration is used with increased reliability due to redundant supply channels. For this DC Zone 2, slow mechanical DCCBs are installed.

The proposed protection arrangement is applied to the MTDC system shown in Fig. 4 and is verified in MATLAB/SIMULINK environment. The MMC converters are modeled as average value models which provide faster simulation speeds. The average MMC model consists of controllable voltage and current sources where additional semiconductor devices are added to replicate the same functionality during normal operation as well as fault event. In this system configuration π model of the cable is considered.

Simulation result

The Station 2 and 4 are assigned to transmit 800 MW and 700 MW power to the DC grid, respectively. Both Station 3 and 6 transmit 400 MW power to their respective AC grids whereas Station 1 and 5 regulate the

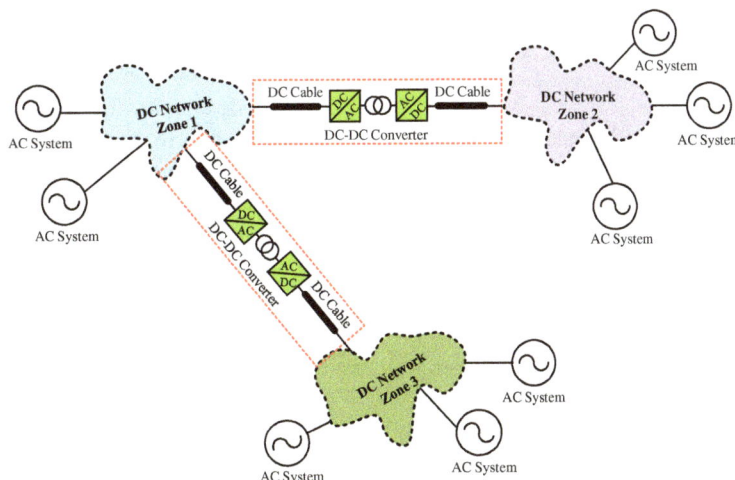

Fig. 3 Possible partitioned MTDC networks using DC-DC converters

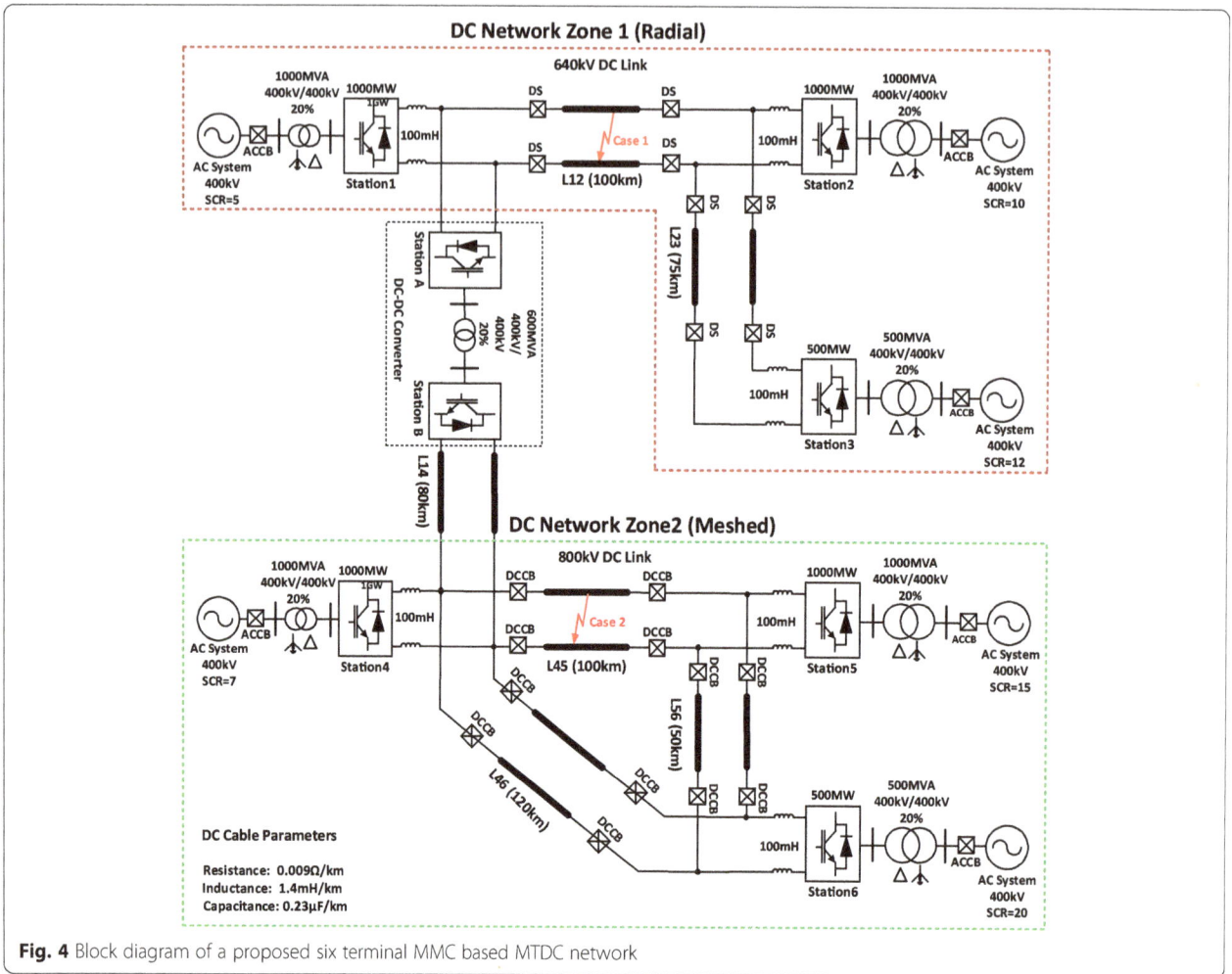

Fig. 4 Block diagram of a proposed six terminal MMC based MTDC network

DC link voltages (640 kV and 800 kV respectively) of DC Zone 1 and 2, respectively. The DC-DC converter's Station B (connected to Zone 2) is designed to transmit 200 MW power from Zone 1 to Zone 2 while Station A (connected to Zone 1) is set for controlling the internal AC source voltage in the DC-DC converter. All converters operate at unity power factor for simplicity.

Case 1-fault in zone 1

When the system reaches its steady state condition, a DC line-to-line fault is applied at the time instant of 1.5 s. The fault is placed at the midpoint of the transmission cable L12 and the protection system is in place throughout the network as required.

The main concept of protection arrangement in DC Zone 1 (radial network) is that, in case of any fault events within Zone 1, the DC-DC converter can quickly isolate the faulty Zone 1 by blocking its converter such that DC network Zone 2 can remain operational all the times. The faulty section in Zone 1 can then be isolated by means of using ACCBs and DC switches. In this case, after isolating the faulty section Station 1 is restarted to

reconnect it to Zone 2 where Station 2 and 3 transmit power among themselves after restarting process.

The obtained results demonstrating the system's behavior are presented in Figs. 5, 6 and 7 for DC voltage, DC currents, and arm currents, respectively. It is obvious from Figs. 5 and 6 that the DC voltages in Zone 1 (all stations) are severely affected after fault initiation leading to a step increase in the DC link current. The DC over current flowing through the DC-DC converter (not shown due to space limitation) is quickly detected resulting an immediate blocking of the DC-DC converter which isolates Zone 1 from Zone 2. Apart from the loose of 200 MW previously transmitted from Zone 1 to Zone 2 through the DC-DC converter which results in the change of the power (DC current) for Station 5 (DC voltage controller) it is evident that there is insignificant impact on the DC voltages in Zone 2 due to the fast blocking of the DC-DC converter.

Figure 7 represents the upper arm currents. In this proposed system, fault is detected in the DC-DC converter and each AC-DC converters located in Zone 1 when their respective arm currents exceed pre-defined

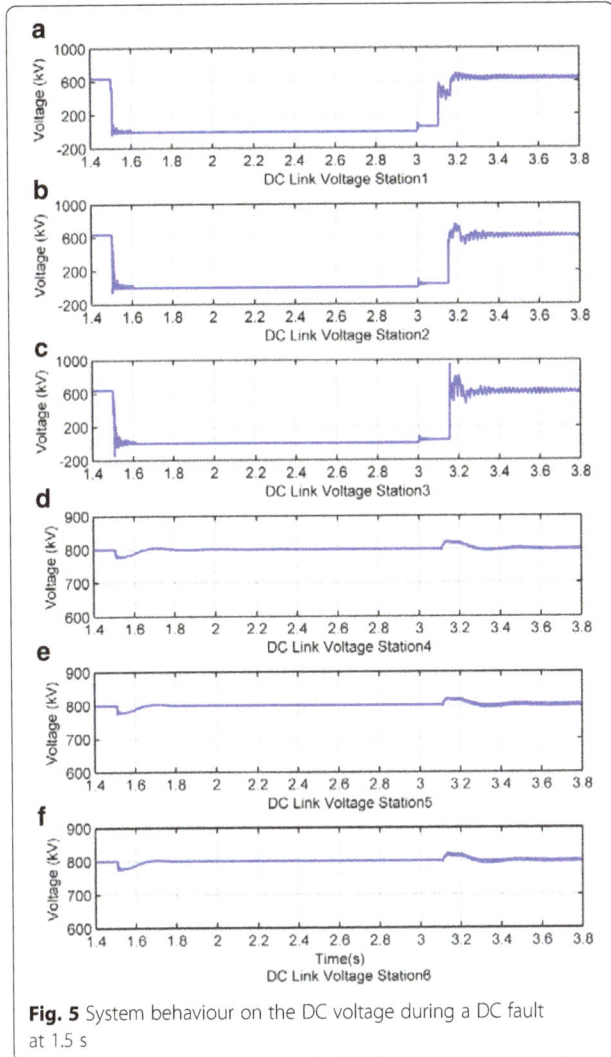

Fig. 5 System behaviour on the DC voltage during a DC fault at 1.5 s

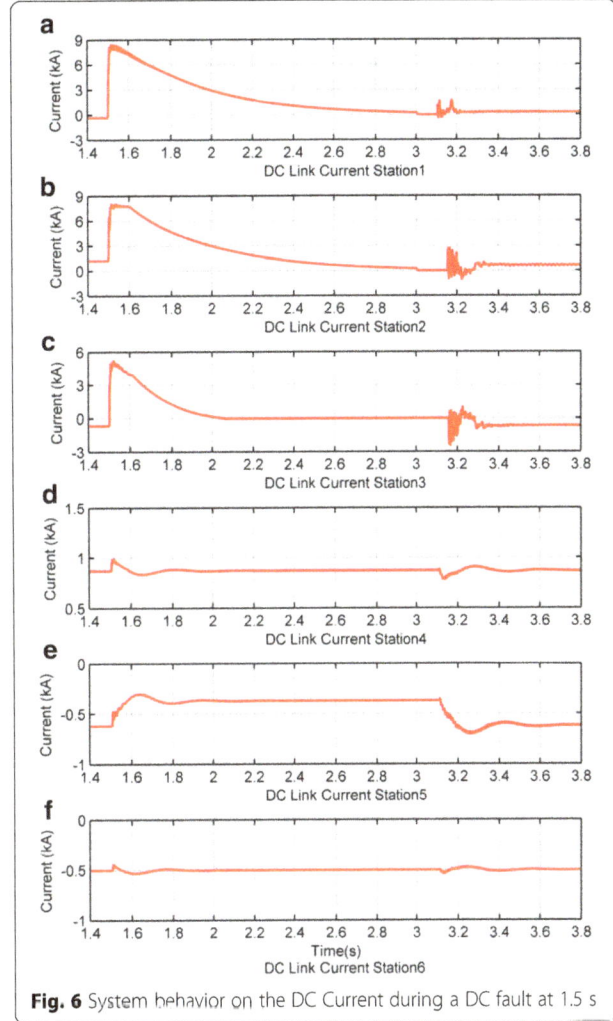

Fig. 6 System behavior on the DC Current during a DC fault at 1.5 s

maximum values. In this simulation study Station 1–3 and DC-DC converter stations are blocked at 7 ms, 4 ms, 8 ms and 7 ms respectively after the fault initiation. After blocking the converter arm currents continue increasing (see Fig. 7) through the freewheeling diodes.

In this study the ACCBs equipped in Station 1–3 are opened at 107 ms, 104 ms and 108 ms respectively, after the fault initiation (including 100 ms delay). Upon opening the ACCBs the converter arm currents in station 1–3 are gradually brought to zero as shown in Fig. 7. The DC current in Zone 1 can take considerable time to decay as evident in Fig. 6 due to the low resistance in the DC cables.

System recovery process is one of the key factors for a large MTDC system. It is worth noted that loss of a transmission line due to fault results in a reduction in overall power capacity of the MTDC network leading to a direct consequence on the remaining healthy lines of the network.

Proper power rescheduling is required to ensure stable system operation during system recovery process. In this simulation, the faulty transmission cable L12 are disconnected by DC switches which are opened when the DC link current of the faulty cable reaches approximately zero. Here DC switches are opened at around 3 s (1.502 ms after fault initiation). Once the faulty part is cleared from Zone 1, all ACCBs are reclosed again for Station 1–3 at 3.057 s, 3.054 s and 3.058 s respectively. Then Station 1 and the DC-DC converter are restarted at 3.1 s to reconnect with DC network Zone 2. The transmitted power from Station 1 through the DC-DC converter to Zone 2 is again set at 200 MW after recovery.

As can be seen in Fig. 6, the DC link current at the DC voltage controller, i.e., Station 5, changes accordingly. Station 2 and 3 are restarted at 3.154 s and 3.158 s respectively to transmit power among the two. Here Station 2 controls the DC link voltage and Station 3 regulates active power at 400 MW. The complete system reaches in steady state within 400-500 ms after the

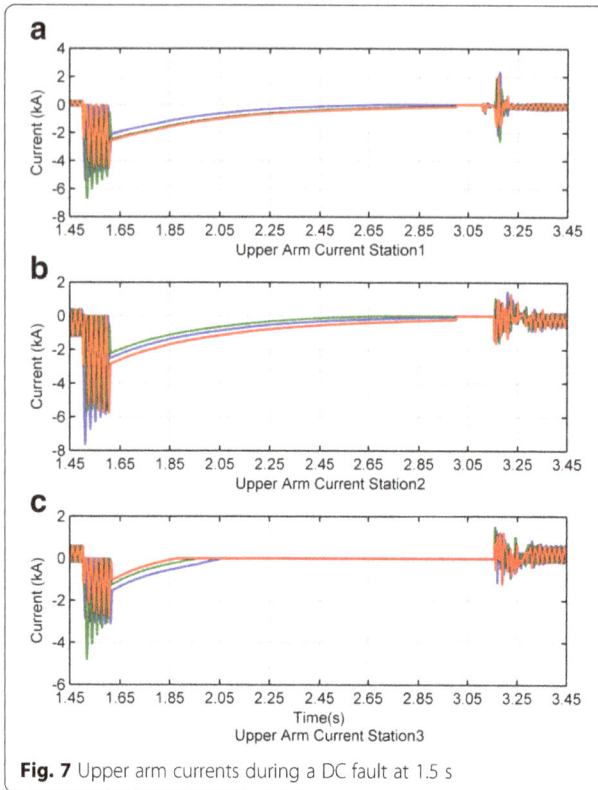

Fig. 7 Upper arm currents during a DC fault at 1.5 s

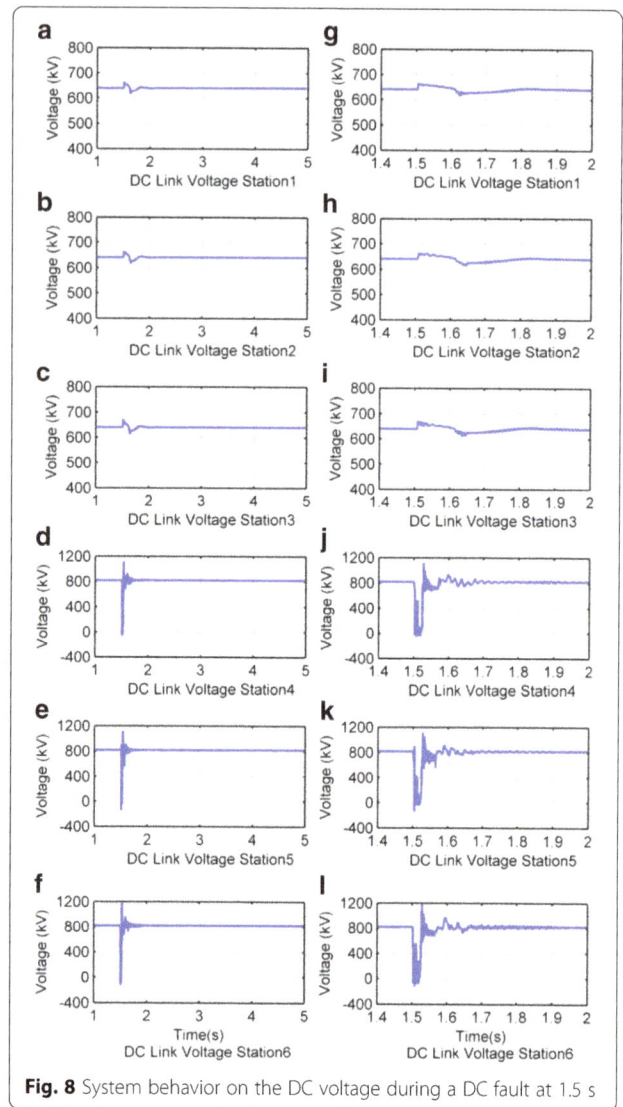

Fig. 8 System behavior on the DC voltage during a DC fault at 1.5 s

recovery process. The obtained results presented in Figs. 5, 6 and 7 clearly show satisfactory protection and restoration process.

Case 2-fault in zone 2

The fault is placed at 1.5 s at the midpoint of the transmission cable L45. The main concept of this protection arrangement in DC network Zone 2 (meshed network) is that, in case of any fault events within Zone 2, the DC-DC converter can quickly isolate the faulty zone by blocking its converter such that the DC network Zone 1 can remain operational all the times. Fault section with the faulty Zone 2 can then be isolated by means of using slow DCCBs. In this case, after isolating the faulty section all stations in Zone 2 are restarted and are reconnected with Zone 1.

The obtained results demonstrating the system's behavior are presented in Figs. 8, 9 and 10 for the DC link voltages, DC link currents, and arm currents, respectively. Long and short duration time-scale waveforms have been presented for ease of analysis. It can be seen from Figs. 8 and 9, that the DC voltages at all station in Zone 2 are severely affected after fault initiation leading to step increases in DC link currents. The DC over current flowing through the DC-DC converter is quickly detected resulting an immediate block of the DC-DC converter isolating Zone 2 from the healthy Zone 1. It is evident that there is little

impact on DC Zone 1 apart from the temporary reschedule of the power flow as Station 1 has to transmit the extra 200 MW previously flowing through the DC-DC converter to Zone 2.

Figure 10 represents the upper arm currents. In the proposed system, faults are detected in each converter located in Zone 2 and the DC-DC converter using automatic arm over-current detection and blocking method. In the simulation Station 4–6 and the DC-DC converter are blocked at 3 ms, 4 ms, 4 ms and 4 ms respectively after the fault initiation.

After blocking the converter, the arm currents continue increasing (see Fig. 10) through the freewheeling diodes. Here Zone 2 is protected using slow mechanical DCCBs. In this simulation study the slow DCCBs are opened with 20 ms mechanical delay after over-current detection and only those DCCBs whose detected over-currents flow into the connected DC cables are opened.

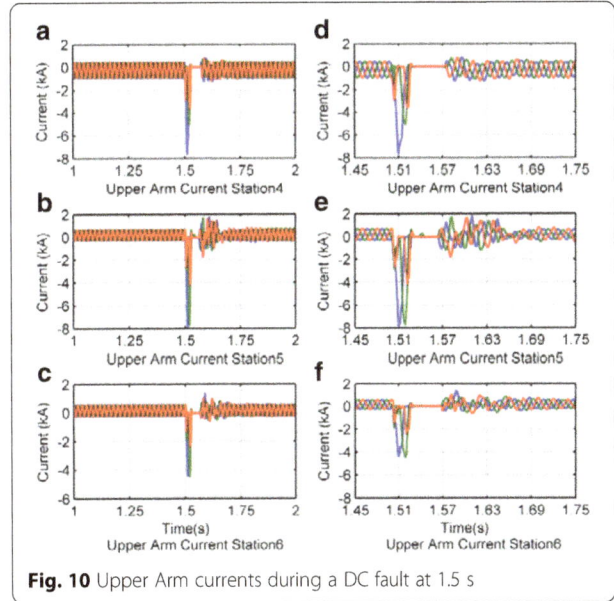

Fig. 9 System behavior on the DC Current during a DC fault at 1.5 s

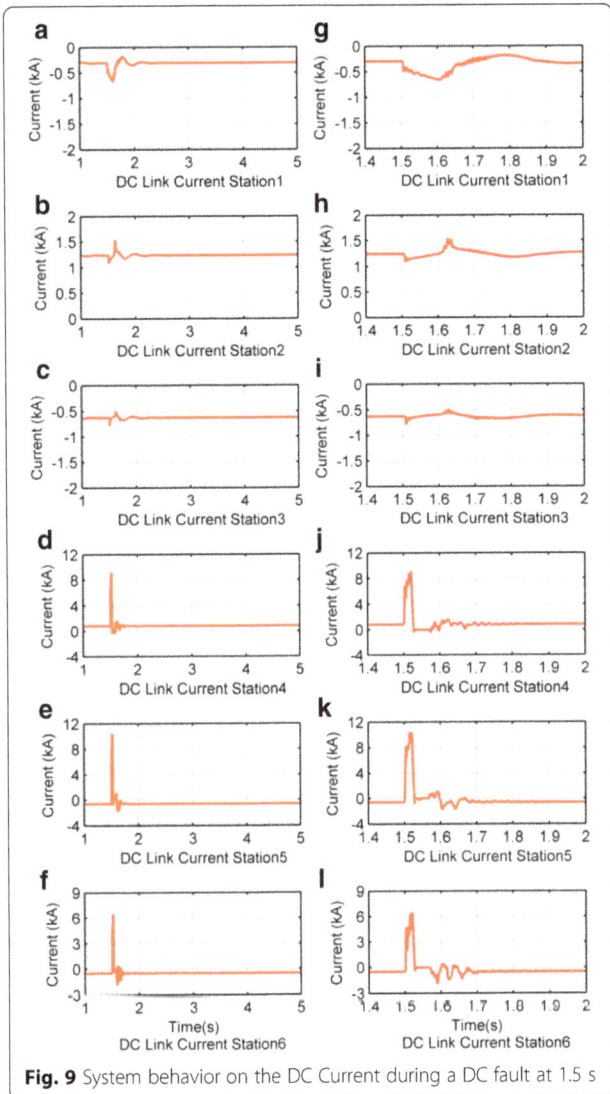

Fig. 10 Upper Arm currents during a DC fault at 1.5 s

As the fault is in cable L45, DCCBs at both ends of L45 will see current flowing into the fault and whereas for other cables (i.e., L46 and L56) only DCCBs at one side of each cable see fault current flowing into the cable. Therefore, L45 will be completely isolated by the DCCBs whereas L46 and L56 only disconnect on one ends. Upon the opening of the DCCBs the DC link currents and converter arm currents in Station 4–6 are quickly brought to zero (see Figs. 9 and 10).

System recovery process for Case 2 is quicker than Case 1 due to the different protection arrangements installed in Zone 2 (DCCBs) compared to that of Zone 1 (ACCBs and DC switches). After the faulty cables L45 are disconnected by opening the relevant mechanical DCCBs all others DCCBs (meaning L46 and L56) are reclosed again for restoration of Station 4–6 at 73 ms, 64 ms and 69 ms respectively. Then the DC-DC converter is restarted at 104 ms for normal operation

reconnecting Zone 1 and 2 where the same pre-fault 200 MW is transmitted from Zone 1 to Zone 2. All Stations in Zone 2 are operated as their pre-fault control modes. The entire MTDC network transmits the same amount of power after losing one cable (L45) due to meshed configuration in Zone 2. The system's restoration process presented in Figs. 8, 9 and 10, gives satisfactory performance. In this simulation study, the system is reached in steady state within 80-100 ms after the recovery process.

Conclusion

Partition of a large MTDC network into different DC network zones is proposed where DC-DC converters installed at strategic locations allowing interconnected network operating with inherent DC fault isolation and 'firewall' between the different DC zones. This proposed protection configuration ensures accurate and robust protection option for the system with low investment in protection cost, and continuous operation of the healthy zones during a fault event in other zone of the MTDC network is achieved. The simulation results corresponding to DC fault protection have been presented for a MTDC network containing one radial DC zone and one meshed DC zone, and give satisfactory results. The proposed concept can be an attractive approach to interconnect various grids to form a large MTDC network in future.

Acknowledgment
This work is supported in part by China Electric Power Research Institute (CEPRI).

Authors' contributions
MHR carried our the simulation studies and drafted the manuscript. LX and LY provided technical leadership to the studies. All authors read and approved the final manuscript.

Competing interests
The authors declare that they have no competing interests.

Author details
[1]University of Strathclyde, Glasgow, UK. [2]China Electric Power Research Institute, Beijing, China.

References

1. Xu, L., & Andersen, B. (2006). Grid integration of large offshore wind farms using HVDC. *Wind Energy, 9,* 371–382.
2. Kirby, N. M., Xu, L., Luckett, M., & Siepmann, W. (2002). HVDC transmission for large offshore wind farms. *IET Power Engineering Journal, 16,* 135–141.
3. MacIver, C., Bell, K. R. W., & Nedic, D. P. (2015). A reliability evaluation of offshore HVDC grid configuration options. *IEEE Transactions on Power Delivery, PP,* 1.
4. Yan, X., Difeng, S., & Shi, Q. (2013). Protection coordination of meshed MMC-MTDC transmission systems under DC faults. In *TENCON 2013–2013 IEEE Region 10 Conference (31194)* (pp. 1–5).
5. Yousefpoor, N., Kim, S., & Bhattacharya, S. (2014). Control of voltage source converter based multi-terminal DC grid under DC fault operating condition. In *Energy Conversion Congress and Exposition (ECCE), 2014 IEEE* (pp. 5703–5708).
6. Tang, L., & Ooi, B.-T. (2002). Protection of VSC-multi-terminal HVDC against DC faults. In *Power Electronics Specialists Conference, 2002. pesc 02. 2002 IEEE 33rd Annual* (Vol. 2, pp. 719–724).
7. Lu, W., & Ooi, B.-T. (2003). DC overvoltage control during loss of converter in multiterminal voltage-source converter-based HVDC (M-VSC-HVDC). *IEEE Transactions on Power Delivery, 18,* 915–920.
8. Yang, J., Fletcher, J. E., & O'Reilly, J. (2010). Multiterminal DC wind farm collection grid internal fault analysis and protection design. *IEEE Transactions on Power Delivery, 25,* 2308–2318.
9. Rafferty, J., Xu, L., & Morrow, D. J. (2015). Analysis of VSC-based high-voltage direct current under DC Line-to-earth fault. *IET Power Electronics, 8*(3), 428–438.
10. Tang, L., & Ooi, B.-T. (2007). Locating and Isolating DC faults in multi-terminal DC systems. *IEEE Transactions on Power Delivery, 22,* 1877–1884.
11. Hajian, M., Jovcic, D., & Bin, W. (2013). Evaluation of semiconductor based methods for fault isolation on high voltage DC grids. *IEEE Transactions on Smart Grid, 4,* 1171–1179.
12. Jovcic, D., Taherbaneh, M., Taisne, J. P., & Nguefeu, S. (2015). Offshore DC Grids as an interconnection of radial systems: protection and control aspects. *IEEE Transactions on Smart Grid, 6,* 903–910.
13. Rahman, M. H., Xu, L., & Bell, K. (2015). DC fault protection of multi-terminal HVDC systems using DC network partition and DC circuit breakers. In *Protection, Automation and Control World Conference(PACWorld2015)* (pp. 1–10).
14. R Li, L Xu, D Holliday, F Page, S Finney, and B Williams. (2016). "Continuous operation of radial multi-terminal HVDC systems under DC Fault". *Power Delivery IEEE Transactions, 31,* 351–361.
15. Marquardt, R. (2011). Modular multilevel converter topologies with DC-short circuit current limitation. In *Power Electronics and ECCE Asia (ICPE & ECCE), 2011 IEEE 8th International Conference* (pp. 1425–1431).
16. Zeng, R., Xu, L., Yao, L., & Williams, B. W. (2015). Design and operation of a hybrid modular multilevel converter. *IEEE Transactions on Power Electronics, 30,* 1137–1146.
17. Rui, L., Adam, G. P., Holliday, D., Fletcher, J. E., & Williams, B. W. (2015). Hybrid cascaded modular multilevel converter with DC fault ride-through capability for the HVDC transmission system. *IEEE Transactions on Power Delivery, 30,* 1853–1862.
18. Franck, C. M. (2011). HVDC circuit breakers: a review identifying future research needs. *IEEE Transactions on Power Delivery, 26,* 998–1007.
19. Tahata, K., El Oukaili, S., Kamei, K., Yoshida, D., Kono, Y., Yamamot, R., et al. (2015). HVDC circuit breakers for HVDC grid applications. In *AC and DC Power Transmission, 11th IET International Conference* (pp. 1–9).
20. F Page, S Finney, and L Xu. "An alternative protection strategy for multi-terminal HVDC". In: 13th Wind Integration Workshop. Energynautics GmbH: Berlin; 2014.
21. Zeng, R., Xu, L., & Liangzhong, Y. (2015). DC/DC converters based on hybrid MMC for HVDC grid interconnection. In *AC and DC Power Transmission, 11th IET International Conference* (pp. 1–6).
22. Jovcic, D., Taherbaneh, M., Taisne, J. P., & Nguefeu, S. (2014). Developing regional, radial DC grids and their interconnection into large DC grids. In *PES General Meeting Conference & Exposition, 2014 IEEE* (pp. 1–5).
23. Chang, B., Cwikowski, O., Barnes, M., & Shuttleworth, R. (2015). Multi-terminal VSC-HVDC pole-to-pole fault analysis and fault recovery study. In *AC and DC Power Transmission, 11th IET International Conference* (pp. 1–8).
24. Rafferty, J., Xu, L., & Morrow, D. J. (2012). DC fault analysis of VSC based multi-terminal HVDC systems. In *AC and DC Power Transmission (ACDC 2012), 10th IET International Conference* (pp. 1–6).
25. Gao, Y., Bazargan, M., Xu, L., & Liang, W. (2013). DC fault analysis of MMC based HVDC system for large offshore wind farm integration. In *Renewable power generation conference (RPG 2013), 2nd IET* (pp. 1–4).
26. Bell, K. R. W., Xu, L., & Houghton, T. (2015). Considerations in design of an offshore network. In *CIGRE Science and Engineering* (pp. 79–92).
27. CD Barker, RS Whitehouse, AG Adamcyzk, and M Boden. "Designing fault tolerant HVDC networks with a limited need for HVDC circuit breaker operation'. Cigre General session: Paris; Paper B4-112: 2014.

8

Adaptive concentric power swing blocker

Jalal Khodaparast and Mojtaba Khederzadeh[*]

Abstract

The main purpose of power swing blocking is to distinguish faults from power swings. However, the faults occur during a power swing should be detected and cleared promptly. This paper proposes an adaptive concentric power swing blocker (PSB) to overcome incapability of traditional concentric PSB in detecting symmetrical fault during power swing. Based on proposed method, two pairs of concentric characteristics are anticipated which the first one is placed in a stationary position (outer of zone3) but the position of the second pair is adjustable. In order to find the position of the second pair of characteristic, Static Phasor Estimation Error (SPEE) of current signal is utilized in this paper. The proposed method detects the abrupt change in SPEE and puts the second pair of characteristic in location of impedance trajectory correspondingly. Second concentric characteristic records travelling time of impedance trajectory between outer and inner zones and compares to threshold value to detect symmetrical fault during power swing. If recorded time is lower than threshold, three-phase fault is detected during power swing. Intensive studies have been performed and the merit of the method is demonstrated by some test signals simulations.

Keywords: Concentric PSB, Phasor estimation error, Power swing, Symmetrical fault during power swing

Introduction

Distance relay malfunction has raised concerns about blackouts in power systems. Distance relays make decisions based on entering of impedance trajectory in protected zones. When a fault occurs in a protected line, the impedance trajectory enters in distance relay zones and the relay operates. However, this impedance penetration may also occur during power swing condition. During a power swing the voltage and current fluctuate simultaneously, causing fluctuation in the measured apparent impedance at the distance relay, which may enter the relay tripping zones. This condition causes relay malfunction and may lead to consecutive events (cascading outages) and even a blackout eventually [1–3].

To avoid this malfunction, Power Swing Blocker (PSB) is installed in modern distance relay [4]. The main task of PSB is discriminating power swing from fault and block distance relay from operating during power swing. Moreover, it should detect any

fault during power swing and unblock distance relay. However, due to the symmetric nature of power swing, detection of symmetrical faults during power swing is more difficult than unsymmetrical faults. Therefore, this issue attracts attentions of many researchers at the moment.

There are various suggestions in the literature as to how to deal with this issue. The most traditional method is utilizing rate of change of impedance for power swing detection [4]. However, this method cannot detect fault during power swing when impedance trajectory crosses concentric characteristics during power swing (it is exemplified in Fig. 4). New methods based on voltage phase angle are presented in [5] and [6] but high resistance and symmetrical faults are not considered in these references. Fault detection based on differential power is another proposed method that makes use of auto-regression technique to predict samples in the future [7]. However this method needs lots of simulations to select appropriate parameters for the auto-

* Correspondence: m_khederzadeh@sbu.ac.ir
Electrical Engineering Department, Shahid Beheshti University, Tehran
165895371, Iran

regression technique. Application of time frequency transforms is another solution. Wavelet transform and S-transform are presented in [8] and [9] respectively, to detect power swing but they require high sampling rate, which is a requirement of most Time-Frequency transforms. Another method based on adaptive neuro-fuzzy system is proposed in [10]. This method requires many simulations in different conditions for training and even retraining in new case. Mathematical morphology is presented in [11] for detecting symmetrical fault during power swing and it is based on monitoring shape of signal. Although this method uses time domain transformation, selection of processing function and its length is difficult. Moving average is a low-pass filter that is presented in [12] to discriminate power swing from fault. The moving average varies periodically during power swing, while it becomes either positive or negative consistently during fault. However, utilization of all three phase currents even in symmetrical fault increases computational burden in this method. In [13], a method based on maximum rate of change of three-phase active and reactive powers is proposed. However, the mathematical demonstration of the proposed index is based on a somewhat impractical hypothesis that considers impedance without resistive component. Combination of Park's transformation and moving data window is presented in [14] to extract power coefficents during fault and power swing. These coefficient are approximately zero during power swing and significant during fault. Computational burden of calculation of power coefficients limits the application of this method. Reference [15] proposes a method based on fundamental frequency component that is created in instantaneous three-phase active power after inception of a symmetrical fault. However, it assumed that the fault resistance is negligible. Reference [16] proposes a technique based on negative sequence component of current and cumulative sum (CUSUM) for detecting three-phase fault during power swing in series compensated line. A new method based on extracting created transient of current signal by least square dynamic phasor estimation is proposed in [17]. The challenge of this method is high computational burden of dynamic phasor estimation.

The purpose of this paper is to modify the traditional concentric PSB to enable it for detecting three-phase fault during power swing. In this paper, Phasor Estimation Error (PEE) is employed as a quantity with high abrupt at fault initiation that helps the proposed method in determining the location of second pair of concentric PSB. According to proposed method, two indices are used in this method for detecting fault during power swing, which these indices complete each other. The first index (IX1) is transient monitor that shows occurrence of transient in signal and determines the location of second pair of characteristic and the second index (IX2) that is the output of second concentric characteristics as final index for detecting three-phase fault during power swing.

Static phasor estimation error
Static Phasor Estimation Error (SPEE) is calculated by static phasor estimation process in every sample and can be used as a quality measure of phasor estimation. In phasor calculation process, windowed signal is utlized for every sample of time. According to Fig. 1, when a transient occurs in the power system, there will be a seri of windows, contain pre and post transient data which are illustrated in shaded box in Fig. 1. It is obvious that the calculated phasors resulted from just pre or just post data of transient periods are accurate, which are illustrated in unshaded box in Fig. 1. The calculated phasors based on the shaded windows (boxes) are not accurate which can be used as a detector of transient in any signal. Therefore SPEE can be formulated as:

$$SPEE_n = \sum_{n=r-N_1}^{n=r} |S_n - \widehat{S_n}| \qquad (1)$$

where r is the first sample of time window, S_n is the real sample, which is measured by relay and $S^\wedge_n$ is recumputed sample of S_n obtained based on static phasor estimation.

Discussion
Limitation of traditional concentric PSB
In normal situation, the measured impedance is far away from the distance relay protection zones. However, when a fault initiates, the measured impedance moves in the complex plane (R, X) rapidly from load point to characteristic of line impedance. As a result of the electrical property of a fault, the rate of change of impedance is very high but it is very slow during power swing a result of the mechanical property of power swing. Traditional concentric PSB utilizes this difference to discriminate power swing from fault. To achieve this goal, two concentric impedance characteristics (outer and inner zones) along with a timer are used in traditional concentric PSB. The required time for impedance movement between outer and inner zones during quickest power swing is considered as threshold value. If the recorded time is lower than

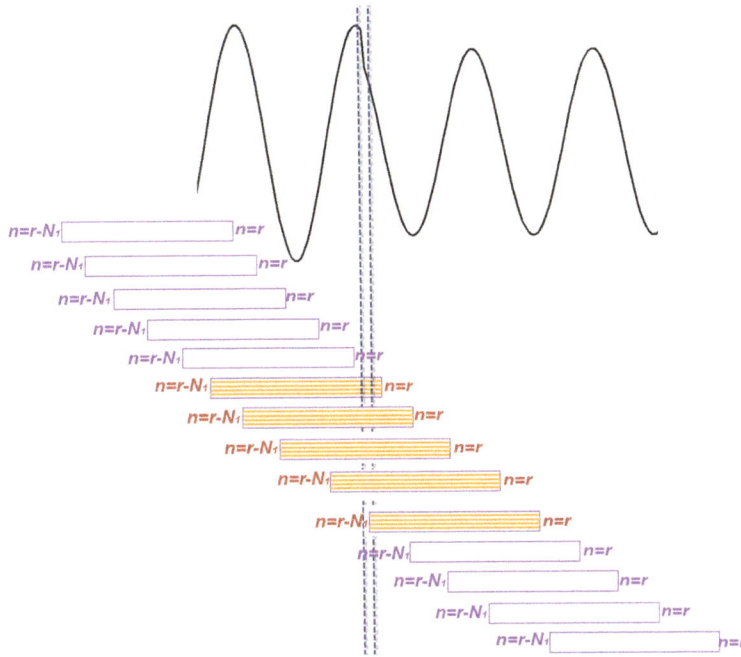

Fig. 1 Concept of *PEE* (N_1 is sample number in one cycle and r is the last sample inserted in window)

the threshold value, it is detected as a fault and in contrast, if the recorded time is higher than threshold, it is detected as power swing.

In order to analyze the performance of traditional concentric PSB in discriminating power swing from fault, a series of tests are carried out on a two-machine equivalent system, shown in Fig. 2. The data of the power system, are: $E_B = 1\angle 0$, $E_A = 1\angle\delta(t)$, $Z_A = 0.25\angle 75^0$, $Z_B = 0.25\angle 75^0$, $Z_{Line} = 0.5\angle 75^0$. The power system frequency is $50\,Hz$ and simulation time step is $500\,\mu s$.

Case1: first test is programmed to examine capability of traditional CPSB in detecting power swing. In order to simulate the power swing, displacement angle of source A is considered as:

$$\delta(t) = \delta_0 + k \cdot e^{-t/\tau} \cdot \sin\left(2\pi \cdot f_{slip} \cdot t\right) \qquad (2)$$

where $k = 5$ is constant scaling coefficient, $\tau = 0.3$ is the damping time constant and $f_{slip} = 1Hz$ is the slip frequency. Impedance trajectory of this case is shown in Fig. 3. According to this figure, power swing starts at $t = 0.4$ s with impedance value *1.88–0.24i*, which is outside of the relay outermost zone. After power swing initiation, impedance starts to move and come near the relay zones. Before entering the relay's outermost zone, two circular concentric characteristics (outer and inner zones) are located to record the travelling time of impedance trajectory between outer and inner zones. This recorded time is compared with threshold value for detecting power swing. Therefore, if threshold value is selected accurately (threshold value is selected based on traveling time in fastest power swing), traditional CPSB can detect power swing in this condition.

Case2: A second test is programmed to show capability of traditional CPSB in detecting symmetrical fault during power swing in special condition. Similar to the previous case, power swing is simulated by displacement angle of

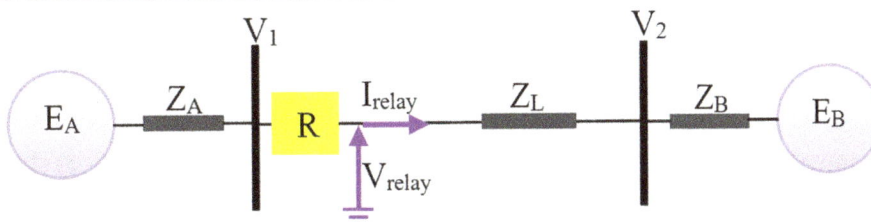

Fig. 2 Simple two-machine-system

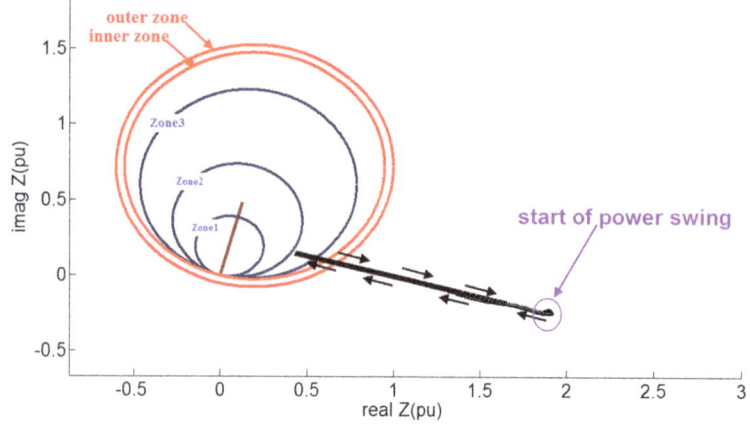

Fig. 3 Capability of concentric PSB in pure power swing detection

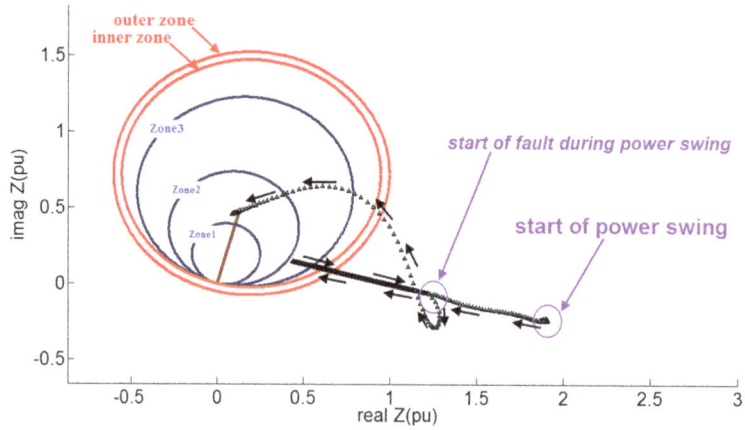

Fig. 4 Capability of concentric PSB in power swing detection and fault during power swing

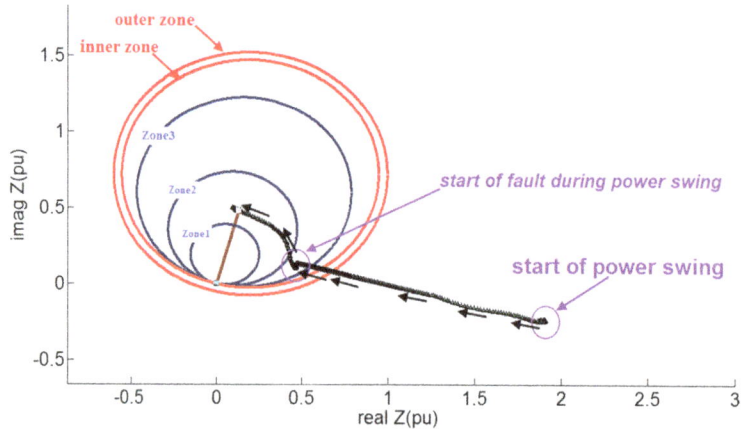

Fig. 5 Incapability of concentric PSB for detecting fault during power swing

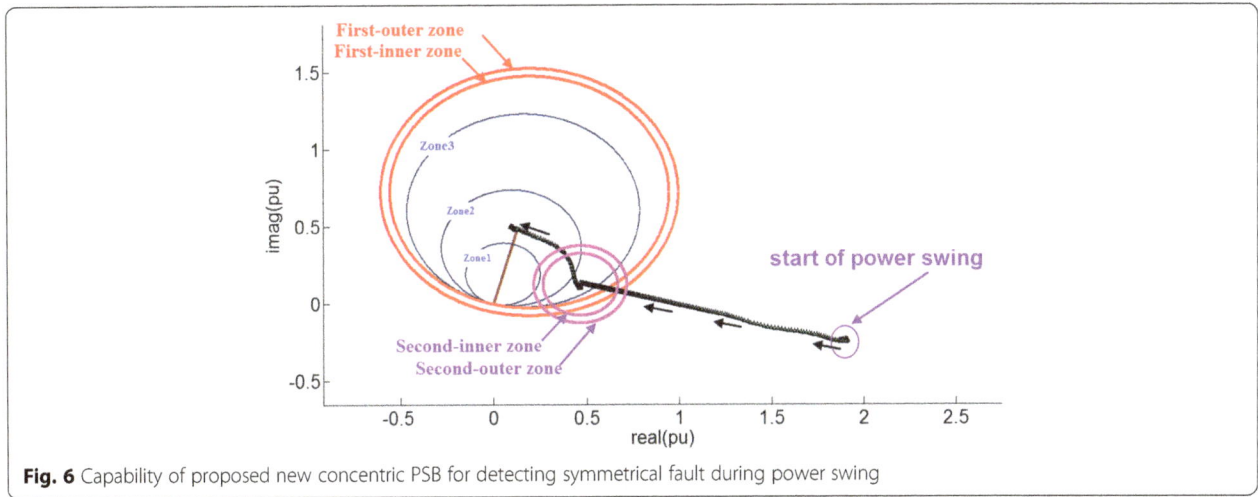

Fig. 6 Capability of proposed new concentric PSB for detecting symmetrical fault during power swing

source A as Eq. (2). A three-phase fault is simulated in right end of the protected line at $t = 0.85\,s$ during power swing.

The impedance trajectory of this case is shown in Fig. 4. According to the figure, after power swing initiation, impedance moves toward distance zones so that it crosses the CPSB for the first time and then the timer records the traveling time between outer and inner zones. Therefore, power swing can be detected by comparing the recorded time with threshold value and then distance relay is blocked. As a consequence of power swing, impedance trajectory moves back and gets away from the distance zones and so leaves the outer zone of CPSB.

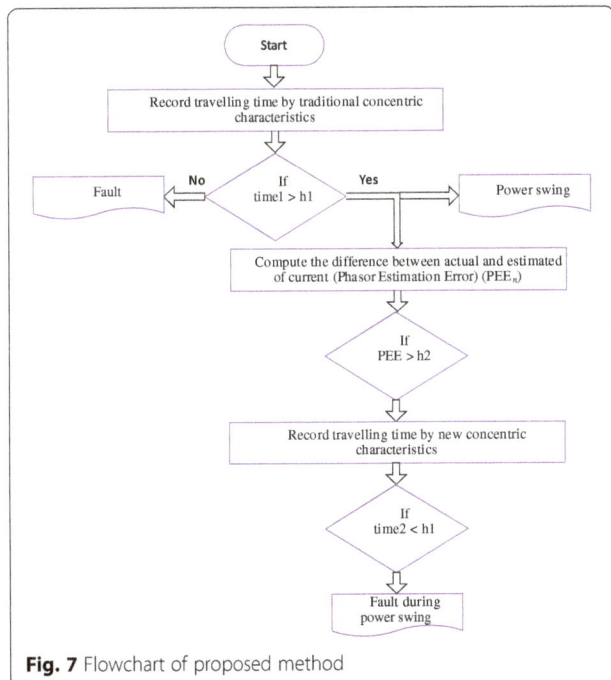

Fig. 7 Flowchart of proposed method

Next a three-phase fault occurs at $t = 0.85\,s$ during power swing. This causes the impedance trajectory crosses CPSB again during fault and so a new travelling time is recorded by timer, which can be used for detecting fault individually. Hence, traditional CPSB can detect both power swing and fault during power swing in this case.

Case3: A third test is simulated to show the condition in which traditional CPSB cannot detect a three-phase fault during power swing. Impedance trajectory of this case is shown in Fig. 5. Power swing is programmed similar to the two previous cases. According to Fig. 5, as a result of power swing, impedance trajectory crosses CPSB for the first time and the travelling time is recorded by timer, which can be used for detecting power swing. However, a fault occurs at $t = 0.55\,s$, when the impedance trajectory is inside the inner zone of CPSB. According to Fig. 5, impedance trajectory does not cross the traditional CPSB again during fault. This condition shows inability of traditional CPSB in three-phase fault detection during power swing.

Although, distance relay can easily detect unsymmetrical faults with various faulted loops by

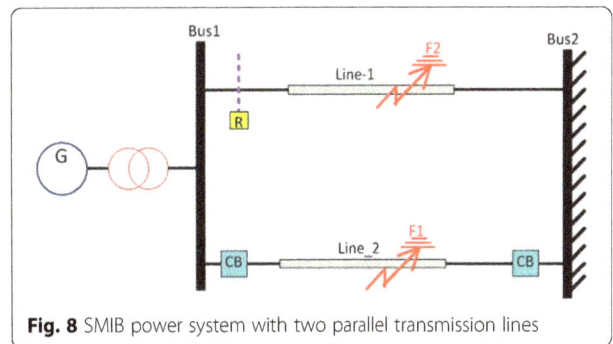

Fig. 8 SMIB power system with two parallel transmission lines

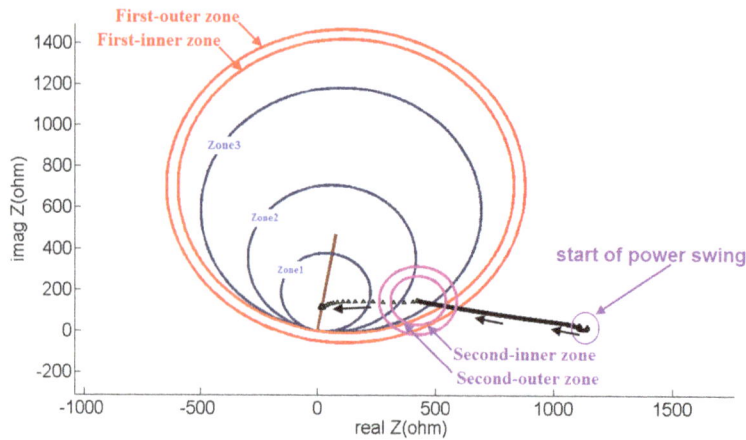

Fig. 9 Impedance trajectory for power swing at =1 sec and three-phase fault at t = 1.9 sec at 25% line during stable power swing in SMIB

assessing the negative sequence of current signal, it is faced by challenge in symmetrical faults during power swing because of inconsiderable amount of negative sequence during three-phase fault.

Methods
Proposed adaptive concentric psb
According to motioned simulations and explanations, traditional concentric PSB has limitation for detecting symmetrical fault during power swing and cannot detect it in special condition. When a symmetrical fault occurs, while impedance trajectory of power swing is inside of inner zone of CPSB, this kind of CPSB cannot detect fault because there is no second cross through zones of the CPSB during fault period.

In order to solve this problem, adaptive CPSB is proposed in this paper. According to proposed method, second pair of CPSB is programmed, which

its location is adapted by PEE, for detecting symmetrical fault during power swing. This idea is shown in Fig. 6. According to this figure, proposed method provides two independent pairs of CPSB for power swing and symmetrical fault during power swing. Therefore, recorded time by second PSB is used for detecting fault during power swing.

Another key point of this proposed method is detection of the location of impedances trajectory (the place in complex plane) for placing the second CPSB. In order to achieve this goal, phasor estimation error (PEE) is employed in this paper. By monitoring PEE during power swing, abrupt change of PEE can be used as primary indicator of symmetrical fault initiation and then the second CPSB is set at corresponding impedance in complex plane.

Hence, the proposed method in this paper includes two steps. In the first step, first CPSB is placed farther zone3 to discriminate power swing from fault. Recorded time by this CPSB is compared to predefined threshold

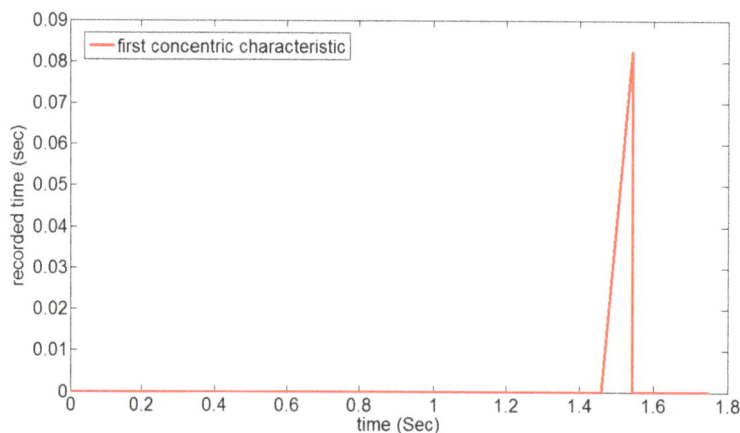

Fig. 10 Recorded time by first concentric characteristic

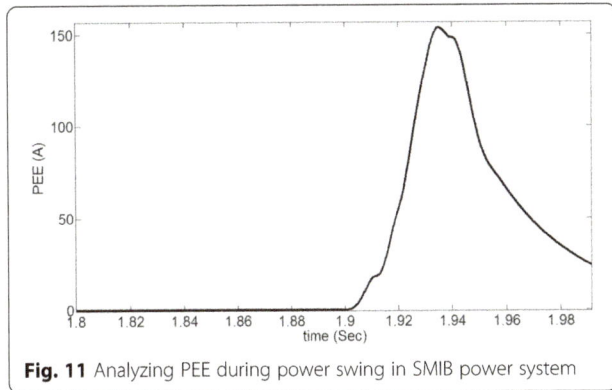

Fig. 11 Analyzing PEE during power swing in SMIB power system

(h1) so that it is detected as power swing if it is higher than threshold otherwise it is detected as fault. The second step of proposed method is employed when power swing is detected by first step. In the second step, PEE is calculated during power swing continuously and analyzed (compare to predefined threshold (h2)) to anticipate three-phase fault during power swing. In order to verify this anticipation, the second CPSB is placed where impedance trajectory presents at this time. Recorded time by the second CPSB is compared to the predefined threshold (h1) so that it is detected as symmetrical fault during power swing if it is lower than threshold value. Therefore combination of these two pairs of CPSB provides a complete method which can detect power swing and three-phase fault during power swing in different situations. Flowchart of the proposed method is shown in Fig. 7.

Results

Simulation part of this paper is divided into three parts. In the first part, the proposed method for detecting three-phase fault during power swing is examined in single machine to infinite bus (SMIB) and in the second part; the performance of the proposed method in three-machine power system is verified and in the last section, the performance of the proposed method is examined in IEE 39-Bus power system.

Simulation results of the proposed method in single machine to infinite bus (SMIB)

In order to validate performance of the proposed method (shown in flowchart (Fig. 7)) in discriminating three-phase fault from power swing, power system shown in Fig. 8 is considered, which its data are presented in [8]. A distance relay is considered at bus 1 in the upper line (line with impedance $76.8 + 469.98i$). A three phase fault (F1) is simulated at the middle of lower line which occurs at $t = 1$ s and is cleared after 0.03 s by opening the breakers at both ends (CB1, CB2). This event causes a stable power swing in the line between buses 1 and 2 and is observed by the relay R. Therefore, distance relay should be blocked by power swing blocker during power swing. Moreover, A three-phase fault (F2) initiates at $t = 1.9$ s (at 25% protected line) during power swing which should be detected by power swing blocker and then distance relay should be unblocked.

Impedance trajectory of this condition is shown in Fig. 9. According to this figure, the stable power swing causes the impedance trajectory enters into protected zone 3, which could lead to malfunction of the distance relay. In order to prevent this malfunction, first pair of CPSB is designed farther zone3 to detect power swing. Travelling time between first-outer and first-inner zone is shown in Fig. 10. According to this figure, impedance

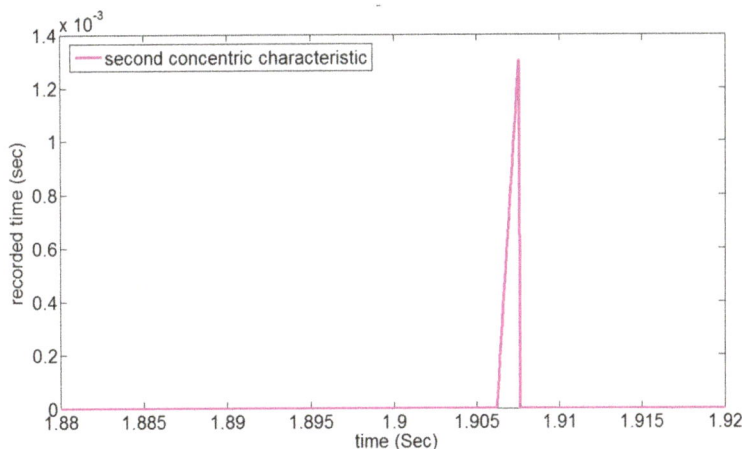

Fig. 12 Recorded time by second concentric characteristic in SMIB

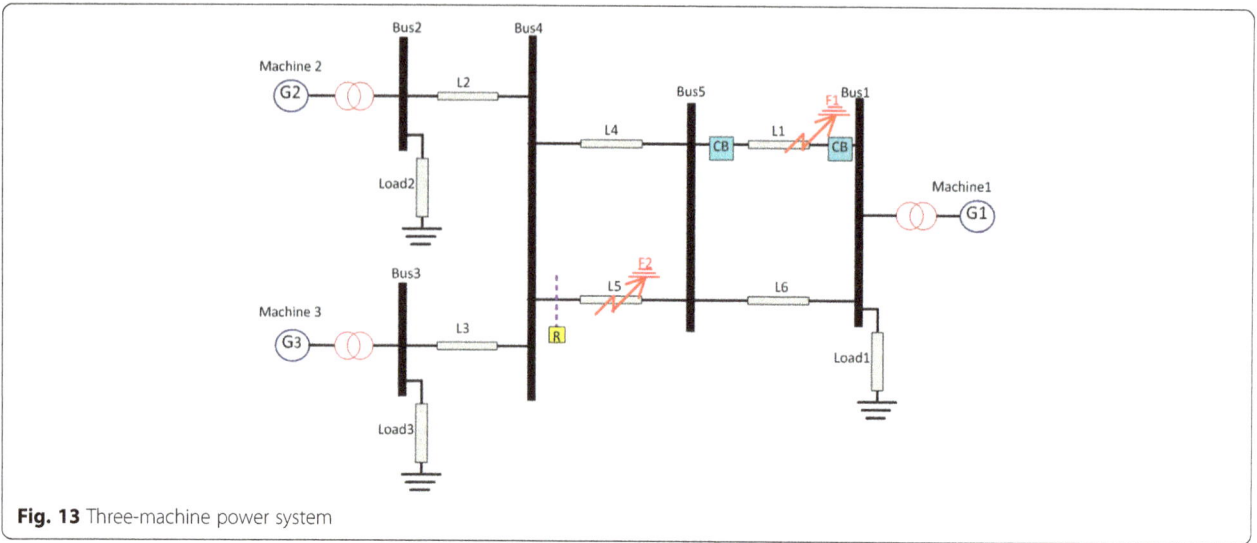

Fig. 13 Three-machine power system

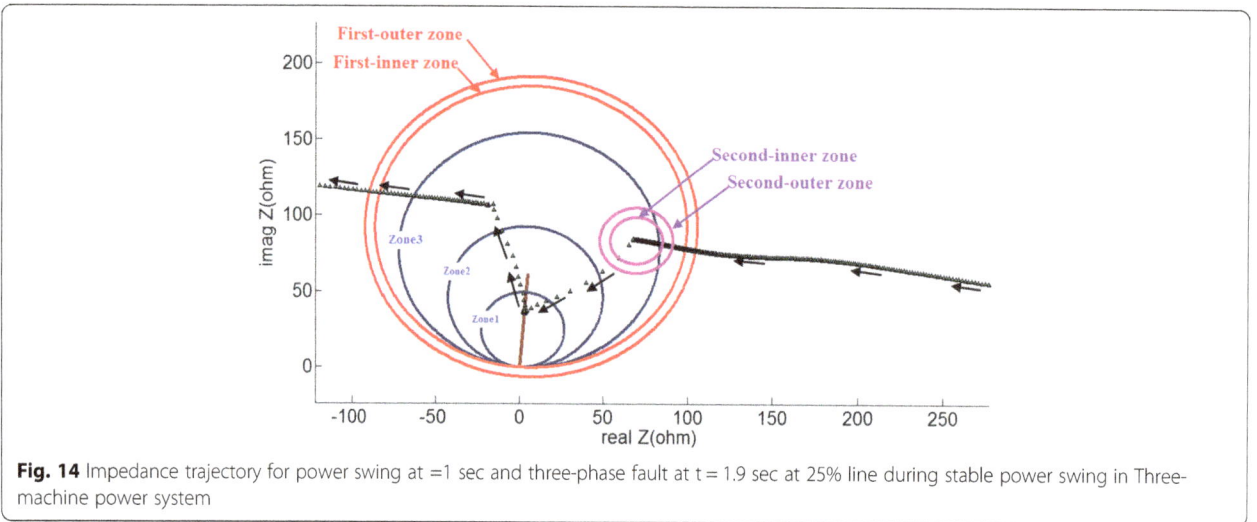

Fig. 14 Impedance trajectory for power swing at =1 sec and three-phase fault at t = 1.9 sec at 25% line during stable power swing in Three-machine power system

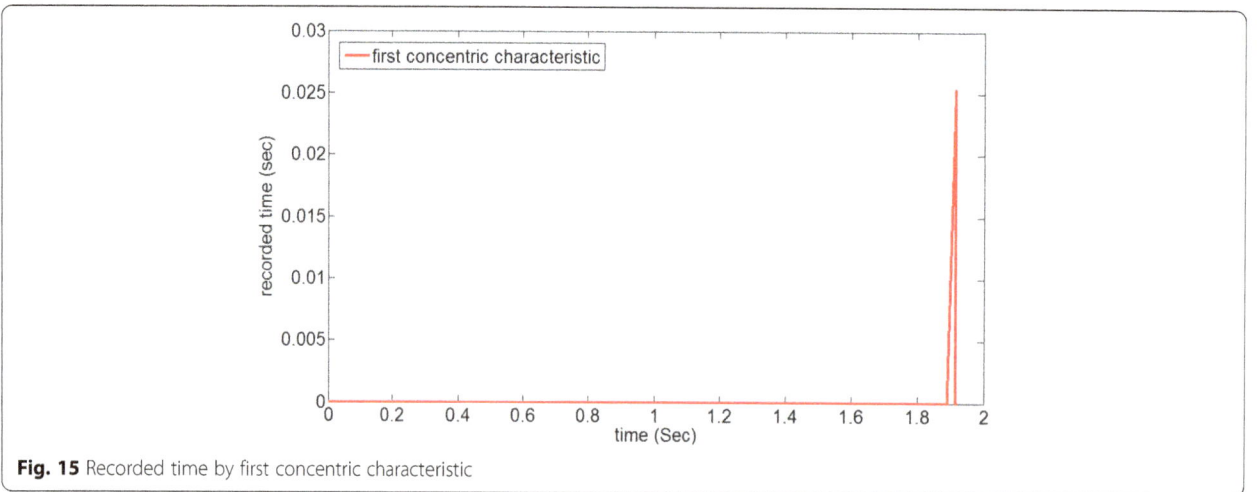

Fig. 15 Recorded time by first concentric characteristic

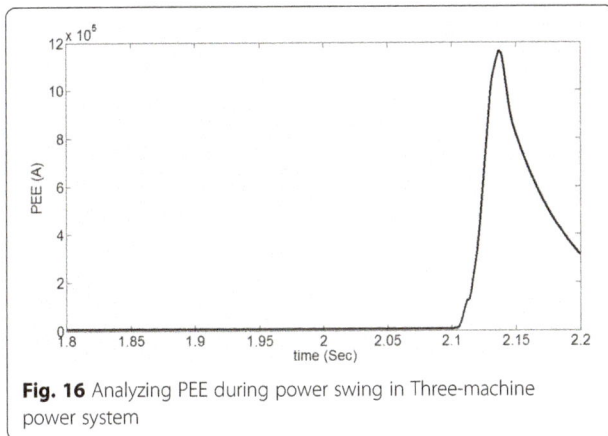

Fig. 16 Analyzing PEE during power swing in Three-machine power system

Simulation results of three-machine power system

In order to examine the proposed method in larger power system, the three-machine power system shown in Fig. 13 is considered [18]. A three-phase fault is simulated at 90% of the line connecting buses 5 and 1. The fault (F1) occurs at $t = 1$ s and is cleared after 0.25 s. This event causes an unstable power swing that is observed by the distance relay. Moreover, another three-phase fault (F2) initiates at $t = 2.1$ s (during unstable power swing) in 57% protected line.

Impedance trajectory of this condition is shown in Fig. 14. According to this figure, the unstable power swing causes the impedance trajectory enters into protected zone 3. First CPSB is designed farther zone3 at the first step of proposed method and the process of recording time by CPSB (between first-outer and first-inner zones) is shown in Fig. 15. According to this figure, impedance trajectory needs 0.026 s for travelling between first-outer and first-inner zones. By comparing the recorded time with threshold value (0.005 sec), it can be understand that this is power swing.

PEE of current signal is monitored during unstable power swing detection. PEE of current signal is shown in Fig. 16. According to this figure, a new transient happens at $t = 2.1$ s which is anticipated to be a three-phase fault. In order to verify this anticipation, the second CPSB is designed (as shown in Fig. 14) and recorded time by this CPSB is shown in Fig. 17. According to this figure, impedance trajectory needs 0.0008 sec for travelling between two zones of second CPSB. By comparing recorded time with threshold value (0.005 sec), it can be understudied that this is symmetrical fault during power swing.

trajectory enters first-outer zone at $t = 1.465$ s and enters first-inner zone at $t = 1.54$ s that results in 0.085 s recorded time by first CPSB. By comparing recorded time with threshold value (0.01 sec), power swing can be detected.

Based on the proposed method, PEE of current signal is monitored continuously during power swing. PEE of the current signal is shown in Fig. 11. According to this figure, a new transient happens at $t = 1.9$ s which is anticipated to be three-phase fault. In order to verify this anticipation, the second CPSB is designed (as shown in Fig. 9) and recorded time by this CPSB is shown in Fig. 12. According to this figure, impedance trajectory enters outer zone at $t = 1.906$ s and enters second-inner zone at $t = 1.907$ s which result in recorded time close to $0.0013se$. By comparing recorded time with threshold value (0.01 sec), it can be detected that this is symmetrical fault during power swing.

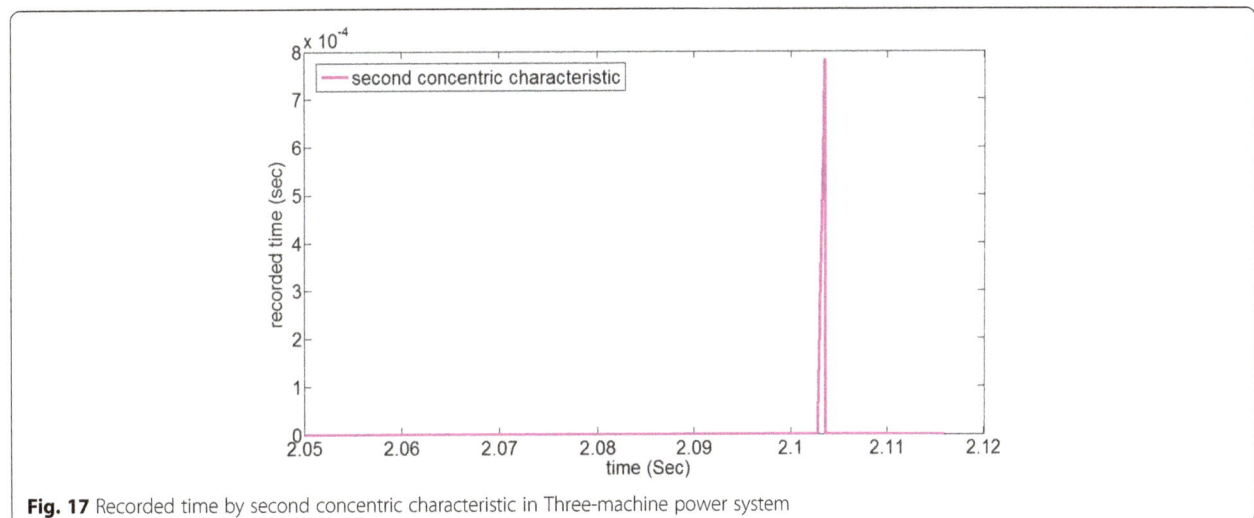

Fig. 17 Recorded time by second concentric characteristic in Three-machine power system

Simulation results of IEEE 39-Bus power system

IEEE 39-Bus power system is examined as a large test system (Fig. 18) in this paper. A three-phase fault (F1) is simulated at 50% of the line connecting buses 10 and 13. The fault occurs at t = 1 s and is cleared after 0.2 s. This event causes an unstable power swing and is observed by the distance relay (R). In order to examine the performance of the proposed CPSB, another three-phase fault (F2) is simulated at 100% of the protected line (line connecting buses 4 and 14) during unstable power swing. The impedance locus for this condition is shown in Fig. 19. According to this figure, at the first the impedance starts to move at t = 1.2 s and enters into the

protective relay's characteristics of distance relay due to the power swing. In this condition, the distance relay is blocked by the first CPSB; meanwhile, another three-phase fault (F2) occurs at t = 1.9 s. So the impedance leaves the power swing locus immediately and reaches to the fault impedance point. As is shown in Fig. 19, since the three-phase fault occurs after the impedance trajectory leaves the first CPSB characteristics (special condition); the first CPSB is not capable of detecting the fault.

According to Fig. 19 and proposed strategy, first CPSB characteristics are designed farther zone3 for a distance relay to discriminate fault from power swing. Travelling time between first-outer and first-

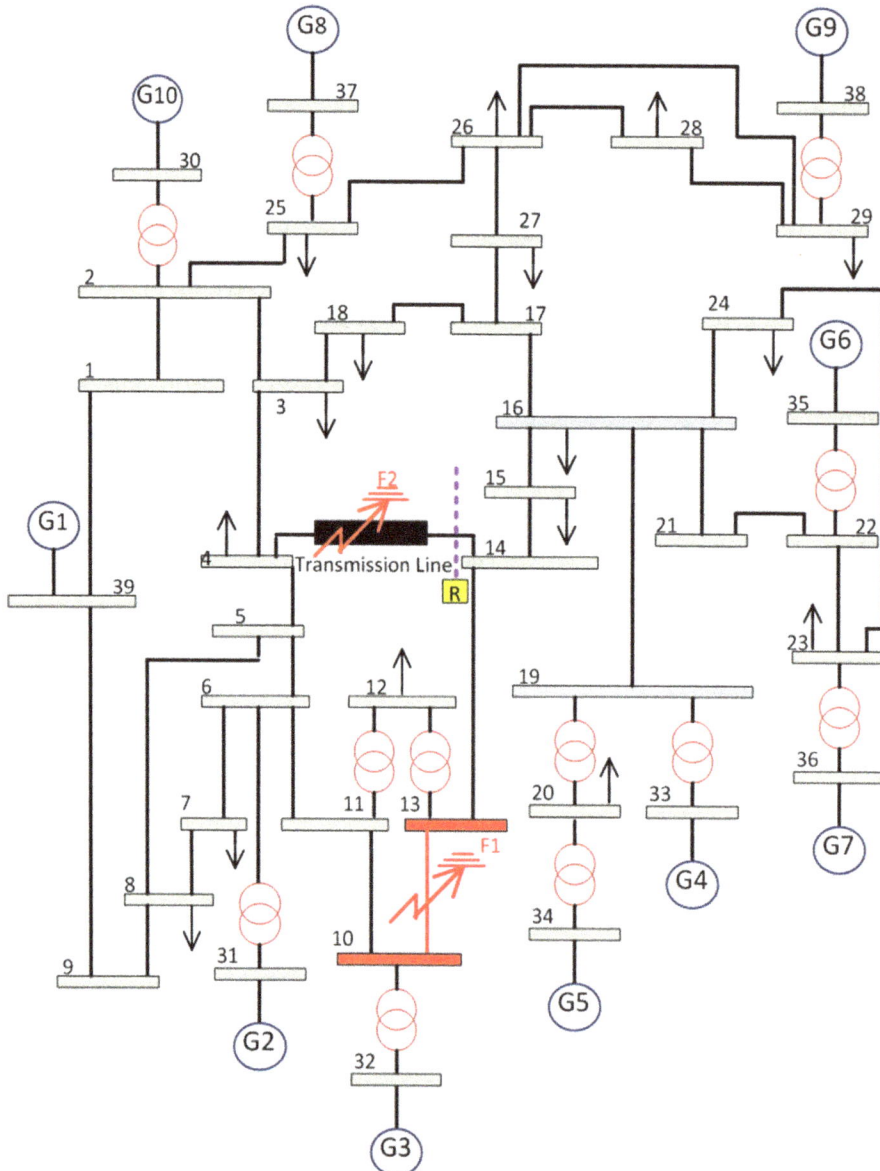

Fig. 18 IEEE 39-Bus power system

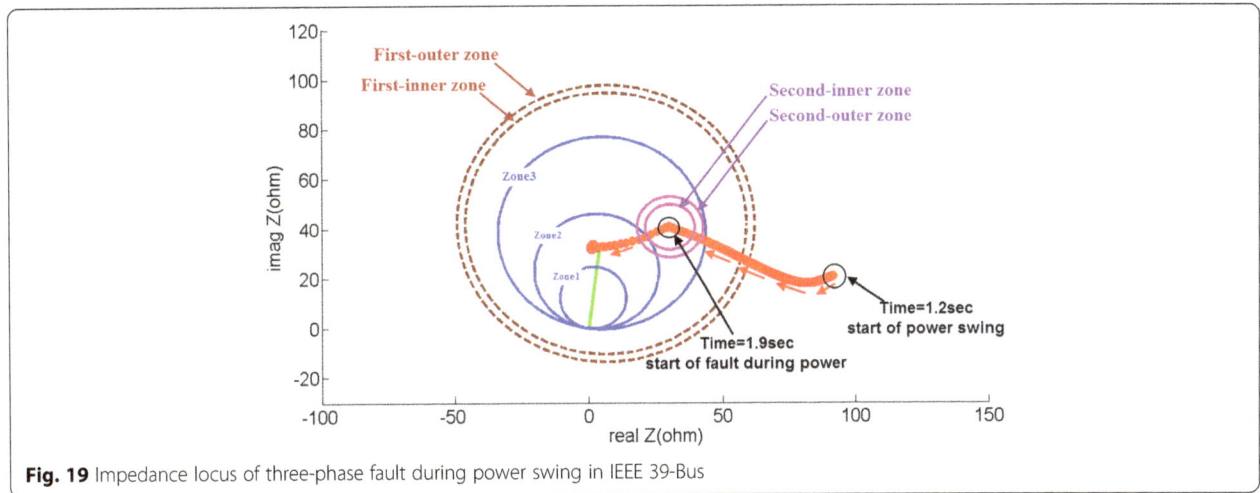

Fig. 19 Impedance locus of three-phase fault during power swing in IEEE 39-Bus

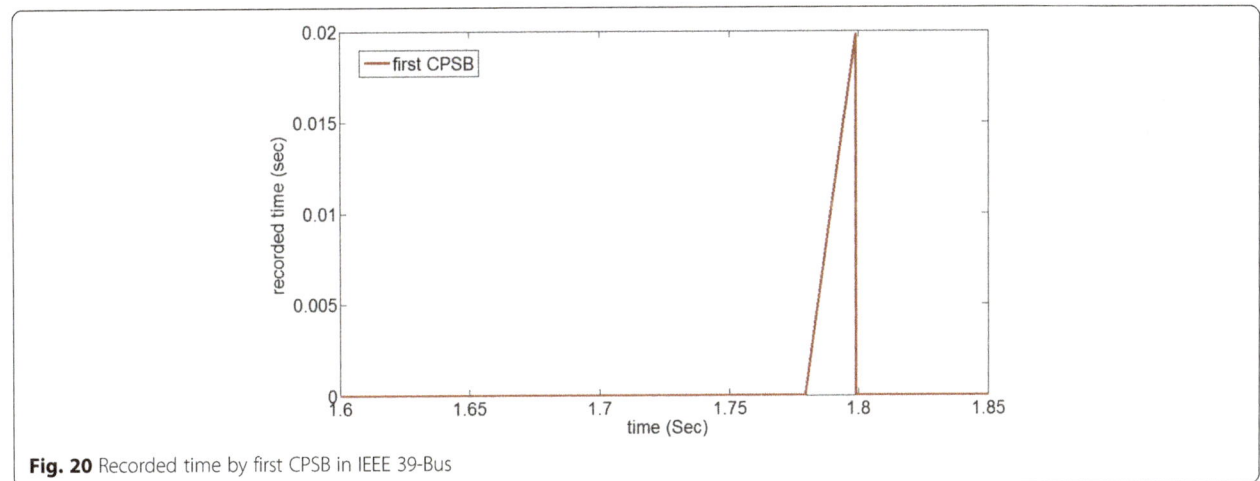

Fig. 20 Recorded time by first CPSB in IEEE 39-Bus

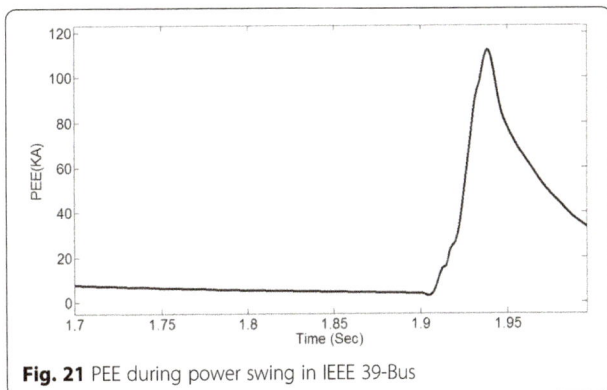

Fig. 21 PEE during power swing in IEEE 39-Bus

inner zone is shown in Fig. 20. According to this figure, impedance trajectory enters outer zone at t = 1.78 s and enters inner zone at t = 1.8 s which result in 0.02 s recorded time by first CPSB. By comparing recorded time with threshold value (0.005 sec), it can be understand that this is power swing. Based on proposed strategy, PEE of current signal is monitored during time after power swing detection. PEE of current signal is shown in Fig. 21. According to this figure, a new transient starts at t = 1.9 s which is anticipated to be a three-phase fault. In order to verify this anticipation, second CPSB is designed (as shown in Fig. 19) and recorded time by this CPSB is shown in Fig. 22. According to this figure, impedance trajectory enters second-outer zone at t = 1.907 s and enters second-inner zone at t = 1.908 s which result in 0.001 s

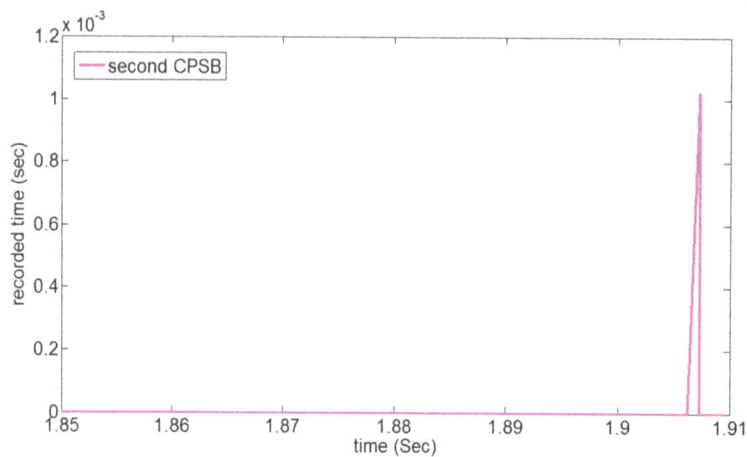

Fig. 22 Recorded time by second CPSB in IEEE 39-Bus

recorded time by second CPSB. By comparing recorded time with threshold value (0.005 sec), it can be understudied that this is symmetrical fault during power swing.

Conclusion

Measured apparent impedance by a distance relay moves into relay operating zones during power swing as a consequence of disturbance in power system that causes malfunction of distance relay. Traditional CPSB is designed inside of distance relay to prevent this malfunction by blocking distance relay during power swing. However, if a fault occurs during power swing, it should be detected and distance relay is blocked. Traditional CPSB is a common method for detecting power swing. However, it has limitation in detecting symmetrical fault during power swing. Therefore, adjustable concept of this method is proposed in this paper to overcome this difficulty. According to the proposed method, two pairs of CPSB are employed; the first CPSB is used for discriminating fault from power swing and the second CPSB is used for detecting symmetrical fault during power swing. According to results, the proposed method demonstrates its ability to unblock distance relay in three-phase fault during power swing.

Authors' contributions

JK, Ph.D. student, brings up the idea of adaptive procedure, performed the primary simulations and drafted the manuscript. MK, JK Ph.D. supervisor, participated in enriching the manuscript (in theoretical idea and simulation section (IEEE 39-Bus power system)) and carried out the revising the manuscript (response to the reviewers and editing grammatical and lexical mistakes). Both authors read and approved the final manuscript.

Competing interests

The authors declare that they have no competing interests.

References

1. Kundu, P., & Pradhan, K. (2014). "Synchrophasor-assisted zone 3 operation". *IEEE Trans Power Del, 29*(2), 660–667.
2. Nayak, P., Pradhan, K., & Bajpai, P. (2015). "Secured zone 3 protection during stressed condition". *IEEE Trans Power Del, 30*(1), 89–96.
3. Horowitz, S., & Phadke, A. (2006). "Third zone revisited". *IEEE Trans Power Del, 21*(1), 23–29.
4. IEEE PSRC WG D6. (2005). "Power swing and out of step considerations on transmission lines", *A report to power system relaying committee of the IEEE power engineering society.* IEEE PES (Power and Energy Society).
5. Mechraoui, A., & Thomas, D. W. P. (1995). *"A New Blocking Principle with Phase and Earth Fault Detection during Fast Power Swings for Distance Protection", IEEE Transactions on Power Delivery, 10*(3).
6. Mechraoui, A., & Thomas, D. W. P. (1997). *"A New Principle for High Resistance Earth Fault Detection during Fast Power Swings for Distance Protection", IEEE Transactions on Power Delivery, 12*(4).
7. Ganeswara, J., & Pradhan, A. (2012). "Differential power based symmetrical fault detection during power swing". *IEEE Trans Power Del, 27*(3), 1557–11564.
8. Brahma, S. (2007). "Distance relay with out of step blocking function using wavelet transform". *IEEE Trans Power Del, 22*(3), 1360–1366.
9. Mohamad, N., Abidin, A., & Musirin, I. (2014). "Intelligent power swing detection scheme to prevent false relay tripping using S_Transform". *International Journal of Emerging Electrical Power systems, 15*(3), 195–311.
10. Zade, H., & Li, Z. (2008). "A novel power swing blocking scheme using adaptive neuro-fuzzy inference system". *Electr Power Syst Res, 78*(7), 1138–1146.
11. Gautam, S., & Brahma, S. (2012). "Out-of-step blocking function in distance relay using mathematical morphology". *IET Generation, Transmission & Distribution, 6*(4), 313–319.
12. Rao, J., & Pradhan, K. (2015). "Power swing detection using moving averaging of current signals". *IEEE Trans Power Del, 30*(1), 368–376.
13. Lin, X., Gao, Y., & Liu, P. (2008). "A novel scheme to identify symmetrical fault occurring during power swings". *IEEE Trans Power Del, 21*(1), 73–78.
14. Andanapalli, K., & Varma, B.R.K. (2014). *"Parks transformation based symmetrical fault detection during power swing",*(pp. 1-5). Guwahati: Power Systems Conference (NPSC) Eighteenth National.
15. Mahamedi, B., & Zhu, J. (2012). "A novel approach to detect symmetrical faults occurring during power swings by using frequency components of instantaneous three phase active power". *IEEE Trans Power Del, 27*(3), 1368–1376.
16. Nayak, P., & Bajpai, P. (2013). "A fault detection technique for the series-compensated line during power swing". *IEEE Trans Power Del, 28*(2), 714–722.
17. Khodapaast, J., & Khederzadeh, M. (2015). "Three-Phase Fault Detection During Power Swing by Transient Monitor". *IEEE Transactions on power system, 30*(5), 2558–2565.
18. Moravej, Z., Pazoki, M., & Khederzadeh, M. (2014). "Impact of UPFC on Power Swing Characteristic and Distance Relay Behavior". *Power Delivery IEEE Transactions on, 29*, 261–268.

9

Development approach of a programmable and open software package for power system frequency response calculation

Yuzheng Xie[1], Hengxu Zhang[1], Changgang Li[1*] and Huadong Sun[2]

Abstract

Dynamic behaviour of frequency is crucial for power system operation and control. Several frequency response models have been proposed to reveal frequency dynamics from different aspects. A comprehensive software package incorporating major frequency response models is needed for analysis and control of power system frequency dynamics. In this paper, an approach for developing a programmable and open software package for frequency response studies is proposed. The framework of the package is extendable with reduced frequency response models. Essential models for frequency response study are included, e.g., generator, load, and under-frequency load shedding (UFLS). The provided application program interfaces (APIs) enable simulation with high-level languages by calling dynamic link library and makes the package programmable. An advanced application module is developed for quantitative assessment of transient frequency deviation. APIs can also be used for model extension and secondary development. To demonstrate the usage of the package, several examples are illustrated to explain how to perform simulations with the package, and to perform advanced applications using scripting with the provided APIs.

Keywords: Frequency response, Frequency control, Software engineering, Under-frequency load shedding, Transient frequency deviation security, Power systems

1 Introduction

As one of the most important electrical parameters of power systems, frequency and its dynamic characteristic is crucial for power system operation and control [1]. Frequency dynamics interacts with many devices in power systems in two ways. First, the performance of many devices are affected by frequency, e.g., speed governor of synchronous generators, induction motors, power system stabilizer (PSS) with frequency as input, and reactance and susceptance of transmission lines or shunt components. Second, the dynamic behaviour of system frequency is also affected by those frequency-dependent devices. Besides, dynamic characteristic of frequency is a key factor influencing power system protection. Generators are protected against abnormal

frequency deviation with over-speeding and under-speeding protective relays. Under-frequency load shedding (UFLS) is an important resort to prevent power system collapse in the event of large generation deficit [2, 3]. The cooperation between the generating unit protective relays and UFLS is important for power system frequency stability [4]. Moreover, investigation into some blackouts indicates that large frequency deviation is a main factor pushing power systems to the edge [5]. In the process of power systems restoration, frequency deviation should be carefully restricted by gradually starting generators and loads to avoid large frequency deviation and subsequent system failure [6]. Furthermore, with large scale of renewable energy integrated into power grids, power fluctuation from renewables will lead to continuous power system frequency fluctuation and deviation [7]. Primary and secondary frequency regulation have important effect on preventing frequency deviation. The participation of renewable energy generation in frequency regulation also plays an important role in frequency security, regulation,

* Correspondence: lichgang@sdu.edu.cn
[1]Key Laboratory of Power System Intelligent Dispatch and Control of the Ministry of Education (Shandong University), 17923 Jingshi Road, Jinan, Shandong 250061, China
Full list of author information is available at the end of the article

and control. Therefore, it is necessary to study the dynamic behaviour of power system frequency for improving the operation performance of modern power systems.

There are two ways to obtain frequency response of power systems. One is the measurement from devices such as phasor measurement units (PMUs) [8] and digital fault recorders from high voltage levels, and frequency disturbance recorder (FDR) [9, 10] and PMU Lights [11] from low voltage levels. The measured frequency reveals the actual dynamic behaviour of power systems. However, without the knowledge of event type and location, system frequency behaviour can hardly be examined with the measured frequency. In most situations, power systems are operated in ambient mode with little frequency deviation. Dynamic behaviour of frequency with large frequency deviation can be rarely observed. Therefore, the frequency dynamic behaviour can hardly be studied using measurement data.

The other way to get frequency response is to perform numerical simulations with mathematical models of devices. It can easily create scenarios with large frequency deviation by setting up appropriate events. It is the most used technique for studying frequency dynamic characteristics and designing proper control strategies. Numerical simulation methods can be classified into two categories: detailed methods and reduced methods. Full time-domain simulation is the widely used detailed method and provides detailed models of the network and dynamic equipment with appropriate simplification. With coupled active power-frequency dynamics and reactive power-voltage dynamics, frequency, voltage, and angle dynamics can be studied at one time. With detailed network, full time-domain simulation can easily reveal the space-time distribution characteristics of frequency dynamics [12]. With area interconnection and integration of large numbers of devices, computational burden of full time-domain simulation is significantly increased. Thus, full time-domain simulation is not suitable for such circumstance as online security evaluation and emergency control. To supplement the study of frequency dynamics, reduced models such as average system frequency model (ASF) [13], single machine model and system frequency response model (SFR) [14] are adopted. Unlike full time-domain simulation, the reduced methods consider only the active power-frequency dynamics to reduce computation burden.

During the long history of power systems research and operation, full time-domain simulation is extensively used in commercial power system analysis software such as PSS/E and DigSILENT Power Factory. The availability of those commercial software package greatly promotes the research and operation of modern power systems. The reduced models, however, have been developed by researchers for specific studies, and there is no comprehensive software package incorporating these reduced models. The aim of this paper is to propose a framework for developing a programmable and open software package for frequency dynamics study with reduced models.

The rest of the paper is organized as follows. A direct current power flow based frequency response model (DFR) is proposed and reduced frequency response models are reviewed in Section 2. A software package is proposed in Section 3 to incorporate major reduced frequency response models. To demonstrate the usage of the package, the IEEE 9-bus model, NPCC 140-bus model, and a 1000-bus model from China are tested in Section 4. The features of the proposed package are summarized and conclusions are drawn in Section 5.

2 Reduced frequency response models
2.1 DFR model
Full time-domain simulations can simulate frequency response of power systems in detail. However, due to the coupled active power-frequency dynamics and reactive power-voltage dynamics, both frequency and voltage need to be considered when studying power system dynamic behaviour. The influence of frequency and voltage can hardly be distinguished. Consequently, the DFR model is proposed in this paper to decouple frequency and voltage dynamics and to consider the influence of the network. In the DFR model, system network is simulated by direct current power flow so the redistribution of imbalanced power between different generators and the space-time distribution characteristics of frequency can be considered. To focus on active power-frequency dynamics, some assumptions are made as follows. (1) Excitation and regulation system is strong enough to hold the generator terminal voltage and thus, the dynamics of the excitation and regulation systems and the PSS can be eliminated for its negligible influence on active power-frequency dynamics. (2) Generators swing equations are reserved while the influence of transient process of the internal windings on system frequency change can be neglected due to the constant generator terminal voltage. Since turbine-governors have significant effect on power system frequency dynamics, details of the turbine-governor are modelled in the DFR model.

In the DFR model, the dynamic behaviour of frequency is only influenced by active power change. With constant bus voltages, direct current power flow is introduced to simulate the network when calculating active power flow under initial operating condition [15]:

$$\mathbf{P}=\mathbf{B}\boldsymbol{\theta} \tag{1}$$

where $\mathbf{P}$ is the active power injection, $\boldsymbol{\theta}$ is the voltage angle of all buses except the slack bus, and $\mathbf{B}$ is the network susceptance matrix.

The assumption of constant voltage leads to the simplification of load models. With constant terminal voltage, reactive power of loads can be ignored and the polynomial load model [16] can be reduced as a static active power load with frequency dependency:

$$P_L = P_0\left(1 + K_{pf}\Delta f\right) \qquad (2)$$

where P_L is the actual load, P_0 is active power of the load under initial condition, K_{pf} is the load regulation coefficient, and Δf is frequency deviation.

Other models can also be simplified with appropriate assumptions. For example, high voltage direct current links (HVDC) can be represented as loads for sending and receiving ends with or without frequency dependency.

With the simplifications of the generating units, network, loads and other equipment, the DFR model can be shown in Fig. 1. Quantities in Fig. 1 are listed as follows. ω_i, δ_i, P_{mi}, and P_{ei} are rotor speed, rotor angle, mechanical power, and electrical power of generating unit i. Δf_j and P_j are the bus frequency and active power of load j.

Similar to full time-domain simulation, the DFR model can be expressed in terms of differential-algebraic equations (DAEs), and can be solved by step-by-step integration such as implicit trapezoid integration. Comparing with full time-domain simulation, the computational burden of the DFR model is greatly reduced and it achieves a better computational efficiency with acceptable accuracy. The DFR model can be used to analyse events of load change, generator tripping, etc. It can be also applied to fast frequency response calculation for active power disturbances and event screening.

With the introduction of direct current power flow, the DFR model is applicable to systems in which the network reactance is significantly greater than the resistance, e.g., high voltage transmission systems. The DFR model is primarily useful for cases where frequency

stability is the main concern and angle stability and voltage stability can be maintained.

2.2 ASF model

In real systems, the frequency difference among buses is trivial if generators remain in synchronism during transient process [13]. Thus frequency at different buses can be treated as uniform and space-time distribution of frequency can be neglected. By neglecting the network, the DFR model reduces to ASF model from which uniform frequency can be achieved. The general diagram of the ASF model is shown in Fig. 2(a) where turbine-governors and loads are modelled explicitly. $P_{m\Sigma}$ and $P_{e\Sigma}$ are total mechanical power and total active power load of the system. $\Delta\omega$ is the uniform frequency of the system which is generated from the equivalent swing equation. In addition, all loads can be aggregated into an equivalent load model to simplify the ASF model. It can be applied in applications such as spinning reserve allocation, load frequency control, etc [17, 18].

The ASF model can be modelled with DAEs and solved with step-by-step integration. With network neglected, the computational burden of the ASF model is much less than that of the DFR model.

2.3 Single machine model

Single machine model can be treated as a special case of the ASF model, as shown in Fig. 2(b). It is obtained by further aggregating all turbine-governors and loads in the ASF model. The nonlinearity of the turbine-governors, such as the valve limits and dead bands, is reserved. The structure of aggregated turbine-governors is usually the same as normal turbine-governors. For example, for stand-alone system with most of electricity generated by thermal generating units, steam turbine-governor is preferred for the aggregated model. Step-by-step integration is also used to solve the nonlinear single machine model.

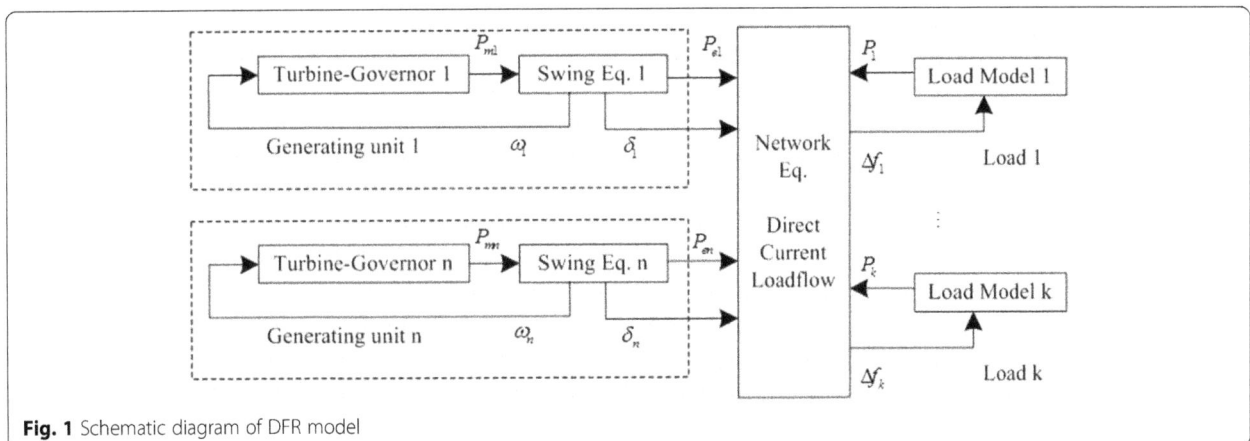

Fig. 1 Schematic diagram of DFR model

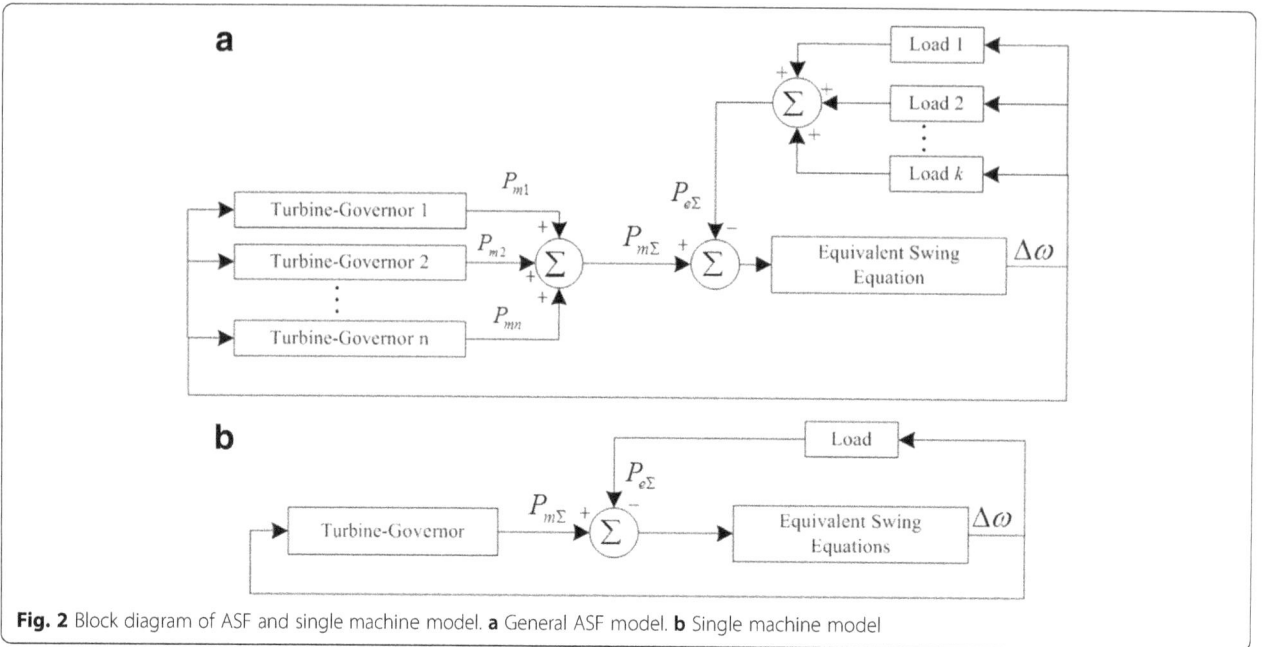

Fig. 2 Block diagram of ASF and single machine model. **a** General ASF model. **b** Single machine model

2.4 SFR model

Nonlinearity of the turbine-governors is considered in the DFR model, ASF model and the single machine model. No analytical expression can be directly obtained and step-by-step integration is the most popular method to get discrete response. By neglecting the nonlinear blocks and small time constants, SFR model was proposed in [14] to derive an analytical expression of frequency dynamics for stand-alone systems, in which the generators are dominated by reheat steam turbines. The block diagram of the SFR model is shown in Fig. 3 where P_d, P_m, H, D, R, F_H, T_R, and K_m are disturbance, mechanical power, inertia, damping, droop, fraction of total power generated by high-pressure turbine, time constant of reheater, and mechanical power gain factor of the aggregated system. Using the analytical expression given in [14], the largest frequency deviation, its corresponding time, and steady frequency under a given active power disturbance can be calculated. Several research adopts SFR model for adjusting UFLS [19, 20].

2.5 Discussion

The frequency dynamic characteristics can be categorized in different ways. For applications depending on the overall dynamic characteristics of frequency, e.g., frequency regulation, uniform frequency is usually assumed and the frequency at different locations is treated as the same. In this case, network can be neglected, and ASF model, single machine model, and SFR model are appropriate. The space-time distribution feature of frequency during event is of most interest for applications such as event location and oscillation detection where the difference between the generators at different locations should be taken into account. In this case, the influence of network should be retained to get the space-time characteristics, and the DFR model is suitable for such applications.

For detailed study of power system dynamic characteristics, the coupling between active power-frequency dynamics and reactive power-voltage dynamics should be included, resulting in the complex full time-domain simulation. However, for cases where frequency dynamic characteristic is of most concern and voltage dynamic is

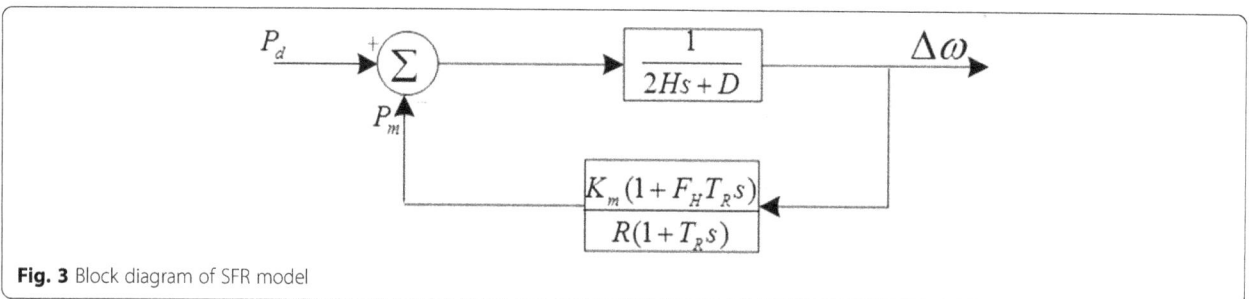

Fig. 3 Block diagram of SFR model

of little interest or voltage can be held at desired levels, the active power-frequency dynamics can be decoupled from the reactive power-voltage dynamics for simplification. It makes active-power the only factor affecting frequency, and the reduced models introduced above are suitable to examine the key impact of active power on frequency.

3 Architecture of software package incorporating reduced frequency response models

3.1 Framework of the software package

The framework of the proposed software package is shown in Fig. 4 with the following modules:

(1) Model library: Models with great impact on the frequency dynamics should be modelled in the package. The models implemented in the package are discussed in the next section in detail.

(2) Data assembler: The data file in supported formats such as the IEEE and PSS/E data formats can be recognized and imported into the memory.

(3) Dynamic equivalence: The function of this module is to supply equivalence calculation for model parameters.

(4) Event library: From the information contained in event library, power system malfunctions or failures, such as generator tripping, load shedding/increasing and continuous loads variation can be set up.

(5) Frequency response calculation: The function of this module is to implement frequency response calculation. The reduced models, DFR, ASF, single machine and SFR model, are implemented in the package using the step-by-step integration method or analytical solution to calculate frequency response.

(6) Frequency security assessment: Security assessment module is implemented for advanced applications.

Details of the frequency security assessment module can be found in Section 3.4.

(7) Application program interfaces (API): The APIs are used to provide interface functions for advanced applications and secondary development.

3.2 Model library

The following models are implemented in the package with appropriate simplifications.

(a) Conventional generator: With decoupling of active power-frequency dynamics and reactive power-voltage dynamics, detailed generator models with damping windings are not required. In DFR model, generators are usually modelled as classical model with swing equation and constant internal voltage behind transient or sub-transient reactance. For ASF and single machine model, no transient or sub-transient reactance is modelled. Only aggregated swing equations are kept in the ASF and single machine model.

(b) Turbine-governor: Turbine-governors provide the mechanical power for generators and are modelled in detail in the package. Typical turbine-governor models are only concerned with active power-frequency dynamics, with reactive power-voltage dynamics ignored. Thus, turbine-governor models can be reserved without simplification.

(c) Load: Considering only active power, two types of load models are implemented in the package: static load model considering frequency dependency and dynamic load model considering induction motors with active power-frequency dynamic response [21].

(d) HVDC: With more and more HVDC projects deployed, the control of HVDC should be modelled for frequency studies, which contains active power

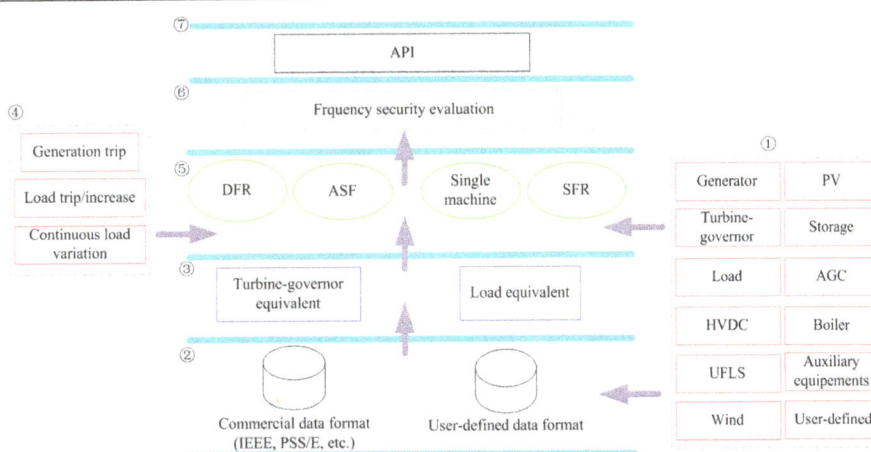

Fig. 4 Framework of the proposed package for frequency response calculation

modulation, dead band of frequency deviation and active power order sub-modules.

(e) UFLS: The control strategies are important for power system frequency stability. Under-frequency load shedding is an important frequency control measure and is modelled in the package in detail.

(f) Wind generator: For wind farm participating into frequency regulation, electrical control is ignored whereas pitch control of the wind turbines is implemented for frequency regulation [22]. Meanwhile, in order to make the wind turbine dynamic process similar to conventional generating units, virtual inertia control can be considered.

(g) Photovoltaic (PV) generation and battery energy storage: With large-scale photovoltaic generation connected to power systems, its fluctuation determines the demands for battery allocation. The PV is usually modelled as negative loads and energy storage is modelled as loads with response to frequency changes.

(h) Protective relay: Generators are equipped with protective relaying devices and their malfunctions can cause serious active power disturbance and large frequency deviation. So the under/over frequency protective relays are included in this package.

(i) Boiler and automatic generation control (AGC): For the purpose of medium and long term simulation, boiler dynamics and AGC should be considered [23, 24].

(j) User-defined model: If other power system models are required, user-defined models can be added in the model library to extend the function of the package.

3.3 Dynamic equivalence module

There are many types of generators in power systems. For the single machine and SFR models, turbine-governors and loads should be aggregated as single turbine-governor and load. Swing equation equivalence [25] and turbine-governor equivalence [26] are implemented in the dynamic equivalence module. For turbine-governor equivalence, two sub-modules are developed for different purposes:

(a) Equivalence of the same type of turbine-governors. When aggregating several turbine-governors of the same type, model structure is retained and the parameters can be generally summed up with weights to give appropriate responses.

(b) Equivalence of different types of turbine-governors. When aggregating several turbine-governors of different types, appropriate model structure should be first selected and then parameters are optimized to match the overall dynamic characteristics.

There are many algorithms to deal with model parameter equivalence, such as particle swarm optimization method

(PSO) [27], dynamic aggregation [28, 29], weighted summation method and least square method [30, 31]. Appropriate algorithms can be implemented for desired applications.

3.4 Frequency security assessment

The output results from the frequency response calculation module can be further analysed with the frequency security assessment module. This module provides transient frequency deviation security (TFDS) assessment and frequency security margin index based on two-element table [32]. According to the requirements of power system operation and control, frequency deviation constraints in extent and duration are given as two-element table $[f_{cr}, t_{cr}]$ where f_{cr} is the deviation extent and t_{cr} is the corresponding maximum duration. For a given frequency trajectory and two-element table $[f_{cr}, t_{cr}]$, the frequency security index can be calculated by considering the cumulative effect of frequency deviation.

Frequency evaluation can be used for further frequency control decision-making. For example, for the setting of UFLS scheme, it can be used to check the feasibility of the scheme and provide guidance for how to optimize.

3.5 API module

The API module is divided into several sub-modules to realize different functional requirements. For example, data assembler APIs are used to read data, and equivalence APIs are used for dynamic parameter equivalence calculation. For security assessment and UFLS setting evaluation, APIs are also implemented for model extension and secondary development.

To improve the programmability, all APIs can be called via dynamic link library (DLL). The advantage of DLL over graphic user interface (GUI) is the freedom to prepare scripts for specific applications. Since almost all high-level languages support loading DLL, the package can be further implemented in other software to extend their functionality of frequency dynamics study.

4 Case study

A programmable and open software package with the framework proposed in Section 3 is implemented in this paper with C++. Dynamic models of PSS/E are supported. The compiled DLL is called in Python modules named *pydfr* for DFR model, *pyasf* for ASF model, *pysm* for single machine model, *pysfr* for SFR model, *pyeqv* for model equivalence, and *pyevl* for transient frequency deviation security evaluation.

4.1 Demonstration of APIs

The following example shows how a simulation with Python codes calling APIs is performed.

```
from pydfr import *   # import DFR module
read("ieee9.raw")     # read power flow data
solve()               # solve DC power-flow
save ()               # save power-flow results
dyre("ieee9.dyr")     # read dynamic data
chan("ieee9.cnl")     # read dynamic channels
dist("ieee9.dis")     # read disturbances
progress("ieee9.log") # set log file
strt()                # dynamics initialization
run()                 # run simulation
```

The API *read*() imports model data (including power flow and dynamic models) into the program. DC power flow is solved with API *solve*(), and power flow results are saved with API *save*(). Prior to running dynamic simulation, dynamic data should be imported with API *dyre*(), and the channels to be exported are set with API *chan*(). Events information can be imported by API *dist*(). If detailed action of the power system is expected, a dynamic action file can be set up with API *progress*(). Before running the simulation with API *run*(), API *strt*() is called to initialize the dynamic models. The simulation results are outputted automatically during simulation.

4.2 Model equivalence

To perform simulations with single machine and SFR models, model equivalence should be performed for multi-machine systems. In this paper, dynamic aggregation and weighted summation method are implemented for model equivalence calculation. The following codes show how an equivalence for the single machine model with typical steam turbine-governor model IEEEG1 is obtained.

```
import pyeqv                  # model equivalence
import pydfr                  # DFR related module
pydfr.read("ieee9.raw")
pydfr.solve()
pydfr.dyre("ieee9.dyr")
pyeqv.aggregate_st()          # aggregate the same type
pyeqv.aggregate_dt("IEEEG1")  #aggregate different types
```

The data imported into the package by the *pydfr* module can be accessed by the *pyeqv* module. To get the proper equivalent model, the same type of turbine-governors are first aggregated by API *aggregate_st*(). The aggregated models can be used for ASF model. If single machine model is to be called, the API *aggregate_dt*() needs to be called to reduce turbine-governors of different types to a single model.

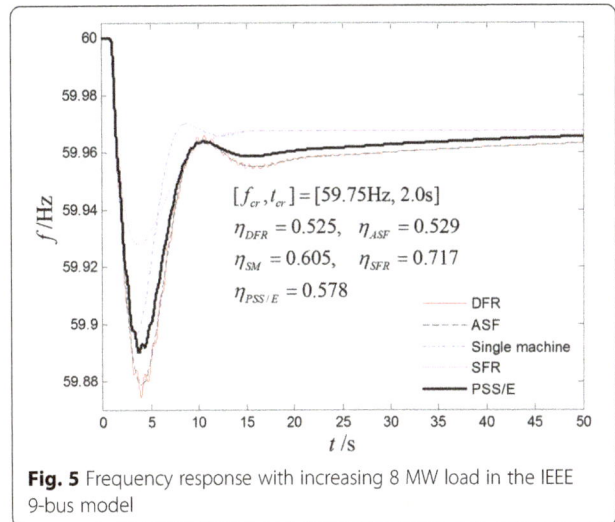

Fig. 5 Frequency response with increasing 8 MW load in the IEEE 9-bus model

4.3 Comparison of different models

Frequency responses of the reduced models are examined in this section with the IEEE 9-bus model and a 1000-bus model from China. Dynamic frequency of the two cases are shown in Figs. 5 and 6, respectively.

For the IEEE 9-bus model, PSS/E is adopted to perform the full time-domain simulation with complete models of generators, exciters, and turbine governors. Loads are modelled as 40% constant impedance load plus 60% constant power load with frequency dependency. In Fig. 5, the response captured by DFR deviates greater than that by PSS/E since the actual load in full time-domain simulation decrease slightly with the drop of voltage which is neglected in the DFR model. The DFR model generally reflects the overall frequency dynamics when voltage dynamics is neglected. For simplified models, the uniform

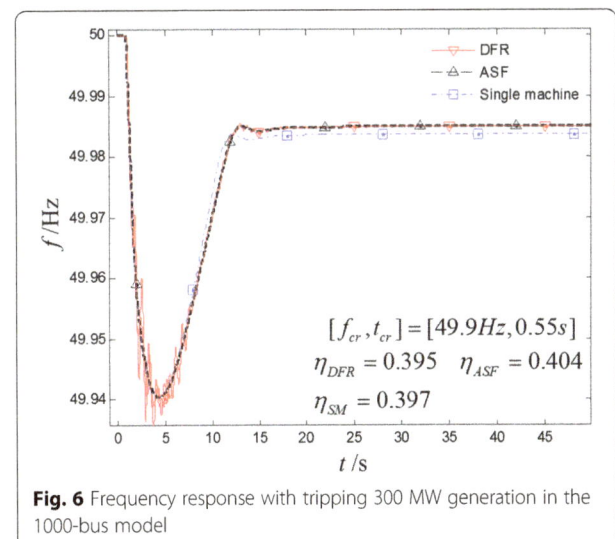

Fig. 6 Frequency response with tripping 300 MW generation in the 1000-bus model

frequency of the ASF model is almost the same as that of the DFR model, and both models can reflect frequency response process with good accuracy. The single machine model has similar tendency as the DFR and ASF models. However, the aggregation of the two steam turbines and one hydraulic turbine in the 9-bus model produces some errors, especially for the overshoot part around 8s. The linear SFR model gives different response from the nonlinear models where spinning reserve is limited. The values of the TFDS index η are calculated by the security assessment module and are shown in Fig. 5 based on a two-element table of [59.75Hz, 2.0s].

For the 1000-bus model from China, similar conclusions can be drawn. Due to the lack of spinning reserve, the frequency response of the equivalent SFR model deviates greatly from the results of other models and is not illustrated in Fig. 6. The two-element table for the 1000-bus model is [49.9Hz, 0.55s].

To compare the computation efficiency between the reduced models, time consumption for the generator tipping event in Fig. 6 is compared with simulation time span of 50s. On a PC with CPU of 2.83GHz, the simulation

time of the DFR model, ASF model, and single machine model are 3.259s, 0.531s, and <1ms, respectively. With the improvement of computational efficiency, the package is suitable for online frequency response analysis.

4.4 UFLS control
To demonstrate the applications of the package in UFLS control, a UFLS scheme is set up for the 1000-bus model. After tripping 5% of total generation, the first step of UFLS is activated when frequency drops beyond 49.25Hz with time delay of 0.2s. Four percent of total load is shed to recover frequency. The dynamics of frequency and total load are shown in Fig. 7.

4.5 Load variation
With large scale integration of renewable generation, the fluctuation of renewables will lead to frequency variation. A load variation model is provided in this package to check the impact of variable loads and renewables (negative load) on frequency dynamics. On the NPCC 140-bus model, perturbation of renewable power is applied with

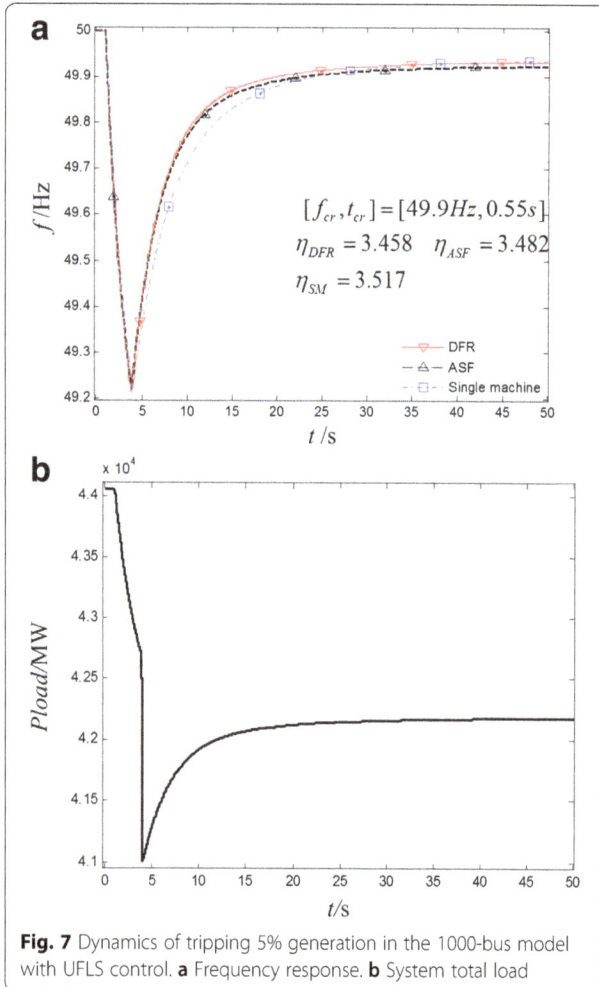

Fig. 7 Dynamics of tripping 5% generation in the 1000-bus model with UFLS control. a Frequency response. b System total load

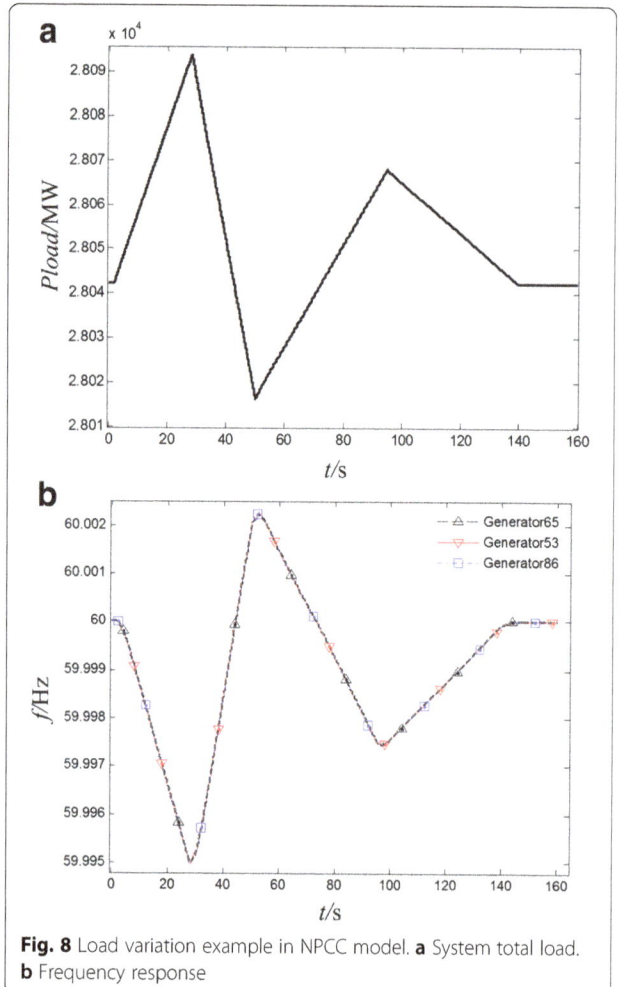

Fig. 8 Load variation example in NPCC model. a System total load. b Frequency response

the load variation shown in Fig. 8 (a). The frequency response is shown in Fig. 8 (b).

4.6 Advanced applications

With the APIs provided, advanced applications can be conducted by calling the APIs with high-level languages. The following codes show the application of searching critical load shedding amount during a sudden generation trip to prevent further activation of UFLS. The general process of simulation and evaluation is re-defined in the new function *sim_evl*().

```
def sim_evl(files,load_shed,ftable):
    read(files[0])
    solve()
    dyre(files[1])
    chan(files[2])
    dist(files[3])
    add_dist("LV 0 0 1.2 1 1.21 "+str(1-load_shed)) #use LV to shed
loads
    run()
    index=tfds_evaluation(ftable)   # perform TFDS evaluation
    free()
    return index

# the following is main searching process
raw_file="test.raw"
dyr_file="test.dyr"
chan_file="test.cnl"
dist_file="test.dis"
ftable=[49.25,0.2]  #threshold to trigger the 1st UFLS step
files=[raw_file,dyr_file,chan_file,dist_file]
load_shed1 = 0         # initial guess, zero control
index1=sim_evl(files,load_shed1,ftable)
load_shed2 = 0.05      # initial guess, 5% load shed
index2=sim_evl(files,load_shed2,ftable)
while 1:
    load_shed3=(load_shed1*index2-load_shed2*index1)/(index2-
index1)
    index3=sim_evl(files,load_shed3,ftable)
    if(abs(index3)<1e-3):
      break
    if(abs(index1)>abs(index2) and abs(index1)>abs(index3)):
      load_shed1 = load_shed3
      index1 = index3
    else(index3<0&&index1<0&&index3>index1):
      load_shed2 = load_shed3
      index2 = index3
```

Figure 9 shows the iteration process to calculate the critical amount of load shedding. The two-element table of the 1st step of UFLS ([49.25Hz, 0.2s]) is used to check the frequency security index. When the index is zero, the 1st step of UFLS is critically activated. With the initial guess (0% of iteration -1, and 5% of iteration 0), the searching progress converges after 6 iterations, and 1.71% loads should be shed 0.2 s after the event to prevent the activation of the 1st step of UFLS. The percentage shown in Fig. 9 are the load shedding amount of each iteration.

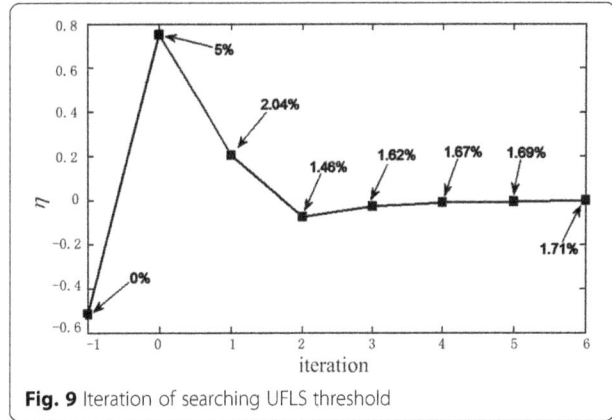

Fig. 9 Iteration of searching UFLS threshold

5 Conclusion

Development approach of a programmable and open software package for power system frequency response calculation is proposed in this paper where reduced frequency response models are incorporated. With the modularized framework, the package can be easily extended by adding new functions to the modules. APIs are provided in DLL to be called by other high-level languages. The programmability makes the package suitable for advanced applications and secondary development. An implementation of the proposed framework is introduced with support of PSS/E data formats. Simulations show that the software package is easy to use and APIs can be reorganized to perform simulations for specific purpose.

Acknowledgement
This work was supported by National Natural Science Foundation of China (No: 51477092).

Authors' contributions
HZ proposed the idea and helped to prepare and revise the draft. CL coded the simulation modules of DFR and ASF in C++, and prepared the draft. YX coded the aggregation module, evaluation module in C++, and the Python modules. Most simulations were conducted by YX. HS helped to revise the draft and reorder the sequence of the draft. The model library part was revised based on HS's suggestion. All authors read and approved the final manuscript.

About the authors
Yuzheng Xie (1991-), M.E. candidate. Major in power system security, stability assessment and control.
Hengxu Zhang (1975-), Ph.D. and professor. Major in power system security and stability assessment, power system monitoring and numerical simulation.
Changgang Li (1984-), Ph.D. and associate research fellow. Major in power system dynamic and control, and wide-area measurement and control.
Huadong Sun (1975-), Ph.D and professor level senior engineer. Major in power system security assessment, stability control and numerical simulation.

reimbursements, fees, funding, or salary from an organization that holds or has applied for patents relating to the content of the comment.

All the authors don't don't have any other financial competing interests.

There aren't any non-financial competing interests (political, personal, religious, ideological, academic, intellectual, commercial or any other) to declare in relation to this comment.

Author details

[1]Key Laboratory of Power System Intelligent Dispatch and Control of the Ministry of Education (Shandong University), 17923 Jingshi Road, Jinan, Shandong 250061, China. [2]China Electric Power Research Institute, 15 Xiaoying East Road, Qinghe, Beijing 100192, China.

References

1. Kundur, P., Paserba, J., Ajjarapu, V., et al. (2004). Definition and classification of power system stability. *IEEE Transactions on Power Systems, 19*(3), 1387–1401.

2. Gu, W., Liu, W., Zhu, J. P., et al. (2014). Adaptive Decentralized Under-Frequency Load Shedding for Islanded Smart Distribution Networks. *IEEE Transactions on Sustainable Energy, 5*(3), 886–895.

3. Larsson M. (2005) An adaptive predictive approach to emergency frequency control in electric power systems, 44th IEEE Conference on Decision Control, pp. 4434-4439.

4. Ahsan, M. Q., Chowdhury, A. H., Ahmed, S. S., et al. (2012). Technique to Develop Auto Load Shedding and Islanding Scheme to Prevent Power System Blackout. *IEEE Transactions on Power Systems, 27*(1), 198–205.

5. Wang, S. P., Chen, A., Liu, C. W., et al. (2015). Efficient Splitting Simulation for Blackout Analysis. *IEEE Transactions on Power Systems, 30*(4), 1775–1783.

6. Medina, D. R., Rappold, E., Sanchez, O., et al. (2016). Fast Assessment of Frequency Response of Cold Load Pickup in Power System Restoration. *IEEE Transactions on Power Systems, 31*(4), 3249–3256.

7. Henneaux, P., Labeau, P. E., & Maun, J. C. (2013). Blackout Probabilistic Risk Assessment and Thermal Effects: Impacts of Changes in Generation. *IEEE Transactions on Power Systems, 28*(4), 4722–4731.

8. Gurusinghe, D. R., Rajapakse, A. D., & Narendra, K. (2014). Testing and Enhancement of the Dynamic Performance of a Phasor Measurement Unit. *IEEE Transactions on Power Delivery, 29*(4), 1551–1560.

9. Lin, Z. Z., Xia, T., Ye, Y. Z., et al. (2013). Application of wide area measurement systems to islanding detection of bulk power systems. *IEEE Transactions on Power Systems, 28*(2), 2006–2015.

10. Liu, Y., Zhan, L. W., Zhang, Y., et al. (2016). Wide-Area-Measurement System Development at the Distribution Level: An FNET/GridEye Example. *IEEE Transactions on Power Delivery, 31*(2), 721–731.

11. Jin Z. S., Zhang H. X. Li C. G. (2015) 'WAMS Light and Its Deployment in China', the 5th International Conference on Electric Utility Deregulation and Restructuring and Power Technologies (DRPT), Changsha, China, pp. 1373-1376

12. Zhang, H. X., Shi, F., Liu, Y. T., et al. (2016). Adaptive Online Disturbance Location Considering Anisotropy of Frequency Propagation Speeds. *IEEE Transactions on Power Systems, 31*(2), 931–941.

13. Chan, M. L., Dunlop, R. D., & Schweppe, F. (1972). Dynamic equivalents for average system frequency behaviour following major disturbances. *IEEE Transactions on Power Apparatus and Systems, PAS-91*(4), 1637–1642.

14. Anderson, P. M., & Mirheydar, M. (1990). A low-order system frequency response model. *IEEE Transactions on Power Systems, 5*(3), 720–729.

15. Stott, B., Jardim, J., & Alsac, O. (2009). DC Power Flow Revisited. *IEEE Transactions on Power Systems, 24*(3), 1290–1300.

16. Wang K., Huang H., Zang C. (2013) 'Research on Time-Sharing ZIP Load Modeling Based on Linear BP Network', The 5th Intelligent Human-Machine Systems and Cybernetics Conference (IHMSC), Hangzhou, China, pp. 37-41

17. Ersdal, A. M., Imsland, L., & Uhlen, K. (2016). Model Predictive Load-Frequency Control. *IEEE Transactions on Power Systems, 31*(1), 777–785.

18. Yousef, H. A., AL-Kharusi, K., Albadi, M. H., et al. (2014). Load Frequency Control of a Multi-Area Power System: An Adaptive Fuzzy Logic Approach. *IEEE Transactions on Power Systems, 29*(4), 1822–1830.

19. Anderson, P. M., & Mirheydar, M. (1992). An adaptive method for setting under-frequency load shedding relays. *IEEE Transactions on Power Systems, 7*(2), 647–655.

20. Denis, L. (2006). A general-order system frequency response model incorporating load shedding: analytic modeling and applications. *IEEE Transactions on Power Systems, 21*(2), 709–717.

21. Aree, P., & Acha, E. (2011). Power flow initialisation of dynamic studies with induction motor loads. *IET Generation, Transmission & Distribution, 5*(4), 417–424.

22. Chang-Chien, L. R., Lin, W. T., & Yin, Y. C. (2011). Enhancing frequency response control by DFIGs in the high wind penetrated power systems. *IEEE Transactions on Power Systems, 26*(2), 710–718.

23. Xie Y. Z., Zhang H. X., Sun H. D., et al. (2015) Frequency response model considering large frequency deviation, Industrial Instrumentation and Control Conference (ICIC), Pune, India, pp. 940–943.

24. De Mello, F. P. (1991). Boiler models for system dynamic performance studies. *IEEE Transactions on Power Systems, 6*(1), 66–74.

25. Zhang B. Z, Zhang Y., Lin L. X., et al. (2012) Study on Two Dynamic Aggregation Algorithms of Coherent Generators, the 4th IEEE Computational Intelligence and Communication Networks Conference (CICN), Mathura, India, pp. 676-680.

26. Ourari, M. L., Dessaint, L. A., & Do, V. Q. (2006). Dynamic equivalent modeling of large power systems using structure preservation technique. *IEEE Transactions on Power Systems, 21*(3), 1284–1295.

27. Voumvoulakis, E. M., & Hatziargyriou, N. D. (2010). A Particle Swarm Optimization Method for Power System Dynamic Security Control. *IEEE Transactions on Power Systems, 25*(2), 1032–1041.

28. Ju, P., Ni, L. Q., & Wu, F. (2004). Dynamic equivalents of power systems with online measurements. Part 1: Theory. *IEE Proceedings-Generation, Transmission and Distribution, 151*(2), 175–178.

29. Ju, P., Li, F., Yang, N. G., et al. (2004). Dynamic equivalents of power systems with online measurements Part 2: Applications. *IEE Proceedings-Generation, Transmission and Distribution, 151*(2), 179–182.

30. Wan, J., & Miu, K. N. (2003). Weighted least squares methods for load estimation in distribution networks. *IEEE Transactions on Power Systems, 18*(4), 1338–1345.

31. D'Antona, G. (2003). The full least-squares method. *IEEE Transactions on Instrumentation and Measurement, 52*(1), 189–196.

32. Zhang, H. X., Li, C. G., & Liu, Y. T. (2015). Quantitative frequency security assessment method considering cumulative effect and its applications in frequency control. *International Journal of Electrical Power & Energy Systems, 65*(65), 12–20.

10

The research and the development of the wide area relaying protection based on fault element identification

Xianggen Yin, Zhe Zhang, Zhenxing Li, Xuanwei Qi, Wenbin Cao[*] and Qian Guo

Abstract

The urgent problem of the relaying protection in the modern AC/DC hybrid connected grid and the development of the wide area communication, the information process and the intelligent technology powerfully promotes the development of the technology of the wide area relaying protection (WARP), which has become a research hotspot that attracts extensive attention. Originated from the basic concept of the wide area relaying protection, this paper analyses the advantages, the effects and the functions of the wide area relaying protection. The two main approaches to realize the wide area protection, which are on-line adaptive setting (OAS) principle and fault element identification (FEI), are introduced in this paper. Aimed at improving the performance of the backup protection, the research content and the technology demand of the wide area protection are proposed, meanwhile, the basic principle and the algorithm of the fault element identification are introduced. At last, the scheme of the limited wide area relaying protection based on the existing pilot channel of the main protection is discussed.

Keywords: Wide area protection, Wide area relaying protection, On-line adaptive setting, Fault element identification, The limited wide area protection

Introduction

Because of the extremely unbalance distribution of the electrical load and the energy in China, the Chinese power grids are required to transport the large-scale electrical power from the west to east over long distances, of which the scale is expanding rapidly, the structure is becoming more and more complicated, the technology continues to improve and the operation is flexible and changeable. Meanwhile, with the intensive construction of the large-scale ultra-high or extra-high voltage HVDC, the national grid is becoming an unprecedented large-scale AC/DC hybrid connected complex power system, of which the system protection is confronted with the severe challenges. If the fault cannot be cut-off in time, there will be a complex and evolutionary accident process in the AC/DC hybrid power system, and what's worse, the HDVC will be forced to quit operation, causing the huge power shortage and severely

threatening the safe and stable operation of the power grid. In order to cut-off the faults rapidly and guarantee the safe operation of the whole power grid, the protection of the AC/DC hybrid system is required to operate faster, more reliably, more accurately and more sensitively, and especially, it is of great significance to increase the operation speed of the backup protection.

The operation of traditional relaying protection depends on the fixed value of off-line setting. However, in complex power system, especially when the operating mode of power system frequency changes, the cooperation and the coordination among the protections is hard to be achieved. For example, the selectivity, speediness and sensitivity of the protection cannot be satisfied at the same time. In the complex power system, because of the difficulty of the setting value calculation of the relaying protection, especially the backup protection, several backup protection configurations are trend to be simplified or abandoned. These conditions will increase the risk of the grid in urgent emergency. For instance, the clearing time of the fault will be prolonged and the isolation region of the fault will be extended, when the

* Correspondence: 1019097704@qq.com
State Key Laboratory of Advanced Electromagnetic Engineering and Technology, Huazhong University of Science and Technology, Wuhan 430074, China

primary protection refuse to operate due to low sensitivity and the breaker fails to trip or loss of protection power, resulting in the local grid disaster. Another example is that, during the large-scale power flow transfers in the power system, the unexpected cascading tripping of the backup protection may occur, which is the main reason for the accident extension and even the large area blackout [1–3].

To improve the performance of traditional protection fundamentally and satisfy the safety protection requirement of the modern AC/DC hybrid complex grid, in recent years, with the development of wide-area information collection technology in power system, the researches on protection and controlling based on wide area measurement information (hereinafter referred to as wide area information) are becoming a focused research hotspot in the global power industry [4–6]. In China, many researchers have put their effort into the studies on the wide area protection. There is progress on the research of the basic principle and the implementation scheme for the wide area protection. Meanwhile, several local industrial wide area protection systems have been put into use in actual power grid.

This paper analyses the advantages, the effects and the functions of the WARP, from the basic concept of the WARP. The two main approaches to realize the wide area protection: the on-line adaptive setting (OAS) principle and fault element identification (FEI) are introduced in this paper. The research content and the technology demand of the wide area backup protection are proposed. The basic principle and the algorithm of the fault element identification are introduced. The scheme of the limited WARP technology with high reliability and speed based on the existing pilot channel of the main protection is discussed.

The wide area relaying protection
The challenge of traditional relaying protection in modern power system

As the primary barrier of the power system, the relaying protection takes the responsibility of the fault clearing. However, with the development of the modern power system, the traditional relaying protection cannot satisfy the requirement for the safe and stable operation of the power system. The main problems are as follows:

The difficulty in calculation and cooperation of the setting value

The relaying protection system is mainly constituted by main protection and backup protection. The multi-stage backup protection mainly consists of the local backup protection and the remote backup protection of the power system, of which the typical mode is three stage protections. Because the grid structure and operation mode of the modern power system is becoming more and more complex and variable, the coordination among the related backup protection is very complex. Under such condition, the backup protection that only utilizes the local measurement and coordinate through the different time delay cannot satisfy the selectivity requirement of the power system protection. Because of the above problem, the backup protection trend to be simplified in the actual engineering application, and what's worse, the zone II backup protection is suggested to be cancelled and the zone III backup protection should be simplified in some situation. It is worth attention that when high resistance fault occurs in the large power system, the dual-configuration main protection still cannot detect the fault sensitively, which will prolong the fault clearing time and expand the scope of tripping. This will undoubtedly increase the risk of the local grid disaster under emergency state.

Long time trip of remote backup protection

The multi-stage cooperation mode will lead to a long delay time of the backup protection, easily resulting in commutation failure in power grid at receiving side, which is harmful to the security of AC/DC hybrid power grid.

The lack of self-adaptive ability

The setting coordination of traditional backup protection is based on the fixed operation mode of the power system. When the grid structure and operation mode of power grid changes, the backup protection may be hard to corporate and coordinate, which may cause the misoperation of the protection and expansion of the fault range.

The potential risk of misoperation

When the grid structure or running condition changes unexpectedly, the large load power flow transferring may occur in a large scale. The zone III distance protection may cascade trip unexpectedly, resulting in the system splitting or large scale blackout.

The major cause of these problems is that the relaying protection operates only based on the local information at the installation position. If more comprehensive information about the present power system can be acquired, for example, if the backup protection can get the information of the present system and the related information from the wider area, the performance of the protection will be greatly improved. For example, if the remote backup protection can get the relevant information from the remote protected device, the fault can be identified more accurately and cleared more quickly. The problems of traditional protection may be eliminated through the wide area information.

Wide area protection and the wide area relaying protection

Wide area protection (WAP) is the protective relay and the security and the stability control of the power system based on the wide area information, which consists of WARP and wide area security control (WASC) [7]. The wide area information is the information from the local and the remote power grid, rather than the information only from the protected device.

The narrow definition for the wide area protection is only related to the security and stability control, emphasizing the protection for the whole power system [3, 8, 9]. There are also several literatures regarding the security control among 0.1 ~ 100 s as wide area protection [10, 11], in order to emphasize the protection for the wide area power system general safety. From the perspective of the history and applications, the stable control function is achieved by using the local information or wide area information. In recent years, the real-time wide area information are more and more used to improve the stable control function [10, 11], which has enriched the concept of wide area stable protection from the aspect of wide area information. In addition, several literatures, call the WARP as wide area back-up protection [3], which indicates that using the wide area information to improve the performance of backup protection is more concerned at present.

The relaying protection and security and stability control can achieve the security emergency control together for the power system. The relaying protection is called as equipment protection, and the security and stability control is often taken as system protection or SPS (Special Protection System or System Protection Scheme), RAS (Remedial Action Scheme) [3, 8, 9]. In the view of above, the relaying protection can be called as fault protection, and the security and stability control can be called as security and stability protection.

The essence of the security emergency control in Chinese power system is the three lines of defense [12, 13]. In the process of emergency control, relaying protection and security control operate independently, perform their own functions and their function cannot replace each other. It is worth attention that there are two basic principles in the security emergency control: the necessity for failure clearing and the priority of failure clearing. This is the duty and responsibilities of relaying protection and WARP. The fault clearing is the premise of effective security control. So the function of backup protection cannot be replaced by security control [3].

The function and characteristics of wide area relaying protection

The important difference between WARP and traditional relaying protection is using the wide area measured information. The traditional relaying protection mainly uses the local information, and at most uses the remote information which is extended from local protected equipment. Please note that the protected line of the remote backup protection is adjacent to the neighbor element and the state of this line, the previous operation result of the other protection (main protection) and the relevant remote backup protection are all related with the remote information. However, the tradition protection only can make judgments just according to the local measurement information. Hence, the information used for the traditional protection is not sufficient for the power system protection, which is the main reason for the existing deficiency of the protection. Therefore, the studies on WARP at present focus on the backup protection, especially the remote backup protection.

The wide area information can effectively improve the performance of the backup protection. The protective information in the faulty region can be acquired by the wide area communication systems. The fault element can be determined by the multi-information fusion. Then the backup protection can make the protection decision based on the fault element, and the coordination between the protections can be simplified. Meanwhile, the real time system structure and the operation mode can be acquired through the wide area measurement and the communication technology. The setting value of the protection can be up-dated based on the current status of the power system. Hence, the sensitivity and the selectivity can be improved.

The farthest protection scope of relaying protection is the next adjacent equipment, which is the region of backup protection. Therefore, the protection scope of the relaying protection is limited. Hence the information of the WARP is limited, which introduces the concept of limited WARP. The most important characteristic of limited WARP is that it only needs the measurement information from the adjacent limited area, which can lighten the burden of wide area communication or information processing, which is beneficial to the realization of WARP. In practical engineering, it is possible to realize the limited-wide-area relaying protection by using the existing optical fiber communication channel of the pilot protection. On the other hand, the system protection is oriented for the safe and stable operation of the entire power system, which needs the information from the wide-range of the whole system. This is another important difference between the protective relay and the system protection.

Besides, in a substation, if not only the information of local protected equipment can be used, but also the information of relevant equipment in the station can be used, the performance of relaying protection can be improved effectively. This is called substation area relaying protection, which uses the substation area measurement information. The substation area protection does not need the remote communication, which is feasible for the industrial application, especially for the digital substation. The sharing and the utilization of the substation area information can improve the performance of the relaying protection and control functions in the substation, integrate the substation relaying protection and simplify the relaying protection configuration, extent the new relaying protection function flexibly, such as the bus-bar protection and the breaker failure protection in the medium or low voltage substation, implement the coordination between the relaying protection and the control and the unified substation protection system. Meanwhile, the substation protection system can access to the wide area protection system as the child-station, and corporate with the wide area protection to achieve the protection and control for the regional power grid. Therefore, some scholars have proposed the concept of the hierarchical protection, which is constructed by the bay layer relaying protection (traditional primary relaying protection), substation area layer relaying protection (substation area protection) and wide area layer relaying protection (WARP), of which the main design is shown in Fig. 1.

In Fig. 1, the bay layer relaying protection is only aimed at single equipment bay. With the information in bay unit, it achieves the protection function of devices in bay unit. It also collects information and executing orders for WARP. Substation area layer relaying protection is aimed at substation. With synthesized information from multiple equipments in the substation, it judges, decides uniformly and achieves protection and controlling function in substation area. It also preprocesses information and provides communication service for WARP. Wide area layer relaying protection is aimed at limited area grid. With synthesized information from multiple substations, it judges, decides uniformly and achieves wide area protection function in wide area.

The basic approach to achieve the wide area relaying protection

At present, there are mainly two different approaches to realize the basic function of WARP. One approach is based on the principle of online adaptive setting (OAS); another is based on fault element identification (FEI) to achieve WARP [7].

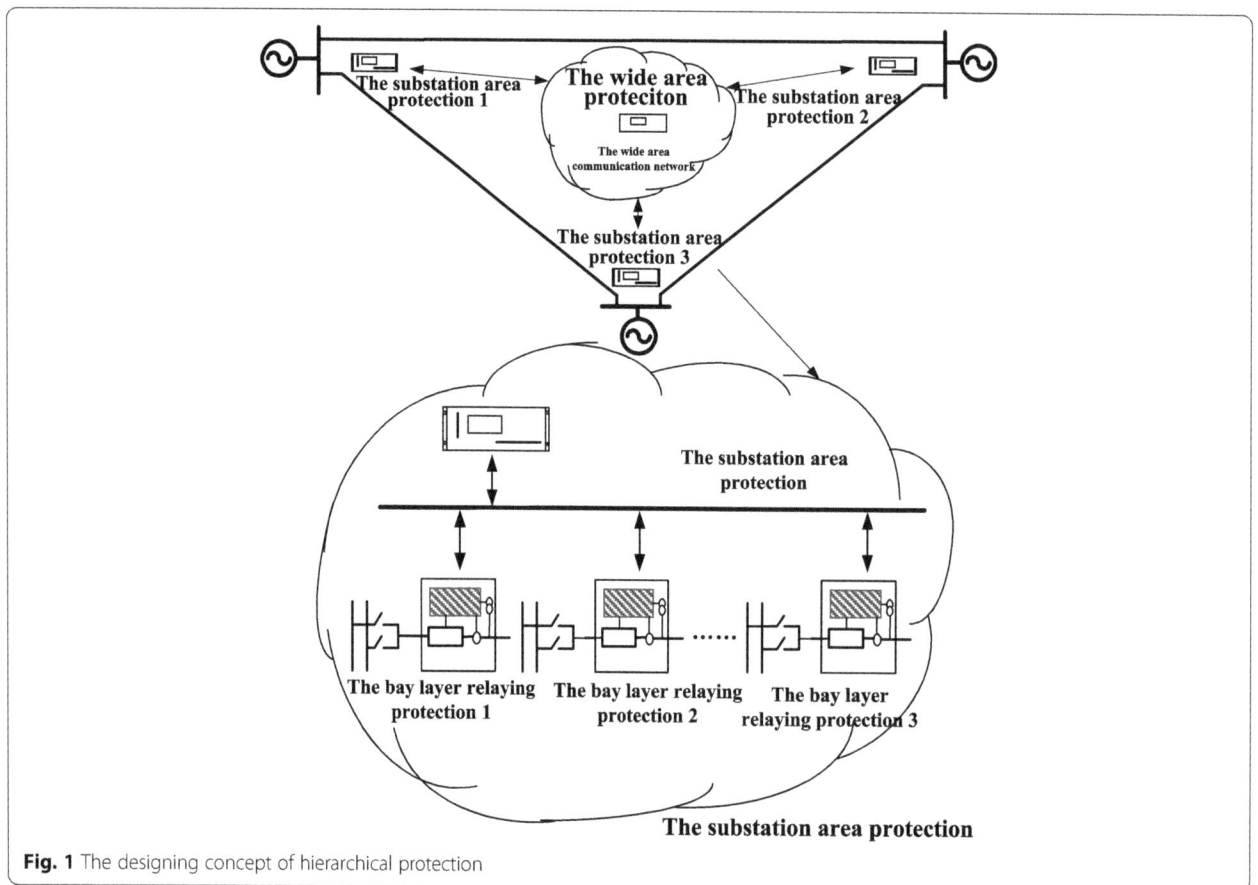

Fig. 1 The designing concept of hierarchical protection

The wide area relaying protection based on on-line adaptive settings

The study on on-line adaptive setting (OAS) started at 1980s and the foreign scholars have defined it as following [14]: An on-line activity that modifies the preferred protective response to a change in system conditions or requirements. It is usually automatic, but can include timely human intervention. And several Chinese scholars express it as following [15]: Taking the event-triggered mode, tracking the change of power system operation mode in time, calculating and on-line regulating the setting value of protection to prevent the protection from losing coordination and improve its sensitivity.

In recent 20 years, the researches on OAS setting approach are mainly around the fault disturbance domain identification, the search of minimum cut point and quick short circuit calculation and so on [15–18]. In China, the protection information system for the setting values online check has been constructed in the power system. Based on the wide area communication system, the protection information system can acquire the in-time operation condition of the power grid, check and reset the setting value, and transfer the updated setting value to each protection device [19–22].

WARP based on OAS has been studied for a long time. There are several progresses achieved. But its application is limited. This is because, although this approach improves the sensitivity and selectivity by adjusting the fixed value online, it still cannot overcome the shortcomings of the traditional backup protection, including the difficulty of the setting value coordination, the long operation time delay and so on. This shortcoming is also the important reason for hidden failure of the backup protection, which leads to the cascading trip and threats the system safety. In addition, how to ensure the timeliness and reliability of the wide area information communication, and how to calculate the setting value quickly after the change of the operation mode are still the questions to be solved.

The wide area relaying protection based on fault element identification

The study on WAPR based on FEI begins at the end of 1990s. For example, Yoshizumi Serizawa came up with the wide area current differential backup protection [23] at 1998, which form the scope of differential protection according to the area of backup protection and it can identify the fault element accurately and determine the operation scope of backup protection.

The WARP based on FEI can be described as following: By using the wide area multi-point measurement information in the power system and taking several fault discriminate mechanism, the position of fault element and the state of fault clearing can be determined. And

then, the cutting point of WARP can be determined. The breakers at the cutting point operate to isolate the fault through bay layer's equipment. Its superiority is that there is no need for setting calculation, the selectivity of backup protection can be guaranteed only by simple coordination of time and logic and the operation time of back up protection can be shorted effectively. For example, no matter where the remote backup protection locates, because it only coordinates with near backup protection (or main protection), the coordination between protection is simplified. Meanwhile, the backup protection can be prevented from cascading tripping during large scale load transferring.

The WARP based on FEI doesn't require the information changes in time of entire power system. Even for the remote back up protection, it only needs the relevant fault information of the nearby equipment which is in the rim of adjacent substation groups (the scope may be expanded appropriately from the perspective of information fault tolerance). Therefore, this is a limited wide area protection and it is beneficial to the achievement of its engineering.

Recently, the WARP based on FEI has drawn the research concern. There are many achievements in the aspect of identification of fault elements, system constitution and the industrial test. However, according to the requirement of engineering application, there are many technical problems remain to be solved. The following will discuss the topic.

Aiming at the potential risk problems that the unexpected cascading operation of backup protection which is probably caused by large scale load transfer, it can use the wide area information to analyze and distinguish the state of the load transferring in power system and take the measures [24, 25] such as, blocking or changing action characteristics for the relevant backup protection and so on, which can avoid the cascading trip of backup protection and guarantee the system security.

The research contents of wide area relaying protection based on fault element identification
System constitution and zone-division technology

WARP system has three basic links: 1) information acquisition and operation; 2) communication networks and information transferring; 3) fault element identification and action decision-making. In the whole power grid, based on the substation distribution situation, different system structures can be formed by reasonably distributing and organizing these three links, which are also accordant with the hierarchical protection structure. The first link is located at the bottom of the protection system which can be implemented by the intensively configured intelligent electric device (IED). The second link is the wide area communication network, which transfers the information between the first and third level. The third link makes the protection decision, of which the configuration form determines the wide-area protection system structure.

The system structure type of FEI-based WARP includes three main types: the centralized structure, the distributed structure and the distributed-centralized hybrid structure [26].

The distribution-centralized hybrid structure divides the protected zone of the power grid, every divided protected zone has a decision-making center which is responsible for the wide-area relaying protection and the information exchanges with the adjacent protected zone. This structure combines the advantages of the former two structures, and has a better research prospect. The zone-division distribution-centralized structure proposed in the article [27] is a typical hybrid structure. Hybrid structure needs to solve the issue of dividing the protected zone of the power grid, including the principle and method of automatic zone-division, the selection of the main substation, the boundary processing technology of the adjacent protected zone and so on [27, 28].

The principle of fault element identification (FEI)

The mechanism of fault element identification is the critical technologies for the FEI-based WARP. Consequently, the principle and the algorithm of fault element identification based on wide area information are the research hotspot in recent years, which mainly includes two aspects.

The first aspect is the research of the basic principle of fault element identification. In early time, several backup protection principles are proposed, such as the wide area current differential protection, wide area direction directional pilot protection [23, 29–32]. The wide area current differential backup protection is a kind of current differential protection utilizing the multi-side current in the protected region, of which the protected zone can cover the region of the local and the remote backup protection. This protection only needs a simple time delay cooperation between each protection, and can effectively guarantee the selectivity and the action speed of the backup protection. The wide area current differential protection has a simple and clear principle, and can detect various types of internal fault. But it relies on high precision synchronous sampling and still has some problems concerning with the threshold value setting method and the sensitivity of the protection, such as the wide area measurement error, the capacitance current compensation, and the high resistance grounding fault. The wide area directional pilot protection needs small communication traffic, and has low requirement for the synchronization sampling. However, the

directional element is highly affected by high resistance grounding, open-phase operation and fault transition and so on. To overcome these problems, several principles of FEI based on fault voltage distribution, wide area integrated impedance or integrating variety of information are presented in recent years.

The second aspect is to improve the reliability of FEI by using multi-source information fusion. The process of distributed collection, decentralized processing and remote communication of the wide area measurement information may cause the information error. On the other hand, the WARP can acquire multi-source information from the protected zone, such as the electric quantity, the state quantity and even the operation results of the protection criterion. This multi-source information will be processed by using the method of artificial intelligence, such as the genetic [33] and the probability identification [34]. The adverse influence of bad data can be avoided and the error-tolerance of decision-making can be realized.

Other technology issues

A brief description of some important research issues are briefly introduced in the following:

Tripping strategy

After the fault elements are determined, the action status of breaker connected to fault element also need to be monitored. Only when a breaker refuse to work, can the wide area backup protection (far backup) trip the remote breakers that is associated with the faulty breaker., The wide area backup protection should cooperate with the other protection and the breakers, and trip the minimum breakers to isolate the fault.

Communication technology

Communication technology of the wide area protection involves two aspects: the communication in the substation area and the wide area communication. With the development of the digital substation technology based on IEC61850, the substation-area communication already has a perfect scheme and will be continuously improved.

As for the wide area protection communication, the region is wider, the distance is longer and the data flow is larger. The communication notes and links are numerous. The wide area protection communication has to bear the different kinds of the information exchange, such as the periodicity, the stochastic and the burst information, and the high-precision clock synchronization is of great importance. Recently, the PTN (Packet Transport Network, PTN) communication technology has become a widely-used communication technology in the telecom and industry, and has been applied in the power system. PTN can achieve the flexible transmission for the non-fixed granular business, overcoming the shortcomings, such as the low frame utilization of the SDH network, the poor safety of the GPS synchronization and the in-adaptation for the burst communications of the protection. PTN can get the target of the high real-time performance, reliability and security for the wide area protection. In addition, some scholars also study on the application of the WAMS (Wide Area Measurement System, WAMS) for the wide area protection [35–37].

Except the above content, the research subjects on the communications for WARP also include the structure model of wide area communication system, the analysis and control of communication traffic, the error information analysis, and the intelligent error-information-tolerant method.

The problem of wide area information synchronization

At present, this problem is solved by global satellite synchronous clock like GPS and Bei Dou. However, due to the natural and technological factors, the performance of this approach still needs to improve. Therefore, the principle and algorithm of wide area protection need to be concerned besides wide-area information synchronization performance.

The interoperability of wide area protection device

In order to implement the interoperability of wide area protection device, the studies and the practice on the IEC61850 standard for wide area communications are necessary in order to realize the plug and play of all wide area protection.

Better configuration of backup protection system

The wide area communication failures and the sensitivity of the wide area backup protection should be taken into account. To achieve a better configuration of backup protection system, the misoperation of the wide area backup protection should be prevented. The remote back-up protection based on local information with the fixed setting value, the high sensitivity and the long time delay, still needs to be reserved for the last level of the backup protection.

Reliability and risk assessment of wide area relaying protection

Since the wide area backup protection using multi-source information from the wide area communication system, its reliability is an important issue. The multi-source information is benefit to improve the reliability of WARP. The Reliability and risk assessment of WARP can provide an effective guide for the actual application of WARP [36].

New principles of fault element identification

The following briefly describes several new principles of FEI. They involve the basic principle and the multi-source information fusion. Several aforementioned problems are considered when designing the protection algorithm.

The principle of fault element identification based on fault voltage distribution

There are several principles of FEI for a single element (for example a transmission line), such as the current differential, the pilot direction and the pilot distance, etc. Obviously, the current differential protection is strict with the synchronous sampling and difficult for the wide area protection. Meanwhile the performance of the pilot direction protection and the pilot distance protection is still non-perfect under the condition of complex faults. The principle of FEI based on the fault voltage distribution can both solve the above-mentioned two problems [35, 37].

This principle uses the measured fault component voltage and current on one side of the transmission line to estimate the fault component voltage on the other side. Hence, both of the measured and the estimated fault component voltages on two sides of the transmission line can be acquired. The measured value and estimated value of fault component voltages on both sides of the line are consistent, when external faults occur. As for internal faults, there exists some difference between the measured value and the estimated value of fault component voltage. So the faulty element can be identified based on the ratio between the measured values and the estimated values. Figure 2 is the distribution diagram of the measured value and estimated value of fault component voltage.

The principle of FEI is applicable to all types of fault components, including positive-, negative-, and zero-sequence fault components. By the comprehensive utilization of six components combination criterion, this FEI method can effectively deal with the single-phase-to-ground fault, the asymmetric interphase fault and three-phase-short-circuit fault.

Because only the amplitude of remote voltage is acquired for this wide area protection, this FEI method has low requirement for the wide area data synchronism, and only needs the synchronization correction based on the fault characteristics on the both sides of the line at the moment that the fault occurs in the line. The simulation calculation shows that this principle can correctly identify the fault line in the conditions of the high resistance grounding, weak feedback, faults on two-phase operation of the line, evolved faults or fault during oscillation, and is not affected by power flow transferring.

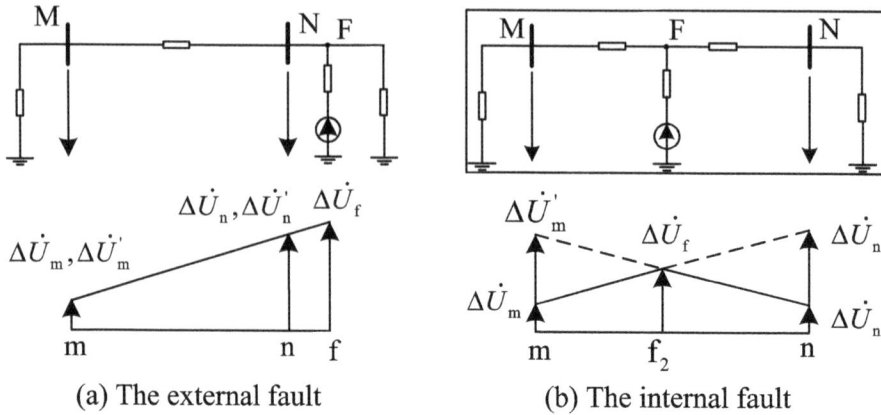

(a) The external fault   (b) The internal fault

Fig. 2 Distribution of fault components of voltages

The principle of fault element identification based on wide area integrated impedance

The wide area current differential protection is more susceptible to the line distributed capacitor than the common current differential protection, resulting in the lower sensitivity. This is because the number of lines under different operation condition is variable in the region of the wide area current differential protection. The distributed capacitance and the capacitance current may change in a wide range. In addition, it is more difficult to estimate and compensate the capacitance current in the wide area region of the differential protection. The pilot protection based on integrated impedance can overcome the influence of distributed capacitance and increase the sensitivity [38]. Introducing the concept of integrated impedance to wide area protection, the principle of FEI based on wide area integrated impedance can be established [39], which can overcome the defects of the wide area current differential protection.

The principle uses the multiport voltage and current in a region to form the integrated impedance, as shown in Fig. 3 . The identification of wide area integrated impedance is given by

$$Z_{cd} = \frac{\dot{U}_{cd}}{\dot{I}_{cd}} = \frac{\sum_i^N \dot{U}_i}{\sum_i^M \dot{I}_i} \quad (1)$$

Where, M and N are respectively the number of current that flow into the wide area relay protection area and the bus number on the region of the wide area relay protection.

When external fault occurs, the wide area integrated impedance reflects the equivalent line capacitive reactance of protection zone, of which the imaginary part value is thousands of ohms and the impedance angle is about -90°. As for the internal fault, the wide area integrated impedance is related to factors, such as the

system impedance, the line impedance and the fault resistance, which appears to be inductive reactance and has small imaginary part. Based on the difference of the integrated impedance when the internal fault and the external fault occur, the characteristic of the proposed protection is shown in Fig. 4 . Simulation verifies that the algorithm has high sensitivity and is capable for the earth fault with high fault resistance. This algorithm is not affected by the capacitive current, two-phase operation of the line, evolved faults and system oscillation, meanwhile, has the advantages of phase selection.

The limited wide area protection

As mentioned before, the limited wider area protection that is based on the limited information can identify the fault element and improve the performance of the backup protection, meanwhile, can also speed up the remote backup protection under the condition of the substation DC power supply failure. The advantage of the limited wide area protection lies in that only the limited information in the adjacent region is needed, reducing the information communication and processing burden. Under the current technology condition, the existing

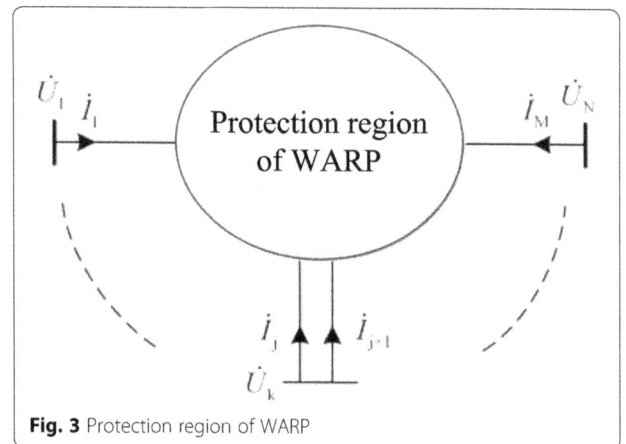

Fig. 3 Protection region of WARP

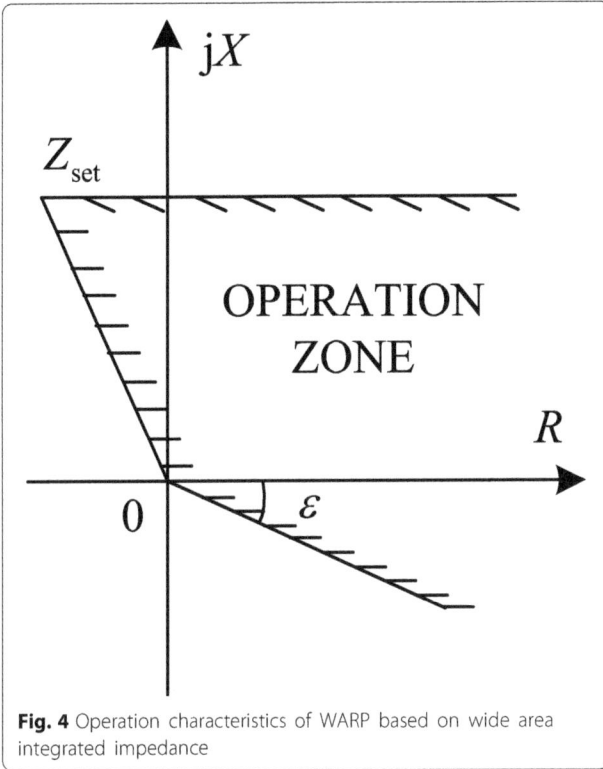

Fig. 4 Operation characteristics of WARP based on wide area integrated impedance

pilot channel of the current differential protection can be applied to transfer the protection information among the neighbor substations. Based on this, the limited wide area protection could be implemented.

Recently, the limited wide area has become a research hotspot due to its extensive application prospects. Some prototyping system has been put into operation in Chinese grid. The following introduces the principle and the application of some typical limited wide area protection, including the backup protection based on the pilot channel and the substation dc power supply failure protection.

The limited wide area backup protection based on pilot protection channel

According to the construction model of mentioned hierarchical protection, the bay layer protection is main protection. In addition, the substation area protection system is constructed inner the substation to measure and collect the internal substation electric information independently, which can achieve the backup protection function of connected lines and components. Combing the station area protection system and inter-substation communication channel, the simplified limited wide area protection can realize the protection information communication of each adjacent substation, which can identify the fault component and improve the action performance of backup protection. In this scheme, the

backup protection (Including the zone-II and zone-III) of each line all takes the setting method that ensure the sensitivity of terminal protection scope, and the selectivity of the protection can be ensured by the inter substation communication. This method can simplify the setting and coordination of the backup protection, which can also accelerate the backup protection operation.

The following takes zone-II distance protection and zone-III distance protection for example which is shown in Fig. 5 to introduce the limited WARP based on inter-substation communication. This protection scheme needs two signal transmission, which respectively achieve the near backup protection and remote backup protection.

The first signal transmission: the zone II distance protection of the fault line immediately startup, and the adjacent normal line zone II distance protection is blocked. When the fault occurs in the line, the substation zone II distance protection detects the fault line and sends the judgement conclusion to the adjacent substation by the inner-substation protection channel: Send the information 'the fault occurs in this line' to the opposite station of the fault line and sent the information 'the fault occurs in the other line' to the other adjacent substations. Only when the substation area protection judges the position of fault in the protection region of the zone-II distance protection and the received the 'the fault occurs in this line' information from adjacent substation, the line can be determined as the fault line which can be cutoff by the zone II distance protection quickly. For example, when the fault occurs in the line L3 in Fig. 5, the distance zone-II of the protection 5 and protection 6 on the both sides of L3 will operate and receive the 'fault occurs in this line' information from the neighbor substation. Hence line L3 is taken for the fault line and can be cutoff fast. The distance protection 3 and 8 may also operate during the fault of line L3 because of the setting method that ensures the sensitivity of terminal protection scope. However, the protection 3 and 8 will receive the information 'the fault occurs in the other line' from the opposite substation, the line L2 and L4 cannot be judged as fault line. Hence, the fault line can be effectively judged based on the information between the adjacent stations, which will ensure the fast operation speed and the selectivity of the zone II distance protection.

The second signal transmission: When the zone-II distance protection operates, the relevant remote backup protection that is adjacent to the fault line is started-up. The signal for the first time identifies the fault line and determines all the relevant remote backup protection of the fault components. When the zone-III distance protection judges the fault occurs in the protection region and receive the startup signal from the neighbor substation, the zone-III distance protection will operate and

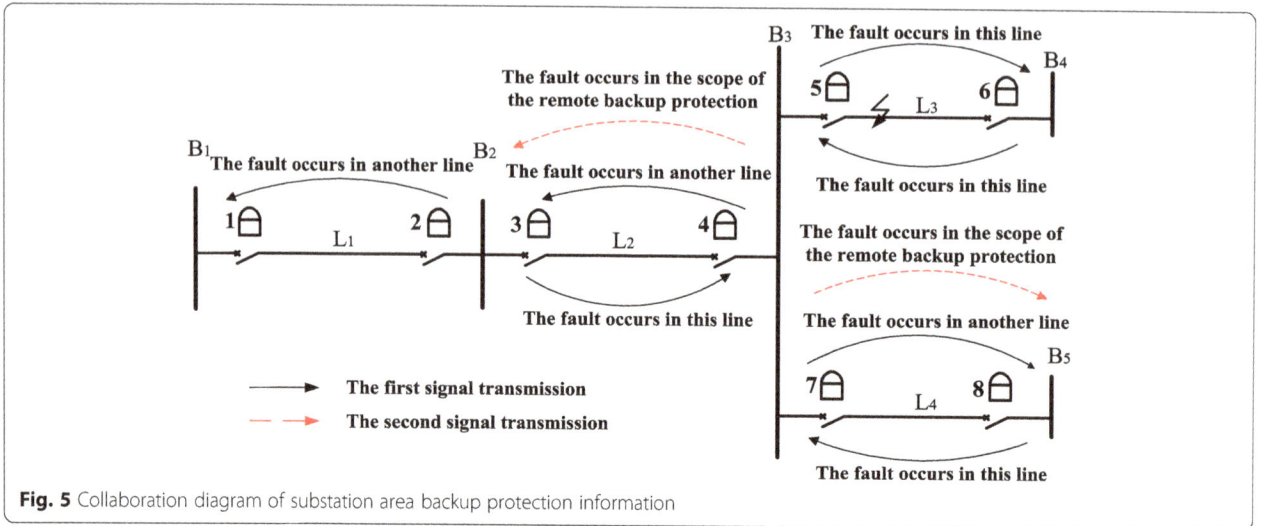

Fig. 5 Collaboration diagram of substation area backup protection information

cutoff the fault fast with short fix time delay. Taking the fault occurs in the line L3 in Fig. 5 as example, the protection 1, 3, and 8 will operate because of the setting mode that ensures the sensitivity of terminal protection scope. But, only the protection 3 and 8 will receive the startup signal and operate and the protection 1 will not misoperate.

Based on the above backup distance protection of inter-substation communication, the backup distance protection can take the invert setting mode to ensure the sensitivity and there is no need to coordinate. The operation time of the zone III distance protection can be set by the fixed time (such as 1 s) delay, without the multi-step time delay, which can speed up the action speed. In addition, when the large scale power flow transfer occurs, because of the lack of the startup signal, the remote backup protection will not misoperate.

In the same way, utilizing the zero-sequence direction element, the limited wide area backup protection can be established based on the pilot channel. The high sensitivity and fast protection for the high resistance grounding fault can be achieved.

This protection scheme takes independent substation protection system to realize the protection function of backup protection. When the bay layer main protection fail or refuse to operate, the scheme can achieve the selective and fast protection of backup protection. For the 110 kV or below 110 kV system which doesn't configure the breaker failure protection, this scheme can achieve the function of fast remote backup protection when the breaker is fail.

The limited wide area protection based on pilot protection channel and the bay layer protection

In the high- or ultrahigh- voltage system, the bay layer protection is configured with the current differential protection and backup protection based on the fiber pilot channel. By using the existing the bay layer protection as well as pilot channel between substations, the function of the limited wide area backup protection shown in Section V. A can also be realized.

The limited wide area backup protection based on the pilot channel and the bay layer protection is made up of the bay layer protection and decision unit of substation area protection. The backup distance protection of the bay layer protection for each line can be directly set based on the sensitivity and operation with the fix short time delay. It is not necessary for each protection to coordinate. The substation layer protection decision unit can acquire the protective results of the layer protection in the substation through the GOOSE network and that in the neighbor substation by the pilot channel. The protection scheme needs two signal transmission to realize the identification of fault element, which can respectively achieve the near backup protection and remote backup protection. Then startup the bay layer backup protection related to the fault element to isolate the fault.

In this scheme, the operation results of the bay layer protection in this substation and neighbor substation are acquired through the existing bay layer protection and pilot channel. The identification of fault element can be realized by the logic information of operation results. So this scheme is feasible for the implementation.

In China, the configuration of the protection in the ultra-high voltage system with the voltage grade of 220 kV and above is completely dual. The signal acquirement system, the power supply of the protection device, the pilot channel and the trip-ping loop of each protection are all dual-configured. Hence, the limited wide area backup protection based on pilot channel has high reliability.

The countermeasure for the failure of substation main power based on the limited wide area protection technology

The substation main power supplies the power for the control and protection devices and the operation loops. The main power failure in a substation occurs, the protection device and operating circuit cannot operate, which is the most severe accident in the substation. Constrained by the construction and operating principle (include the battery system), the operation statistics indicates that the failure of substation main power frequently occurred. This fault is especially severe in the substation supplied by single main power. The fault which occurs in substation and its line can be cleared only by the remote backup protection in neighbor substation. Therefore, it is of great importance for avoiding the expanding of accident to effectively determine whether the main power of substation is failing and to accelerate the remote backup protection when faults occur in the no-power substation.

The basic principle of substation dc power supply failure protection with the help of communication

In the substation, there is usually another communication power used for the remote communication except main power. These two kinds power is independent and backup of each other. Hence the substation area protection system can be supplied by both the main power and communication power. The system can monitor the running state. If the main power is lost, substation area protection system can be supplied by communication power and send the signal of power failure. When faults occur in the no-power substation, substation area protection will send tripping signal to adjacent substation, accelerating the speed of remote backup protection in adjacent substation. So the fault can be quickly isolated.

Example of substation dc power supply failure protection with the inner substation communication

Based on the above principle, a power supply wiring of substation area protection is shown in Fig. 6. This protection device is supplied by main power and communication power. A low-voltage relay is installed in the main power source to monitor the voltage. Once the main power failure occurs, the low-voltage relay operates and the signal will be sent to the substation area protection for the DC power failure judgement. In addition, because of the failure of the main power, the other protection device in the substation will not work. Protection device in the adjacent substation will send channel abnormity signal to the no-power substation because of the communication interrupt. Therefore, substation area protection in the substation without main power can comprehensively judge the loss of main power by collecting channel abnormity signal in adjacent substation and the operation signal of DC power low-voltage relay.

When the main power fails, the substation area protection device supplied by communication power can still work normally. The differential current protection of transforms and bus in the no-power substation still can work and identify the fault. But the breaks in no-power cannot operate to cut-off the fault. In this case, the substation area protection will send tripping signal to adjacent substation to accelerate the isolation of the faults.

Because of some special reasons, such as the voltage measurement is lost, the substation area protection in the no-power substation cannot identify the line fault. The line fault can be identified by collecting the operation performance of pilot impedance component at another side of the line. If the pilot impedance component of the adjacent substation operates meanwhile the breaker is in off-state, the no-power substation will determine that fault occurs in this line, and send tripping signal to adjacent substation to remove fault promptly.

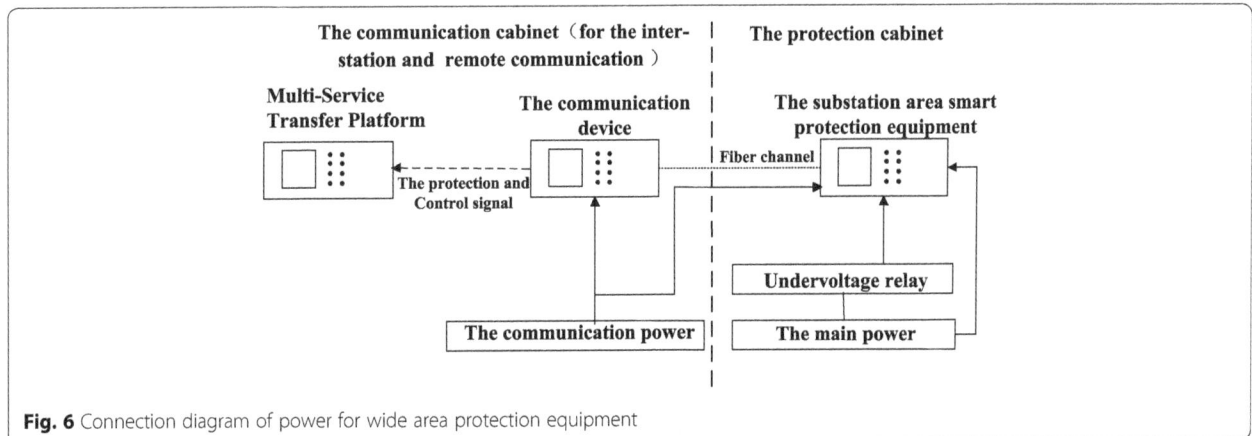

Fig. 6 Connection diagram of power for wide area protection equipment

Other special issues

In the digital substation, the power of the SV and GOOSE network is supplied by the main power of the substation. Under the condition of the main power failure, the substation protection cannot acquire the protection information, and the fault in the no-power substation cannot be detected. Under such condition, the communication between the substations around the no-power substation can achieve the fault element identification, enhancing the performance the remote backup protection. This problem will be discussed in other article.

Conclusion

The urgent problem of the relaying protection in the modern AC/DC hybrid grid and the development of the wide area communication, the information process and the intelligent technology powerfully promotes the development of the technology of the wide area protection. Wide area protection includes wide area relaying protection (WAPR) and wide area security protection (WASP). At present, the research of WARP is focus mainly on wide area backup protection. WARP, used to clean fault, belongs to the first defense of three defense of security emergency control, which cannot be replaced by security emergency control. The farthest region of the wide area backup protection is the range of remote backup protection. Hence wide area backup protection has limited zone in respects of the information and action.

There are two main approaches to realize WARP, which are based on on-line adaptive setting (OAS) principle and fault element identification (FEI) principle. Some achievements have been obtained in these aspects currently, and several test systems have been put into use in the actual grid.

The paper discusses the research of wide area backup relaying protection, including the system constitution and zone-division technology, the principle of FEI, tripping strategy, wide area communication and so on. Then from the view of engineering application, several issues need to be solved when researching the principle of FEI is proposed.

The paper introduces several new judging principles of FEI. These methods have excellent performance in reducing the dependence on wide area synchronous sampling, improving the sensitivity of backup protection.

This paper presents the limited wide area backup protection based on the inter-substation communication channel such as the pilot channel. This protection scheme can identify the fault element correctly simplify the setting and the coordination of the backup protection, and improve the tripping speed. In addition, the limited wide area backup protection can speed up the tripping of remote protection under the condition that

the main power supply of the next substation protected is lost and achieve the fast fault isolation remotely.

Funding

This work is supported by National Natural Science Foundation of China Science Foundation of China (No. 50377031 and No.50837002).

Authors' contributions

XGY contributed to the conception of the study. ZZ contributed significantly to analysis and manuscript preparation; ZXL helped perform the analysis with constructive discussions. XWQ performed the analyses and wrote the manuscript; WBC conceived of the study, and participated in its design and coordination and helped to draft the manuscript. QG revised the manuscript. All authors read and approved the final manuscript.

Competing interests

The authors declare that they have no competing interests.

References

1. Novosel, D., Bartok, G., Henneberg, G., et al. (2010). IEEE PSRC report on performance of relaying during wide-area stressed conditions [J]. *IEEE Transactions on Power Delivery, 25*(1), 3–16.
2. Chen, D. (2004). Preliminary research on security protection technology of large-scale power grid [J]. *Power System Technology, 28*(9), 14–17. 27(in Chinese).
3. Working group c-6, system protection subcommittee. (2002). Wide Area Protection and Emergency Control [R]. IEEE PES power system relaying committee.
4. Xiang-gen, Y. I. N., Yang, W. A. N. G., & Zhe, Z. H. A. N. G. (2010). *Zone-division and tripping strategy for limited wide area protection adapting to smart grid [J]* (Proceedings of the CSEE, pp. 1–7). in Chinese.
5. Xin-zhou, D. O. N. G., & Lei, D. I. N. G. (2009). Research on design of digital integrated protection and control system [J]. *Power System Protection and Control, 01*, 1–5. in Chinese.
6. Zhen-xing, G., Chuang-xin, G., Bin, Y. U., et al. (2011). Study of a fault diagnosis approach for power grid with information fusion based on multi-data resources [J]. *Power System Protection And Control, 39*(6), 17–23. in Chinese.
7. The editorial board. (2014). China Electric Power Encyclopedia [M]. Beijing: China Electric Power Press.
8. Cholley P, Crossley P, Van Acker V, et al. (2001). System Protection Schemes in Power Networks [J]. Cigre Technical Brochure.
9. Ji-xiu, Y. (2007). *Wide area protection and emergency control to prevent large scale blackout [M]*. beijing: China electric power press.
10. Cai, Y., Wang, L., Morison, K., et al. (2004). Current status and prospect of wide-area protection (dynamic stability control) technologies [J]. *Power System Technology, 28*(8), 20–25. in Chinese.
11. Jun, Y. I., & Xiao-xin, Z. (2006). A survey on power system wide-area protection and control [J]. *Power System Technology, 30*(8), 7–12. 30(in Chinese).
12. Standardization administration of China. (2011). GB/T-26399-2 Technical guide for electric power system security and stability control [S]. SAC.
13. Jixiu, Y. (1999). Planning and application of power system security and stability control [J]. *Electric Power, 32*(5), 29–32. in Chinese.
14. Rockefeller, G. D., Wagner, C. L., Linders, J. R., et al. (1988). Adaptive transmission relaying concepts for improved performance [J]. *Power Delivery IEEE Transactions on, 3*(4), 1446–1458.
15. Xian-zhong, D., Zeng-li, Y., & Xiao, C. (2005). Performance analysis of relay settings determined according to Off-line calculation and on-line calculation [J]. *Automation Of Electric Power Systems, 29*(19), 58–61. in Chinese.
16. Cao, G., Cai, G., & Wang, H. (2003). Problems and solutions in relay setting and coordination [J]. *Proceedings Of The Chinese Society For Electrical Engineering, 23*(10), 51–56. in Chinese.
17. Orduna, E., Garces, F., & Handschin, E. (2003). Algorithmic-knowledge based adaptive coordination in transmission protection [J]. *IEEE Power Engineering Review, 22*(12), 63.

18. Ying, L. U., & Boming, Z. (2007). Online relay setting check based on computer cluster [J]. *Automation Of Electric Power Systems, 31*(14), 12–16. 106(in Chinese).
19. Songtao, Q., Chao, H., Falin, Z., et al. (2013). Development of the on-line verification system for Guangxi power grid relay settings [J]. *Southern Power System Technology, 4,* 83–87. in Chinese.
20. Tejun, Z., Jian, Q., Chunyi, W., et al. (2015). Application of visualization technology based on SVG in on-line relay settings verification system [J]. *Power System Protection and Control, 16,* 112–117. in Chinese.
21. Jun, X., Dongyuan, S., Zengli, Y., et al. (2007). An on-line verification of relay settings and early warning system of protective relaying based on MAS [J]. *Automation Of Electric Power Systems, 31*(13), 77–82. in Chinese.
22. Youhuai, W., Zengli, Y., Hubing, Z., et al. (2015). Development and application of online verification and early-warning system for protective relay [J]. *Proceedings of the CSU-EPSA, 27*(6), 91–97. in Chinese.
23. Serizawa, Y., Myoujin, M., Kitamura, K., et al. (1998). Wide-area current differential backup protection employing broadband communications and time transfer systems [J]. *IEEE Transactions on Power Delivery, 13*(4), 1046–1052.
24. Hui-ming, X. U., Tian-shu, B. I., Shao-feng, H., et al. (2007). Study on wide area measurement system based control strategy to prevent cascading trips [J]. *Proceedings Of The Chinese Society For Electrical Engineering, 27*(19), 32–38. in Chinese.
25. Lim, S. I., Liu, C. C., Lee, S. J., et al. (2008). Blocking of zone 3 relays to prevent cascaded events [J]. *IEEE Transactions on Power Systems, 23*(2), 747–754.
26. Tian-qi, X. U., Xiang-gen, Y., Da-hai, Y., et al. (2009). Analysis on functionality and feasible structure of wide area protection system [J]. *Power System Protection And Control, 37*(3), 93–97. in Chinese.
27. Li, Z., Yin, X., Zhang, Z., et al. (2011). Study on system architecture and fault identification of zone-division wide area protection [J]. *Proceedings of the CSEE, 31*(28), 95–103. in Chinese.
28. Zhenxing, L. I., Xianggen, Y., Zhe, Z., et al. (2010). Zone division and implementation on limited wide area protection system [J]. *Automation Of Electric Power Systems, 34*(19), 48–52. in Chinese.
29. Wei, C., Zhen-cun, P., Jian-guo, Z., et al. (2006). A wide area protective relaying system based on current differential protection principle [J]. *Power System Technology, 30*(5), 91–95. 110(in Chinese).
30. Zhao-hui, C., Man-yong, Z., Hong-yang, Z., et al. (2009). Research and implementation of network protection system based on integrated and wide area information [J]. *Power System Protection And Control, 37*(24), 106–108. 113(in Chinese).
31. Yang, Z., Shi, D., & Duan, X. (2008). Wide-area protection system based on direction comparison principle [J]. *Proceedings Of The Chinese Society For Electrical Engineering, 28*(22), 87–93. in Chinese.
32. Lin, X., Li, Z., Wu, K., et al. (2009). Principles and implementations of hierarchical region defensive systems of power grid [J]. *Power Delivery IEEE Transactions on, 24*(1), 30–37.
33. Yang, W., Xianggen, Y., Zhe, Z., et al. (2010). Wide area protection based on genetic information fusion technology [J]. *Transactions Of China Electrotechnical Society, 25*(8), 174–179. in Chinese.
34. Zhen-xing, L. I., Xiang-gen, Y. I. N., & Zhe, Z. H. A. N. G. (2011). A new algorithm of wide area protection on multi-information fusion [J]. *Automation of Electric Power Systems, 09,* 14–18. in Chinese.
35. He, Z., Zhang, Z., Chen, W., et al. (2011). Wide-area backup protection algorithm based on fault component voltage distribution [J]. *IEEE Transactions on Power Delivery, 26*(4), 2752–2760.
36. Zhi-hui, D., & Zeng-ping, W. (2010). Overview of research on protection reliability [J]. *Power System Protection And Control, 38*(15), 161–167 (in Chinese).
37. Hua, W. A. N. G., Zhe, Z. H. A. N. G., & Xiang-gen, Y. I. N. (2011). A wide area protection algorithm based on fault voltage distribution [J]. *Automation of Electric Power Systems, 07,* 48–52. in Chinese.
38. Jiale, S., Kai, L., Xiaohua, S. U., et al. (2008). Novel transmission line pilot protection based on integrated impedance [J]. *Automation Of Electric Power Systems, 32*(3), 36–41. in Chinese.
39. Zhen-xing, L. I., Xiang-gen, Y. I. N., & Zhe, Z. H. A. N. G. (2010). A study of wide-area protection algorithm based on integrated impedance comparison [J]. *Transactions of China Electrotechnical Society, 08,* 1–5. in Chinese.

11

An optimized compensation strategy of DVR for micro-grid voltage sag

Zhengming Li[*], Wenwen Li and Tianhong Pan

Abstract

Introduction: This paper uses a dynamic voltage restorer (DVR) to improve the voltage quality from voltage sags. It is difficult to satisfy various of compensation quality and time of the voltage sag by using single compensation method. Furthermore, high-power consumption of the phase jump compensation increases the size and cost of a dynamic voltage restorer (DVR).

Methods & Results: In order to improve the compensating efficiency of DVR, an optimized compensation strategy is proposed for voltage sag of micro-grid caused by interconnection and sensitive loads. The proposed compensation strategy increases the supporting time for long voltage sags.

Discussion: Firstly, the power flow and the maximum compensation time of DVR are analyzed using three basic compensation strategies. Then, the phase jump is corrected by pre-sag compensation. And a quadratic transition curve, which involves the injected voltage phases of pre-sag strategy and minimum energy strategy, is used to transform pre-sag compensation to minimum energy compensation of DVR.

Conclusions: The transition utilizes the storage system to reduce the rate of discharge. As a result, the proposed strategy increases the supporting time for long voltage sags. The analytical study shows that the presented method significantly increases compensation time of DVR. The simulation results performed by MATLAB/SIMULINK also confirm the effectiveness of the proposed method.

Keywords: Micro-grid, Voltage sag, Dynamic voltage restorer, Compensation time, Optimized compensation strategy

Introduction

Since micro grid can integrate various kinds of renewable energy resources, it has been widely investigated in recent years [1]. However, there also exists some power quality problems that cannot be ignored, such as voltage sag, swell, and harmonic [2]. Furthermore, micro grid contains a variety of powers, sensitive loads and non-linear power electronic devices. As a result, the frequent start-up and shutdown operation of renewable power generators will lead to voltage fluctuations, especially sag. The voltage sag can damage the power quality, may even cause huge economic losses [3]. To improve and manage the power quality of micro grid, the flexible alternating current transmission technology (FACT) has been introduced, such as SVC, D - STATCOM, DVR and UPQC [4–6].

Among these devices, DVR is one of the most powerful apparatus that can protect voltage-sensitive equipment from voltage sag, swell and flicker. In some cases, it can also suppress the voltage harmonics and unbalanced three-phase voltage effectively [7].

DVR needs some power exchange when it is used in compensation. In order to reduce power exchange during the effective compensation of voltage drop, the compensation strategy of DVR should be optimized [8]. The present research and application of DVR compensation strategies for voltage sag are pre-sag compensation, in-phase compensation and the minimum energy compensation [9–12]. In-phase compensation method can compensate the sagged voltage with minimum amplitude [9]. The minimum energy method adjusts the injected active and reactive power of DVR to reduce consumption of the dc stored energy [13]. However, both can't correct the phase jump, even lead to trip of sensitive loads. The pre-sag compensation can

* Correspondence: lzming@ujs.edu.cn
School of Electrical and Information Engineering, Jiangsu University,
Zhenjiang 212000, Jiangsu Province, China

fully restore the amplitude and phase angle of the sagged load voltage [14]. Though pre-sag compensation can guarantee the continuity of grid voltage, the compensation of phase jump needs a large amount of active power in an energy storage system.

Taking advantage of three basic compensation strategies, some comprehensive compensation strategies [15, 16] were proposed. The comprehensive strategy of pre-sag and in-phase compensation [15] reduced the injected voltage while extending compensation time, but it sustainably used the energy of dc unit. The control that combined pre-sag method and minimal energy method [16] were lacked of the time control though it made good use of dc energy. To avoid these drawbacks, an optimized strategy based on the strategy of paper [16] is discussed in this paper. Firstly, the DVR compensating voltage and power flow for three basic strategies is analyzed and the maximum compensation time is derived from the analysis. Then, pre-sag strategy is used to correct phase jump at the beginning of compensating. According to a quadratic transition curve, the injected phase angle varied from pre-sag to the minimum energy compensation is rotated in multi-step. The rest process is completed by the minimum energy strategy. Owing to the introduction of the transition process, the compensation time of the comprehensive strategy is significantly increased. The performance of the proposed method is validated by using MATLAB/Simulink simulation.

Discussion

Three basic compensation strategies

As shown in Fig. 1, DVR system consists of the capacitor (storage device), inverter circuit, LC filter circuit, series transformer. When the voltage sag happens in grid, the dc voltage in energy storage system is converted to ac voltage by the inverter. Then, the ac voltage with a certain amplitude and phase is injected into grid feeder through filter and transformer to restore the sagged load voltage. These operations ensure normal operation of sensitive loads.

As shown in Fig. 2, phase diagrams are applied to analyze characteristics of three basic compensation methods. The Fig. 2(a) shows the in-phase compensation, the Fig. 2(b) and 2(c) shows the minimum energy compensation. The Fig. 2(d) shows the pre-sag compensation. The amplitude and phase angle of DVR compensating voltage are given in the diagrams respectively. In Fig. 2, U_G and U_S are the system voltages before and after the sag, U_L and U_A are the load voltages before and after the sag, U_{DVR} is the DVR output compensating voltage, P_{DVR} and Q_{DVR} are the DVR active and reactive power injected to system, δ is the phase jump of sagged voltage, θ_L is the angle of load power factor. I_L is the load current which is set as the reference vector. ΔU is set as voltage sag depth, so $\Delta U = (U_L - U_S)/U_L$.

In-phase compensation

For in-phase compensation strategy, the DVR compensating voltage has the same phase with the sagged grid voltage. The amplitude of compensating voltage is equal to the difference between the reference voltage of load and the grid voltage (shown in Fig. 2(a)). This compensation strategy only needs to measure instantaneous voltage in the grid, so it has high speed. Though DVR injects the minimum compensating voltage amplitude, it cannot correct phase jump of voltage sag. In this case, it may lead to interruption of load voltage. The DVR compensating voltage amplitude and phase angle injected to the power system are:

$$U_{DVR} = \sqrt{\frac{2}{3}} U_L \Delta U \tag{1}$$

$$\angle U_{DVR} = \theta_L \tag{2}$$

Minimum energy compensation

In reactive power compensation, the injected voltage of DVR is orthogonal to load current. This strategy only provides reactive power to compensate voltage sag without

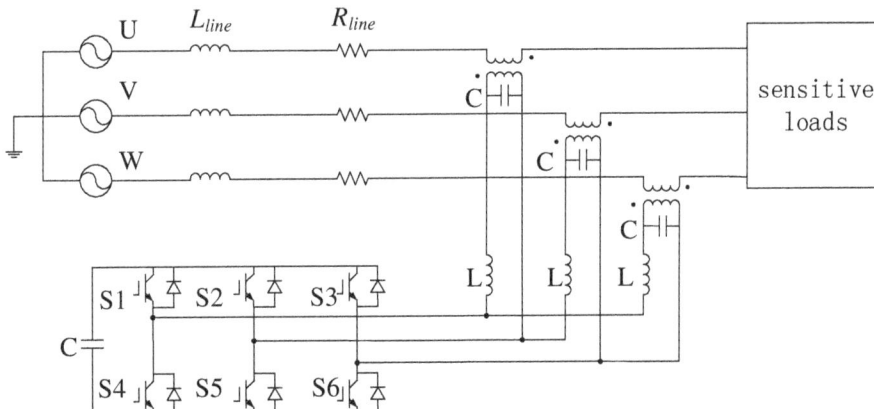

Fig. 1 Topology of DVR

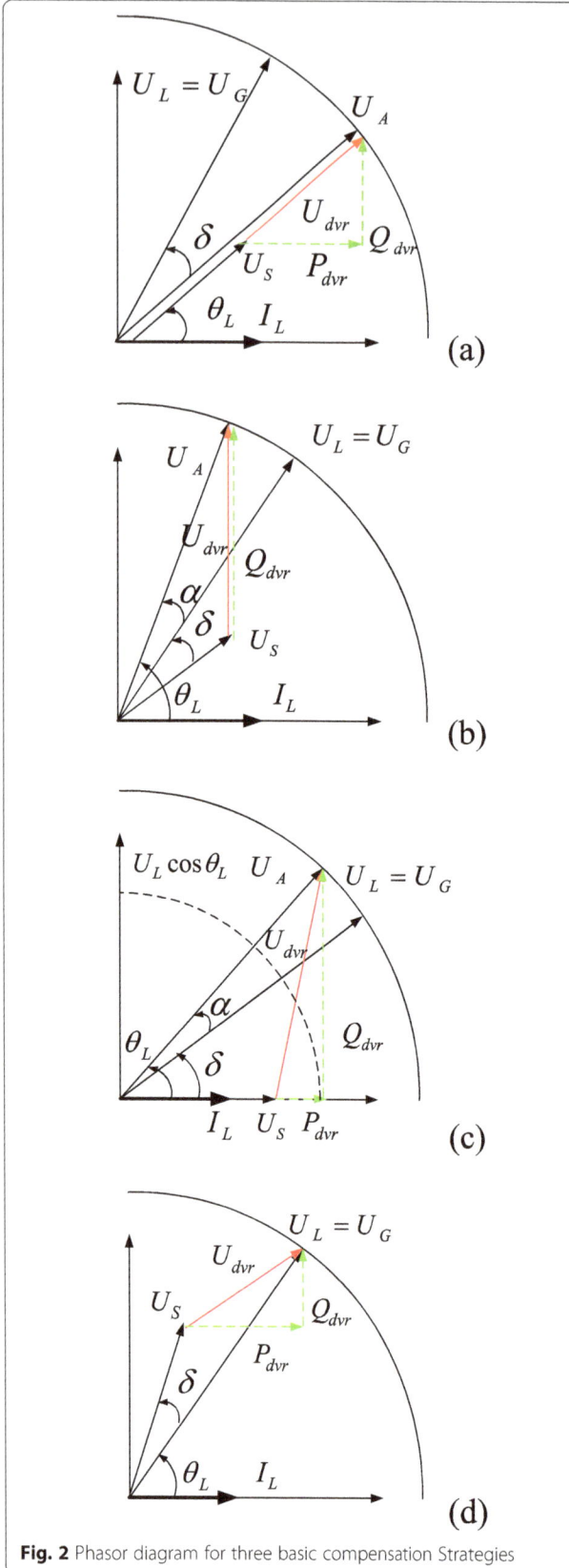

Fig. 2 Phasor diagram for three basic compensation Strategies

active power consumption. As shown in Fig. 2(b), the DVR injected voltage amplitude and phase in the system are:

$$U_{DVR} = \sqrt{\frac{2}{3}}U_L\sqrt{1-2(1-\Delta U)\cos(\alpha+\delta)+(1-\Delta U)^2}$$

$$(3)$$

$$\angle U_{DVR} = \frac{\pi}{2} \qquad (4)$$

where α is the phase change caused by DVR reactive compensation. The maximum sag depth in reactive power compensation, ΔU_{max}, is closely related to the load power factor:

$$\Delta U_{max} \leq (1-\cos\theta_L) \qquad (5)$$

When the sagged voltage of grid U_S is in phase with load current I_L, the compensating voltage reaches the maximum value, i.e:

$$U_{max} = \frac{U_S}{1-\Delta U_{max}}\sin\theta_L \qquad (6)$$

When the voltage sag depth is over the boundary of Eq. (5), DVR must inject some active power into the system of grid to maintain compensation. In this situation, the minimum energy method is put forward to enhance the performance of reactive power compensation. As shown in Fig. 2(c), the amplitude and phase of DVR injected voltage are:

$$U_{DVR} = \sqrt{\frac{2}{3}}U_L\sqrt{1-2(1-\Delta U)\cos\theta_L+(1-\Delta U)^2} \quad (7)$$

$$\angle U_{DVR} = \tan^{-1}\left(\frac{U_L\sin\theta_L}{U_L\cos\theta_L-U_S}\right) \qquad (8)$$

Pre-sag compensation
When the voltage sag happens in the grid system, the phase jump often accompanies with the decrease of voltage amplitude. These methods mentioned above can't compensate the phase jump. Pre-sag compensation ensures that the amplitude and phase of compensating voltage are completely in accord with the pre-sag voltage. However, the storage device has to provide abundant energy in pre-sag compensation. As shown in Fig. 2(d), the amplitude and phase of DVR injected voltage are:

$$U_{DVR} = \sqrt{\frac{2}{3}}U_L\sqrt{1-2(1-\Delta U)\cos\delta+(1-\Delta U)^2} \quad (9)$$

$$\angle U_{DVR} = \tan^{-1}\left(\frac{U_L\sin\theta_L-U_L\sin(\theta_L+\delta)}{U_L\cos\theta_L-U_S\cos(\theta_L+\delta)}\right) \quad (10)$$

Control requirements
The DVR active power injected to the system is equal to the difference between the active power in the grid and

load. From Fig. 2(a), (c) and (d), the active power of pre-sag compensation (P_{pre}), in-phase compensation ($P_{in-phase}$) and the minimum energy compensation (P_{opt}) can be expressed as:

$$P_{pre} = \sqrt{3U_L I_L}(\cos\theta_L - (1-\Delta U)\cos(\theta_L - \delta)) \tag{11}$$

$$P_{in-phase} = \sqrt{3U_L I_L \Delta U \cos\theta_L} \tag{12}$$

$$P_{opt} = \sqrt{3U_L I_L(\cos\theta_L - 1 + \Delta U)} \tag{13}$$

The relation between the DVR active power and other variables (sag depth, phase jump, the load power factor) can be obtained from the three equations. It can be seen that the pre-sag strategy has the highest power consumption among three compensation methods. The operations of the pre-sag compensation require a plenty of energy in dc energy-storage capacitor. In practice, the energy storage of DVR is limited. When the output power reaches a threshold, the compensating voltage will drop. In order to ensure appropriate operations of DVR, it should be satisfied:

$$\frac{U_{DVR}}{n_t} \le \frac{m_{i\,max} U_{dc}}{2} \tag{14}$$

where n_t is the turns ratio of series transformer, m_{imax} is the maximum modulation index of the DVR inverter, U_{dc} is the DVR dc voltage. Once the dc voltage decreases to the minimum threshold value, that is, over the boundary in Eq. (14), DVR will stop compensating process to avoid harmonic pollution of load voltage. The initial energy stored in dc capacitor is:

$$W_{dc} = \frac{1}{2}C_{dc}U_{dc}^2 \tag{15}$$

After a period of time Δt, the value of voltage in dc container decreases from U_{dc} to U_{dcf}. The variation of voltage is ΔU_{dc}. The final energy stored in the capacitor at this time is:

$$W_{dcf} = \frac{1}{2}C_{dc}U_{dcf}^2 \tag{16}$$

During Δt, the total active power provided by capacitor in steady state is:

$$P_{dc} = \frac{W_{dc} - W_{dcf}}{\Delta t} = \frac{1}{2}C_{dc}\frac{d}{dt}U_{dc}^2$$
$$= \frac{1}{2}C_{dc}\left(\frac{U_{dc}^2 - U_{dcf}^2}{\Delta t}\right) \tag{17}$$

In an ideal system, the DVR dc power in Eq. (11) is equal to the ac power in Eq. (17), so the capacity of capacitor C can be obtained from the relation. However, the dc voltage will decrease with the flow of power, so as the output compensating voltage. Accordingly, the DVR

output active power is also limited by the capacity of dc capacitor and the minimal dc voltage, which is sufficient for a proper restoration of the load voltage. In addition, the dc voltage gradient dU_{dc}/dt is proportional to the DVR injected active power P_{DVR} directly. The less P_{DVR} is, the smaller the slope of dc voltage is. Hence, the compensating time is extended. It can be known that improving the rate of dc voltage drop can prolong the compensation time.

According to Eq. (14) and Eq. (17), the maximum compensation time can be expressed as:

$$t_{max} = \frac{C * \left[U_{dc}^2 - \left(\frac{2*U_{DVR}}{m_{i\,max}*n_t}\right)^2\right]}{2 * P_{DVR}} \tag{18}$$

The maximum compensation time t_{max} directly reflects the utilization level of stored energy in dc capacitor.

Methods
The optimized compensation strategy
The dc voltage gradient is controlled by adjusting the magnitude of DVR injected active power to extend the DVR compensating time. Firstly, pre-sag compensation is used to restore the amplitude and phase angle of sagged voltage. Then a transition process is introduced to shift pre-sag strategy to the minimum energy strategy gradually. The transition avoids direct jump of phase angle at running point. This process is usually implemented in one to two cycles, and the sagged voltage can be restored to stabilization in a short time. The compensating control process is described in detail as follows.

The pre-sag compensation
In order to reduce the voltage distortions at the load side, the sagged voltage should be restored to the normal value by the pre-sag compensation. After a certain time, the modulation index will increase as a result of discharging of the dc capacitor. If there is no sufficient energy to support the compensation, DVR will not be able to compensate properly. Therefore, as soon as a certain modulation index is reached, a smooth transition from pre-sag compensation to the minimum energy compensation is initiated to reduce consumption of energy.

To detect phase jump, two phase lock loops (PLLs) are adopted at the load side and the grid side respectively. Once the voltage sag is detected, the initial injected angle of DVR should be determined to compensate the load phase jump. On one hand, the pre-sag voltage angle is locked by freezing the PLL at the load side. On the other hand, the phase angle of the sagged grid voltage is locked through the PLL at the grid side. The two angles are equal when the system is steady. When the sag happens, the difference between the two phases is equal to the phase angle jump of sagged voltage. Hence, the initial phase angle of DVR injected into the system is:

$$\theta_{first} = \tan^{-1}\left(\frac{U_L\sin\theta_L - U_s\sin(\theta_L + \delta)}{U_L\cos\theta_L - U_s\cos(\theta_L + \delta)}\right) \qquad (19)$$

The voltage sag can be detected according to space vector method. Therefore, the sag depth is regarded as the absolute difference between the load reference voltage (1p.u.) and the actual grid voltage (p.u.). Under the synchronous reference frame, the sag depth can be expressed as:

$$\Delta U = \left|1 - \sqrt{U_{Gd}^2 + U_{Gq}^2}\right| \qquad (20)$$

The transition process

After the sagged voltage recovers to normal, pre-sag compensation will be smoothly transformed to the minimum energy compensation in one to two cycles.

From section 1.2, it can be known that the minimum energy method works in two situations. For shallow sag, DVR will make reactive power compensation, in which DVR works in self-supporting mode. It hardly absorbs active power from the grid, or only consumes a little active power to overcome losses in system and maintain the constant voltage in dc capacitor. For deep sag, DVR must carry on the minimum energy compensation. DVR will inject a certain amount of active power to grid, so the energy and voltage in the capacitor will be reduced. In order to maintain the needed output voltage of the inverter, the modulation index will increase until it reaches the maximum value in Eq. (14). Hence, the final injected angle of DVR θ_{final} is:

$$\theta_{final} = \begin{cases} \dfrac{\pi}{2} + \varepsilon, & \Delta U \leq (1 - \cos\theta_L) \\[2ex] \pi - \tan^{-1}\left(\dfrac{U_L\sin\theta_L}{U_L\cos\theta_L - U_S}\right), & \Delta U > (1 - \cos\theta_L) \end{cases} \qquad (21)$$

where ε is the angle change caused by component loss, which can be obtained by PI controller.

The smooth transition of the load voltage will be done by a quadratic function curve to reduce the influence of phase jump to sensitive loads. The curve between the initial and terminal phase is given as:

$$\theta_{tran} = \frac{\theta_{first} - \theta_{final}}{\Delta t_{tran}^2}\left(t - \Delta t_{tran}\right)^2 + \theta_{final} \qquad (22)$$

where Δt_{tran} is the transitional time.

The operation process

In Fig. 3, Phase diagrams of the compound control strategy are given. Figure 3(a) shows the voltage amplitude and phase of the pre-sag strategy, and 3(c) shows the voltage amplitude and phase of the minimum energy strategy, (3b) shows the transition process. As shown in Fig. 3, the sag depth is over the limit in Eq. (5), and it produces a positive phase jump with voltage sag. Here, pre-sag strategy is used to recover the amplitude and phase of the sagged voltage. Once the dc voltage or the modulation index reaches the limited value, a transition process is introduced to avoid direct phase jump at the operating point. The compensating voltage will be gradually transformed to the value of the minimum energy compensation. In the transition process, the reactive power of DVR increases gradually, so as the amplitude and phase angle

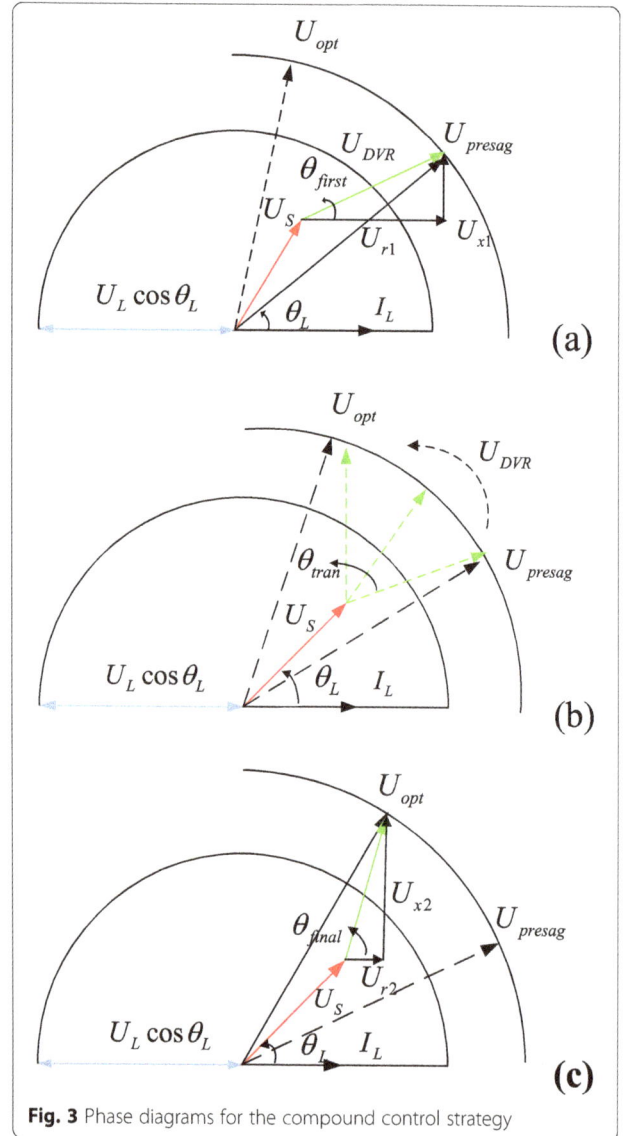

Fig. 3 Phase diagrams for the compound control strategy

(a) The maximum compensation time and sag depth

(b) The maximum compensation time and jump phase

Fig. 4 The maximum compensation time for different methods

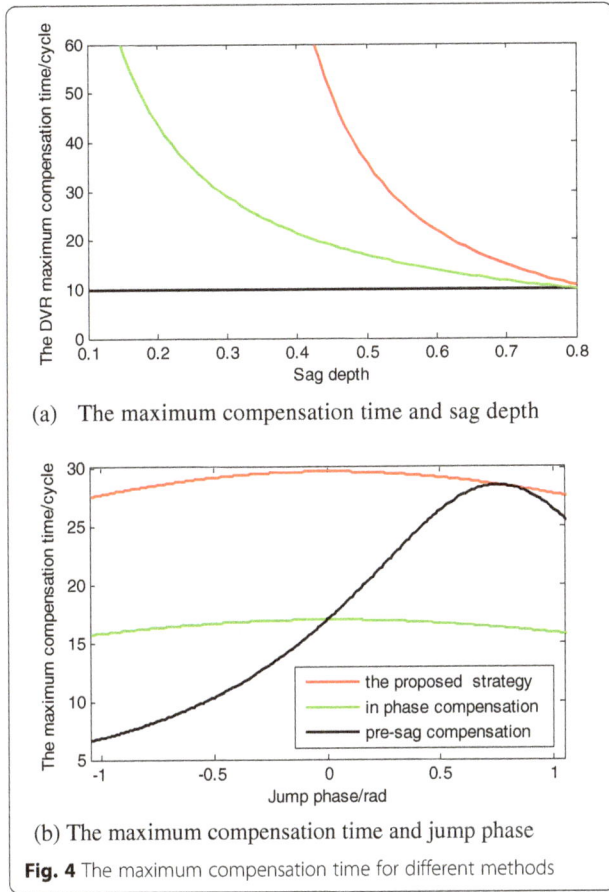

of the compensating voltage. Finally, the $U_{pre\text{-}sag}$ reaches to $U_{A\text{-}opt}$.

Results

Comparisons

As mentioned above, the maximum compensation time plays an important role in reflecting the performance of DVR. It is directly connected to the amount of energy needed for the compensation and the needed capacitance. Based on the mentioned analytic equations, a comparative study is presented to compare the DVR maximum compensating time for different sags and jump phases by using the mentioned strategies (i.e. pre-sag strategy, in-phase strategy, and the proposed strategy.).

As shown in Fig. 1, a medium-voltage DVR system connected with a constant DC-link capacitor. In this system, the nominal grid voltage is 380 V, the apparent load power is 9kVA, the turn ratio is 1, the maximum modulation index is set as 1, the power factor is kept at 0.7 (lag), the phase jump is 45° (positive) with a 50 % given sag depth, and the line frequency is 50Hz while the cycle is 20 ms. Then the obtained capacitance of dc link is 7mF. Using Eq. (1), (12), (7), (13), (9), (11) and (18), As shown in Fig. 4, the maximum compensation time for the proposed strategy, in-phase compensation and pre-sag compensation in different situation are compared. Figure 4(a) shows the maximum compensation time of different strategies with the change of sag depth. When the sag depth is given, Fig. 4(b) shows the maximum time with respect to the different jump angles by using different compensation strategies.

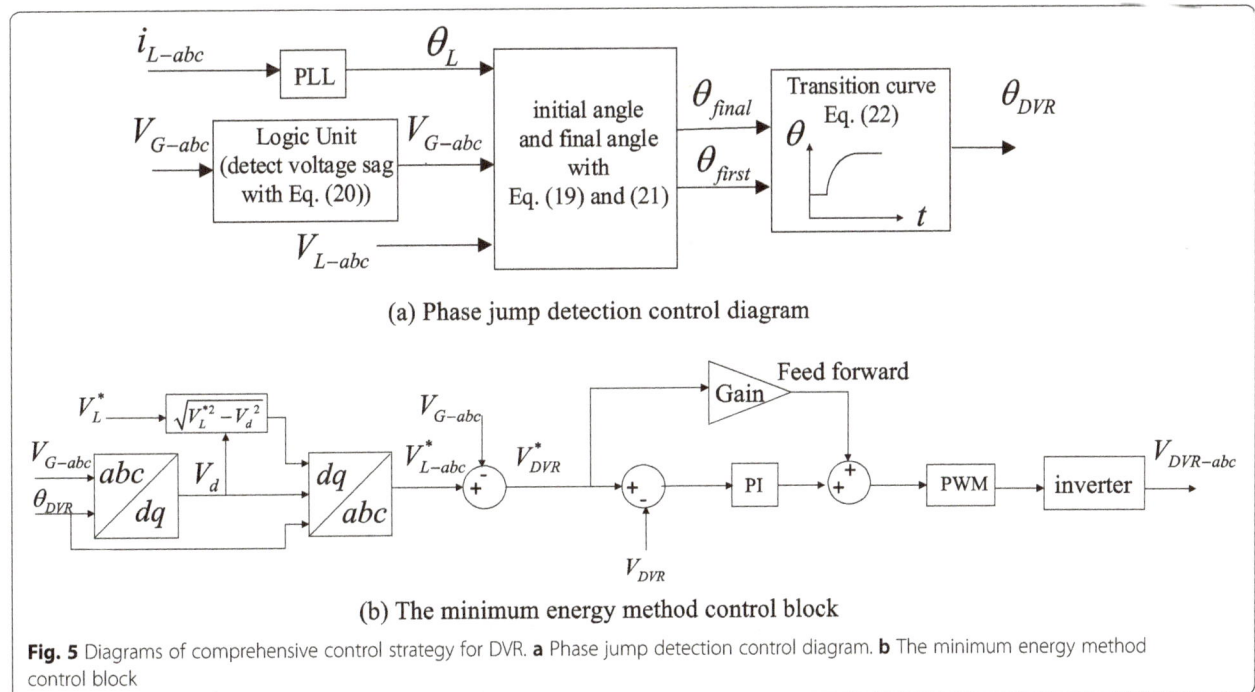

(a) Phase jump detection control diagram

(b) The minimum energy method control block

Fig. 5 Diagrams of comprehensive control strategy for DVR. **a** Phase jump detection control diagram. **b** The minimum energy method control block

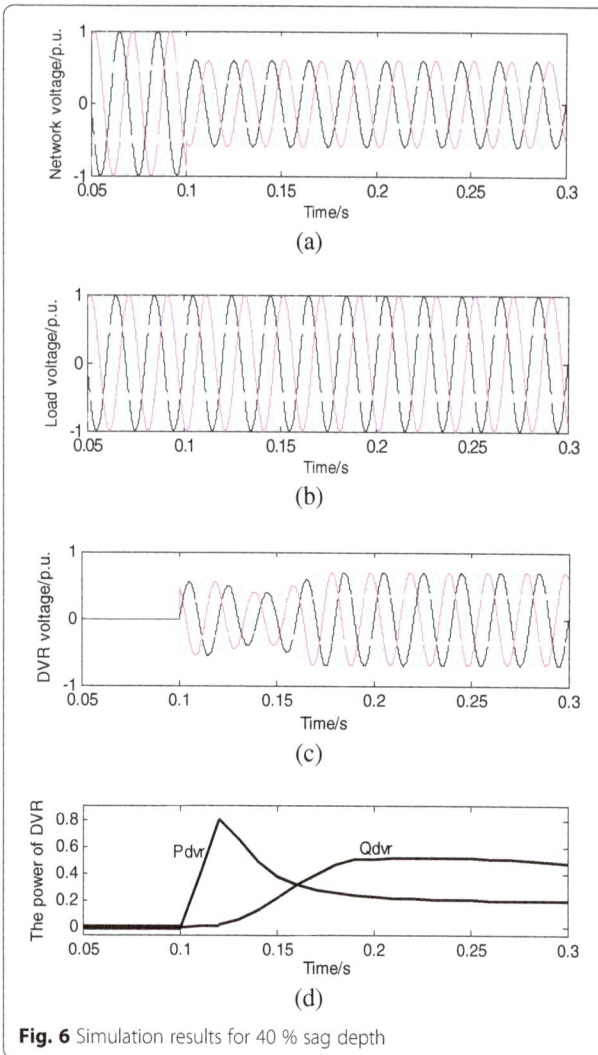

Fig. 6 Simulation results for 40 % sag depth

Simulation study
Block diagram
The simulation complete setup is depicted in Fig. 1. The block diagrams of the comprehensive compensation strategy for DVR are shown in Fig. 5 in detail. Take a 380 V radial micro grid as an example, the sensitive loads is set as R-L loads. Three phase symmetrical voltage drop with 40 % sag depth happens in the grid. Among it, the phase jump is considered as 25°. The frequency is 50Hz, so the cycle is 20 ms. The rest parameters of DVR are the same as that mentioned in section 4.

According to Eq. (20), the voltage sag in grid is monitored by logical unit. Control system includes two sub-modules, one is the calculation of the injected initial angle and final angle of DVR, and another is the transition to the minimum energy compensation. The line current, which is acted as a reference, can be gotten from PLL to realize the decoupling control of active and reactive power. For the phase angle detection module, PLL is used to calculate the DVR initial angle of pre-sag compensation and the final angle of the minimum energy compensation. After the transition, the reference voltage is given by the minimum energy module, i.e. $U_{L-abc}^* = U_{opt}$. The reference voltage U_{DVR}^* is generated by DVR. U_{DVR}^* is compared with the actual voltage U_{DVR} in the stationary reference frame. The difference of U_{DVR} and U_{DVR}^* is tracked by the PI controller accurately. A feed forward control with gain is added to compensate the loss of DVR system.

Simulation result
The Δt_{tran} in transition is taken as 40 ms. From these, it can be obtained that the DVR initial injected angle is 16° while the final angle is 81.5°. The simulation results are given in Fig. 6. Figure 6(a) shows the voltage sag happens at 0.1s. Figure 6(b) shows the amplitude and phase of load voltage return to normal after the recovery of DVR. Figure 6(c) shows the injected voltage of DVR during the sag. Figure 6(d) shows the power change of DVR during compensation.

As shown in Fig. 6, the sag depth was 40 %, and the start time of sag was 0.1 s. The DVR injected voltage and power were shown in the last two graphs of Fig. 6. In the power graph, it was noticed that DVR injected a large number of active power to compensate the phase jump in first cycle of initial stage. After two cycles, pre-sag compensation was gradually shifted to the minimum energy compensation mode. Accordingly, the injected voltage made a smooth transition, which can be seen from the third graph. In the last process, the injected active power was gradually reduced and became steady. After the recovery by DVR, the amplitude and phase of load voltage returned to normal.

Conclusion
To manage the power quality in a micro grid, an optimized compensation strategy of DVR has been proposed based

The sag depth is varied from 10 % to 80 % of the rated voltage. As shown in Fig. 4(a), the proposed method has the longest supporting time among three strategies at the same sag depth. Taking 50 % sag depth for example, the compensation time of the in-phase compensation is longer than that of the pre-sag compensation. The proposed method further increases the compensation time to 20 cycles based on in-phase compensation. When the sag depth is lower than 30 %, the proposed strategy works in the self-supporting mode. The minimize utilization of dc active power makes the slowest discharge of capacitor rate among three compensation methods, so the compensation time can be infinitely prolonged. Figure 4(b) depicts that the phase jump is varied from -60° to 60° for a 50 % sag depth when other conditions don't change. The compensating time of in-phase compensation and the proposed compensation have no obvious fluctuation with the change of phase jump, while pre-sag compensation is opposite. The proposed strategy has the longest compensation time among three strategies.

on three basic strategies in the paper. The proposed strategy protects sensitive loads against the grid voltage sags accompanied with the phase jump. At the beginning of voltage sag, pre-sag compensation is adopted to detect and compensate phase jump. The full compensation of voltage amplitude and phase is achieved. After two cycles, pre-sag compensation is shifted to the minimum energy compensation to reduce the consumption of DVR active power. To mitigate the influence of phase jump on sensitive loads, the voltage phase is converted smoothly according to a quadratic transition curve. Although the presented method has good compensation effects to three-phase symmetrical voltage sags with different characteristics, a more accurate mathematical model needs to be established for unbalanced three-phase voltage drop. Furthermore, the compensation time of DVR under pre-sag compensation can be significantly increased when applying the proposed optimized strategy. In practical application, the proposed method can meet the efficient and accurate requirements of DVR.

Acknowledgment

This work is supported by National Nature Science Foundation under Grant 51477070, and the Priority Academic Program Development of Jiangsu Higher Education Institutions (PAPD).

Authors' contributions

ZML proposed the optimized compensation strategy of DVR for voltage sag, carried out a quadratic transition which involves the injected voltage phases of pre-sag strategy and minimum energy strategy, finished the related calculation of voltage amplitude and phase. WWL carried out the DVR compensating strategy studies, derived the maximum compensation time of DVR by control requirements, participated in the comparison of the maximum compensation time for different methods and drafted the manuscript. THP established the model and carried out the simulation study of the proposed compensation of DVR, gave the voltage and power change during the compensation, made the summary. All authors read and approved the final manuscript.

About the Authors

W.W. Li, received her B.S degree from Jiangsu University in 2013. Now she has been a postgraduate student in School of Electrical and Information Engineering, Jiangsu University, Zhenjiang, China. Her current research interests include micro-grid, power electronic devices.

Z. M. Li, received his B.S. degree from zhenjiang agricultural machinery institute and received his M.S degree from Xi'an Jiao Tong University in 1987. Now he has been a professor in School of Electrical and Information Engineering, Jiangsu University, Zhenjiang, China. His current research interests include industrial computer network, process control, power system remote monitoring etc. He published over 100 papers, more than 20 articles were included by SCI and EI. He obtained second prize for technological progress of the original ministry of machine and the second prize of scientific and technological progress of Jiangsu province, and outstanding teaching achievement prizes of Jiangsu province.

T.H. Pan, received his B.S. degree from Anhui Agriculture University and M.S. degree from Gansu University of Technology in 1997 and 2000 respectively. And he received his Ph.D. degree in control theory and control engineering from Shanghai Jiao Tong University in 2007. Now he has been a professor in School of Electrical & Information Engineering, Jiangsu University, Zhenjiang, China. His current research interests include multiple model approach and its application, machine learning, virtual metrology, predictive control and Run-to-Run control theory and practice, etc.

Competing interests

The authors declare that they have no competing interests.

References

1. Wang, C., & Li, P. (2010). Development and challenges of distributed generation, the micro-grid and smart distribution system [J]. *Automation of Electric Power Systems, 34*(2), 10–14.
2. Wang, C., & Wang, S. (2008). Study on some key problems related to distributed generation systems [J]. *Automation of Electric Power Systems, 32*(20), 1–4.
3. Wang, C., Wu, Z., & Li, P. (2014). Research on key technologies of microgrid [J]. *Transactions of China Electrotechnical Society, 29*(2), 1–4.
4. Kaniewski, J., Fedyczak, Z., & Benysek, G. (2014). AC voltage sag/swell compensator based on three-phase hybrid transformer with buck–boost matrix-reactance chopper. *IEEE Transactions on Industrial Electronics, 61*(8), 3835–3846.
5. Castilla, M., Miret, J., Camacho, A., Matas, J., & de Vicuna, L. (2014). Voltage support control strategies for static synchronous compensators under unbalanced voltage sags. *IEEE Transactions on Industrial Electronics, 61*(2), 808–820.
6. Kumar, C., & Mishra, M. (2014). A multifunctional DSTATCOM operating under stiff source. *IEEE Transactions on Industrial Electronics, 61*(7), 3131–3136.
7. Wang, T., Xue, Y., & Choi, S. S. (2007). Review of dynamic voltage restorer [J]. *Automation of Electric Power Systems, 31*(9), 101–107.
8. Liu, Y.-y., Xiao, X.-n., & XU, Y.-h. (2010). Characteristics analysis on energy steady compensation for dynamic voltage restorer [J]. *Proceedings of the CSEE, 30*(13), 69–74.
9. Sadigh, A. K., & Smedley, K. M. (2012). Review of voltage compensation methods in dynamic voltage restorer (DVR). In *Proc. IEEE Power Energy Soc. Gen. Meet* (pp. 1–8).
10. Feng, X.-m., & Yang, R.-g. (2004). Analysis of voltage compensation strategies for dynamic voltage restorer (DVR) [J]. *Automation of Electric Power Systems, 28*(6), 68–72.
11. Wang, J., Xu, A.-q., Weng, G.-q., et al. (2010). A survey on control strategy of DVR [J]. *Power System Protection and Control, 38*(1), 145–151.
12. Al-Hadidi, H. K., Gole, A. M., & Jacobson, D. A. (2008). A novel configuration for a cascade inverter based dynamic voltage restorer with reduced energy storage requirements. *IEEE Transactions on Power Delivery, 23*(2), 881–888.
13. Sun, Z., Guo, C., Xiao, X., et al. (2010). Analysis method of DVR compensation strategy based on load voltage and minimum energy control[J]. *Proceedings of the CSEE, 30*(31), 43–49.
14. Xiao, X.- n., Xu, Y.- h., & Liu, L.- g. (2002). Research on mitigation methods of voltage sag with phase - angle jump [J]. *Proceedings of the CSEE, 22*(1), 64–69.
15. Meyer, C., Doncker, R. W., Li, Y. W., & Blaabjerg, F. (2008). Optimized control strategy for a medium-voltage DVR—Theoretical investigations and experimental results. *IEEE Transactions on Power Electronics, 23*(6), 2746–2754.
16. Ke, C.-b., & Li, Y.-l. (2012). Study on voltage sags compensation strategy for dynamic voltage restorer [J]. *Power System Protection and Control, 40*(17), 94–99.

12

Experimental studies on impedance based fault location for long transmission lines

Saeed Roostaee[*] ⓘ, Mini S. Thomas and Shabana Mehfuz

Abstract

In long transmission lines, the charging current caused by the shunt capacitance decreases the accuracy in impedance based fault location. To improve the accuracy of fault location, this paper presents a novel scheme, where two Digital Fault Recorders (DFRs) are installed in a line. They can send the transient data of the faults to the both ends of a line. To estimate the distance of a fault, impedance based fault location methods are applied with transient fault data of both ends protection relays and both DFRs installed in a line. To evaluate the proposed scheme, a laboratory setup has been developed. In the lab, several faults have been simulated and associated voltages and currents are injected to a relay IED to compare experimental results.

Keywords: Transmission line fault location, Impedance based fault location

1 Introduction

Accurate fault location can expedite the repair of the faulted components, speed-up restoration, reduce outage time. Therefore, it can improve power system reliability [1, 2]. Many methods and techniques have been introduced on fault location estimation so far. Among them, impedance based fault location methods are most used by utilities [3, 4]. Based on the input data of fault location algorithms, they can be categorized as:

- One-end [5, 6]
- Two-end algorithms [7–9]

One-end impedance fault location method is economic and simple as compared to other fault location methods [5]. One end impedance fault locators calculate the location of a fault based on the impedance from one end of a transmission line. However, this technique is subjected to several sources of error, such as the zero sequence mutual effects, the uncertainty of parameter in transmission line, unbalanced load flow, the influence of facts devices, the accuracy of transmission line model, and measurement errors [10].

Two-end algorithms estimate the location of fault using voltage and current form both the ends of a line. Based on this technique, transient fault data must be collected and synchronized from both ends of a transmission line. The transient data faults can be reported by the protection relays or Digital Fault Recorders (DFRs) which is installed in substations. In order to have an accurate fault location, the data from both ends of a transmission line should be analyzed [8]. Therefore, in this technique, microprocessor relays (or digital fault recorders), communication facility, and analysis software are required.

Two-end algorithms are divided into two categories. One of them makes use of synchronized data, and the other one utilizes unsynchronized data. In the first category, the global positioning system (GPS) is required. The measured values form the both ends are synchronized by GPS clock [7, 8, 11, 12]. The other hand, in other categories, the GPS system is not needed [8, 9]. Thereafter, the main advantage and disadvantage of this class is lower cost of implementation and lower accuracy respectively.

The common characteristics of the fault location methods mentioned in the literature are that the most of them are applied at local and remote end values of a transmission line. However, far too little attention has been paid to the values in the transmission lines. This paper proposes a novel scheme into the fault

* Correspondence: saeed61850@yahoo.com
Department of Electrical Engineering, Faculty of Engineering and
Technology, Jamia Millia Islamia, New Delhi, India

location based the values at both ends as well as in transmission line. The proposed scheme has been compared with one end and two end algorithms with experimental results. The fault location algorithm, hardware and software requirement, experimental setup, and experimental results are presented in this paper.

The overall structure of the study takes the form of five sections, including this introductory section. Section two reviews impedance based fault location technique. The third and fourth sections describe and test the proposed scheme. Finally, the conclusion gives a brief summary of the findings.

2 Impedance based fault location

Impedance based fault location algorithms are used to estimate the location of faults on a transmission line. These methods are commonly used by utilities because of simplicity. They require current and voltage signals along with sequence impedance (Positive, negative and zero) to estimate the distance to fault location. Depending on the availability of the input signals, these algorithms can be categorized into one-end and two-end methods. Following subsection are discussed these two methods.

2.1 One-end method

In this technique, the ground phase voltage and current are needed. Impedance based fault location from one end data can be categorized based on the source at one end and source at two ends. Figure 1 shows the source at one end where $I_F = I_A$ and the estimated distance can be calculated from (2) [1].

$$Z_F = \frac{U_A}{I_A} = dZ_L + R_F \tag{1}$$

$$d = \frac{\mathrm{Im}(Z_F)}{\mathrm{Im}(Z_L)} \tag{2}$$

Where Z_F is the measured impedance from substation A, Z_L is the impedance of the line, R_F is the resistance of fault.

Figure 2 shows impedance fault location based on the source at two ends, where $I_F = I_A + I_B$. In these

types of lines, estimated distance can be obtained from (4).

$$U_A - dZ_L I_A - R_F I_F = 0 \tag{3}$$

$$d = \frac{\mathrm{Im}(U_A \Delta I_A^*)}{\mathrm{Im}(Z_L I_A \Delta I_A^*)} \tag{4}$$

Where star denotes conjugation and ΔI_A can be obtained with the following equation [1]:

$$\Delta I_A = \frac{(1-d)Z_L + Z_B}{Z_A + Z_L + Z_B} I_F \tag{5}$$

2.2 Two-end method

In two ends impedance based fault location, as the same fault happened in a line, the voltage in fault point should be equal. In the other words:

$$U_{Fi}^A = U_{Fi}^B$$

The voltage of fault from each end can be obtained from (6) [1].

$$\begin{aligned} U_{Fi}^A &= \cosh(\gamma_i dl) U_{Ai} - \sinh(\gamma_i dl) Z_{Ci} I_{Ai} \\ U_{Fi}^B &= \cosh(\gamma_i (1-d)l) U_{Bi} - \sinh(\gamma_i (1-d)l) Z_{Ci} I_{Bi} \end{aligned} \tag{6}$$

Where U_{Ai} and U_{Bi} are the ith symmetrical components of voltage in A and B respectively, I_{Ai} and I_{Bi} are the ith symmetrical components of current in substation A and substation B respectively, where can be obtained in the both ends of a line. To simplify equations, following trigonometric identities are applied.

$$\begin{aligned} \cosh(\gamma_i (1-d)l) &= \cosh(\gamma_i l)\cosh(\gamma_i dl) - \sinh(\gamma_i l)\sinh(\gamma_i dl) \\ \sinh(\gamma_i (1-d)l) &= \sinh(\gamma_i l)\cosh(\gamma_i dl) - \cosh(\gamma_i l)\sinh(\gamma_i dl) \end{aligned} \tag{7}$$

To estimate the location of the fault based on two end values, following formula can be derived from (6) and (7) [1].

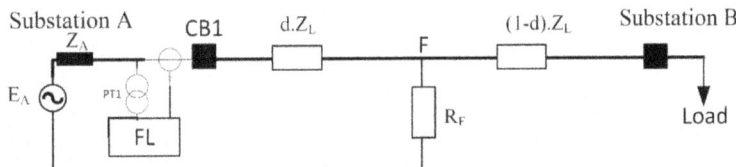

Fig. 1 Impedance based fault location with the source at one end

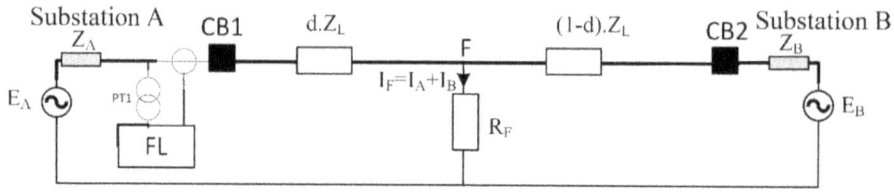

Fig. 2 Impedance based fault location with the source at two ends

$$d = \frac{1}{\gamma_i l} \tanh^{-1} \left(\frac{\cosh(\gamma_i l) U_{Bi} - Z_{Ci} \sinh(\gamma_i l) I_{Bi} - U_{Ai}}{\sinh(\gamma_i l) U_{Bi} - Z_{Ci} \cosh(\gamma_i l) I_{Bi} - Z_{Ci} I_{Ai}} \right)$$

(8)

Where γ_i is propagation constant of the line for the i-th symmetrical component, Z_{Ci} is the characteristic impedance of the line for the i-th symmetrical component.

3 Proposed scheme

Generally, a fault location algorithm uses the recorded transient data from the protection relays at the both ends of a line. In this study, to analysis fault location, we have proposed the use of transient data of the protection relays in substations as well as transient data of the DFRs. As it is shown in Fig. 3, the DFRs are installed in a transmission line and they can send data to the both substations through a communication channel.

There is now available fibre-optic current and voltage transducers (OCT & OVT). Where they can be combined in a light-weight and compact single phase unit with an optical metering unit (OMU). The reduced size and increased accuracy of these optical technology make the design particularly well suited to the proposed scheme. As the transducers have small size and light-weight, they can be mounted between two insulator strings in a tower. And OMU and DFR can be mount in the tower, other mounting options are possible. Under

these conditions, low-cost fault location has become feasible.

Figure 4 illustrates the experimental setup for fault location. In this setup, we can simulate faults in a transmission line in PSCAD; after that, the transient data for the fault is sent to a secondary test kit to generate three phase's fault voltage and current. These three phase signals are connected to a protection relay throughout wiring. Therefore, in this setup, several faults can be simulated and the signals can be generated to provide real values for fault location functions.

Based on this setup, various types of faults in different locations of a transmission line can be simulated via PSCAD (Fig. 5). These faults generate transient data. Transient data can describe the voltage and current signals both before and after the fault. There is a component in PSCAD which can record the transient data in the COMTRADE format. This format is an IEEE standard format for recording transient data and it can be used to different fault analysis tools [13]. The CMC 256 is a test set which can play transient data. The Trans View play is software which can load COMTRADE file and generate corresponding three phase signals (Fig. 6). The signals can be utilized by protection relays to test a fault location algorithm.

4 Results and Discussion

To evaluate the performance of the proposed scheme, various fault scenarios are carried out by utilizing the

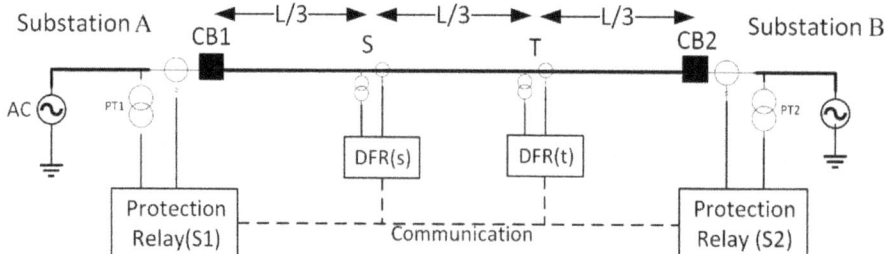

Fig. 3 Proposed scheme

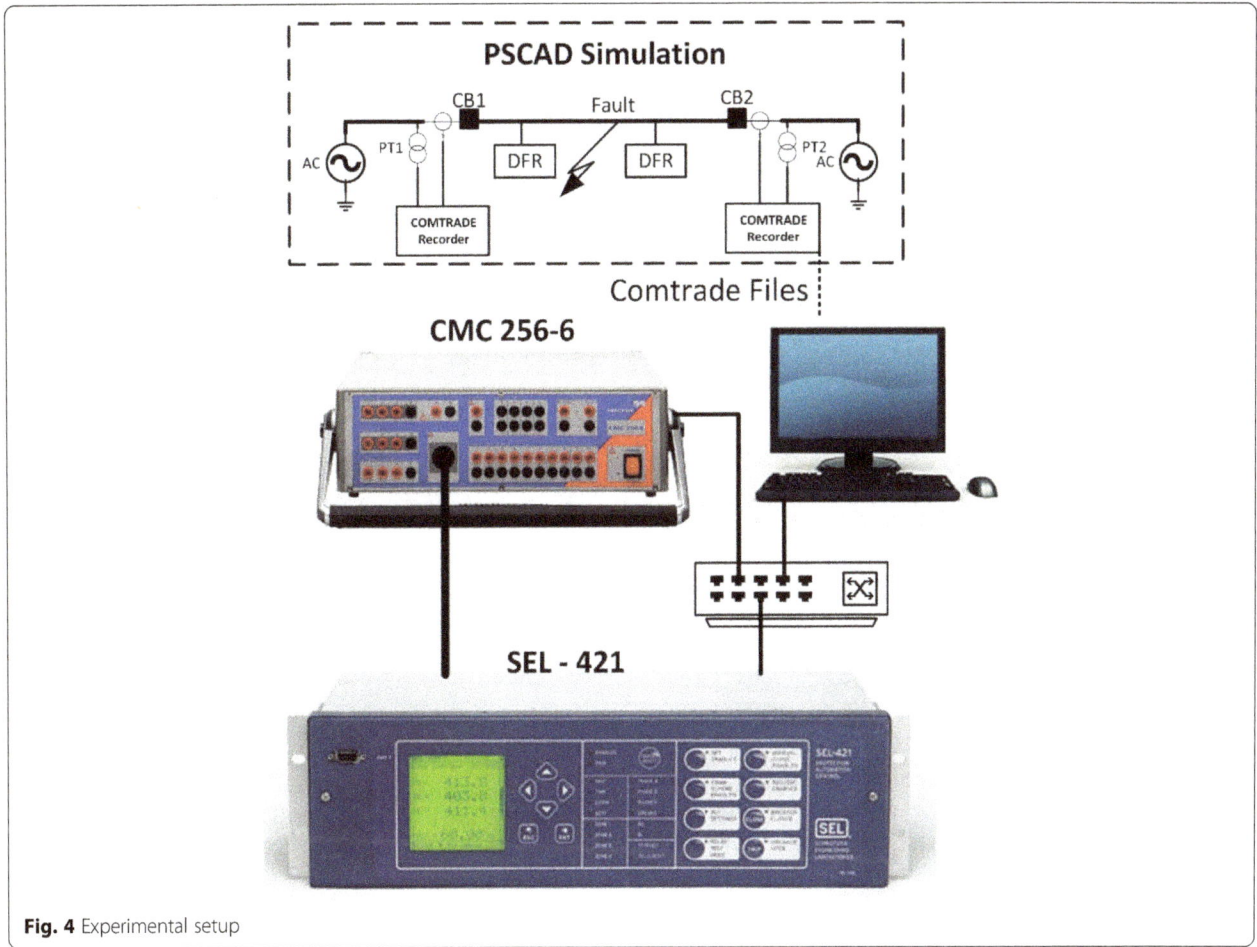

Fig. 4 Experimental setup

hardware implementation in a laboratory environment. To test the proposed scheme, we simulate a long transmission line between substation 1 and substation 2 which is a homogeneous and untransposed line. In this study, standard tower, conductor, and ground wire for 400Kv are selected. Figure 7 shows the main feature of the HS10 tower which is applied in this paper. To transfer the power, each phase contains four conductors in symmetrical bundle mode with 45.72-cm space between two connected conductors. To protect the line against lightening as well as communication purpose, OPGW-24B1/70 is selected as the ground wire. The main feature of conductor and ground wire are summarized in Table 1.

Modelling of a transmission line is very important in fault location algorithms. Impedance based fault location technique estimates the location of a fault by comparison of impedance with the line impedance. Therefore, the accuracy of this technique highly depends on the accuracy of line parameter model. In this study, we used distributed transmission line model. Transmission line parameters can be calculated based on the Carson's'

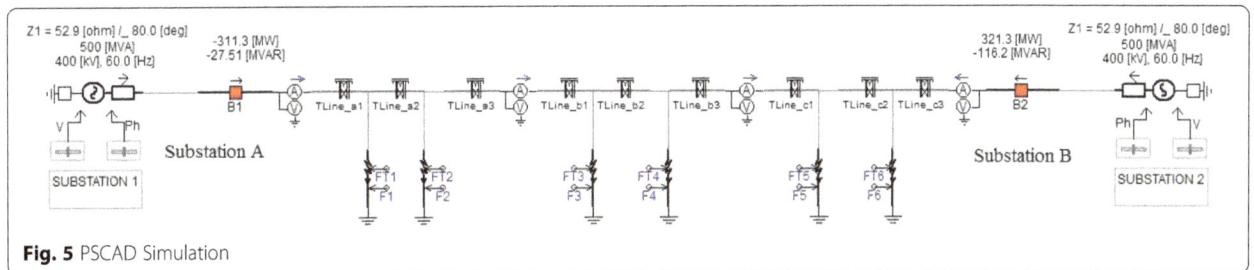

Fig. 5 PSCAD Simulation

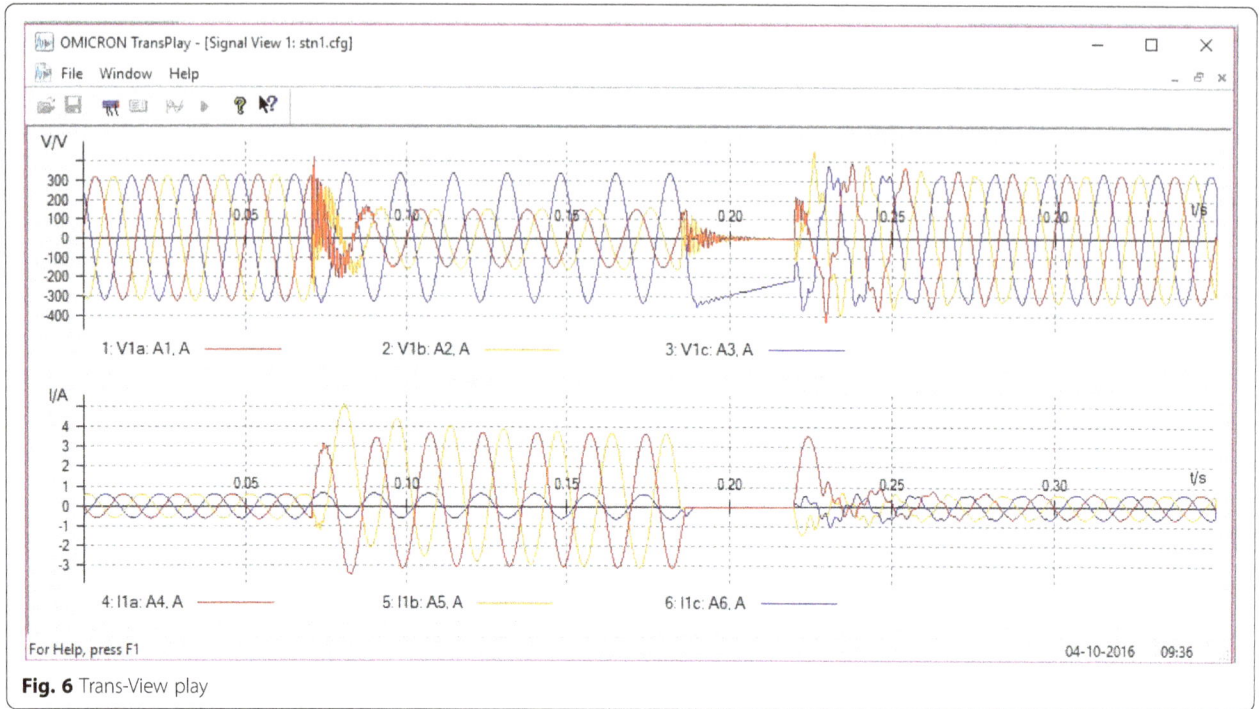

Fig. 6 Trans-View play

equations [14] and the characteristics of the tower, conductor, and ground wire. Following RLC as well as positive and zero sequences of resistance and inductance were calculated in our study as below:

$$R = \begin{pmatrix} 0.038906 & 0.02464 & 0.024644 \\ 0.02464 & 0.038614 & 0.02464 \\ 0.024644 & 0.02464 & 0.038906 \end{pmatrix}$$
$$\Omega/km$$

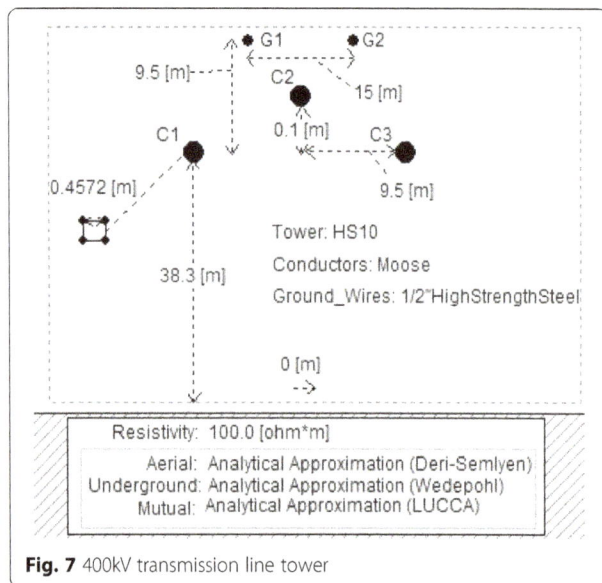

Fig. 7 400kV transmission line tower

$$L = \begin{pmatrix} 0.0013674 & 0.00053991 & 0.00041175 \\ 0.00053991 & 0.0013513 & 0.00053991 \\ 0.00041175 & 0.00053991 & 0.0013674 \end{pmatrix}$$
$$H/km$$

$$C = \begin{pmatrix} 1.0871e{-}8 & -2.734e{-}9 & -1.0124e{-}9 \\ -2.734e{-}9 & 1.1534e{-}8 & -2.734e{-}9 \\ -1.0124e{-}9 & -2.734e{-}9 & 1.0871e{-}8 \end{pmatrix}$$
$$F/km$$

$$[R1 \quad R0] = [0.014167 \quad 0.088091] \quad \Omega/km$$

$$[L1 \quad L0] = [0.00086486 \quad 0.0023564] \quad H/km$$

Following subsections contain the result under different conditions.

4.1 Fault near mid-point

This subsection discusses the faults between the S and T points in the line (Fig. 3). Therefore, a transient fault with following features is implemented in PSCAD.

- Fault Location: 125 km
- Fault type: AB-g (A-B-G)

Table 1 Conductor and ground wire

Name	Application	Diameter mm	RDC (20 °C) Ω/km
ACSR- Moose	Conductor	31.77	0.0547
OPGW-24B1/70	Ground cable	12	0.79

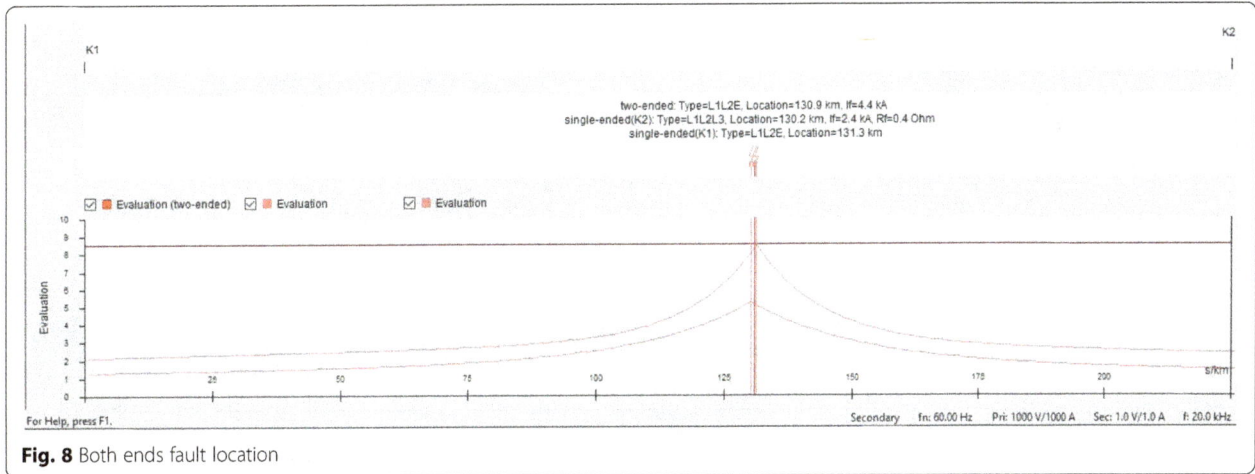

Fig. 8 Both ends fault location

- Fault Start: 0.22 s
- Fault duration: 0.15 s
- Fault Resistance: 0.001 Ω

- Frequency: 60 Hz
- Line Length: 225
- Positive Impedance Magnitude: 73.4
- Positive Impedance Angle: 87.51
- Zero Impedance: 200.76
- Zero Impedance Angle: 84.3

The transient data for the above fault is analysed in OMICRON-Trans View software. As shown in Fig. 8, a screen capture of the software, the estimation of 130.9 km far from substation 1 is the estimation of the location of the fault.

Online relay fault location is also utilized to estimate the fault location. Three phase voltage and current signals associated with the fault are generated by the CMC 256-6. These signals are connected to the SEL-421 relay. The SEL-421 is a transmission line relay and it has five zones of phase and ground. This relay has online fault location and after any fault in a transmission line, the estimation of fault location is shown in relay' LCD as well as, the transient fault data along with estimated fault location are recorded as an event in the relay. The setting of the relay is important in fault location function. Based on the data of the transmission line following settings had been set in the relay:

After injection of transient data of the fault to the relay by CMC 256, the relay issues a trip signal and record the transient data of the fault (Fig. 9). The relay detects an AB-g fault in zone one. Thereafter, online relay fault location estimates a fault on 135.38 km far from substation 1 (Fig. 10).

In the proposed scheme, recorded event of DFR1 and DFR2 are utilized to estimate the fault location. Based on the proposed scheme which is described in section 2, a transmission line is divided into three sections. Section one is between substation A and DFR1. Section two is between DFR1 and DFR2. And the last section in between DFR2 and substation B. For the forehead mentioned fault, only section two has detected the fault. The estimation of the location of the fault is 49.8 km far from

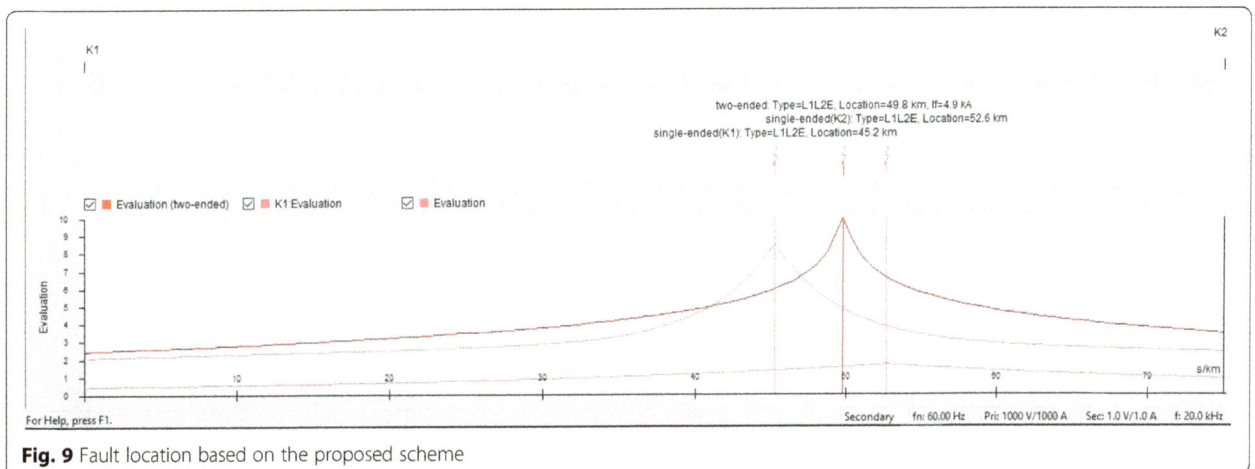

Fig. 9 Fault location based on the proposed scheme

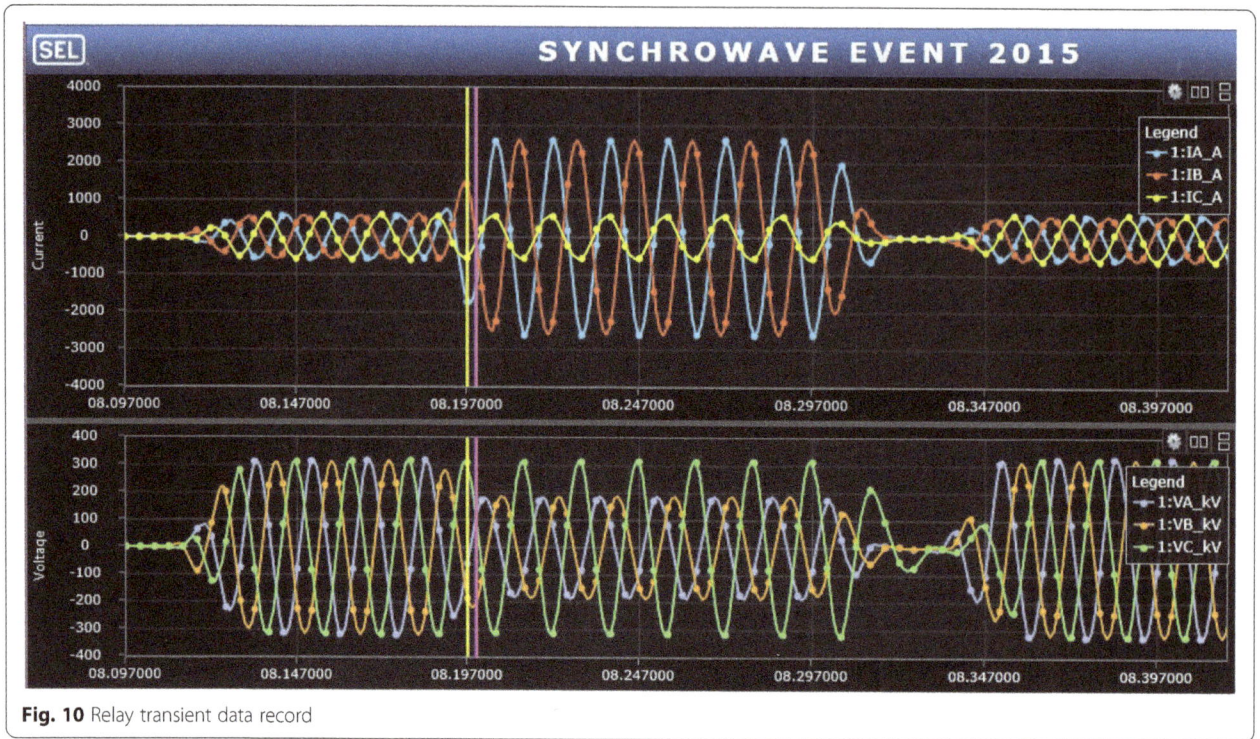

Fig. 10 Relay transient data record

DFR1 (Fig. 11). In the other words, the estimated location is 124.8 km far from the substation one.

To compare the accuracy of fault location based on different methods, the percentage error is determined as:

$$\%Error = \frac{L_{Estimation} - L_{Actual}}{L_{Line}} * 100 \qquad (9)$$

Where $L_{Estimation}$ is the estimated location given by fault location algorithm, L_{Actual} is the actual location of the fault and L_{Line} is the length of the line.

Different factors affecting the accuracy of fault location. Generally, the main factors considered as: fault type and location; fault resistance; power flow; source impedance; line parameter; transient and steady errors of measurement system (including CT saturation), etc.

To improve the fault-location estimation, it is important to reduce errors based on the analysed method. As different input signals are used in the proposed scheme, the accuracy of measuring systems and CT saturation appear important. Therefore, it is suggested to apply optical measurement transformers where they have high accuracy for high voltage system. And regarding to the CTs saturation, the elimination of this source of errors can be achieved by using such a set of fibre optic measurements that are inherently free of magnetic saturation, making them ideal for capturing fast transient currents, and short circuit currents. Otherwise, the sophisticated compensation algorithms aimed at faithful reproduction of the CT primary current have to be applied. The accuracy of different techniques for the fault which is simulated in this subsection can be summarized in Table 2.

Based on this experiment, it is observed that the proposed technique is 29 and 51 times more accurate than two end method and relay function based fault location method respectively. A similar observation can be archived for other faults, which is summarized in the Table 3.

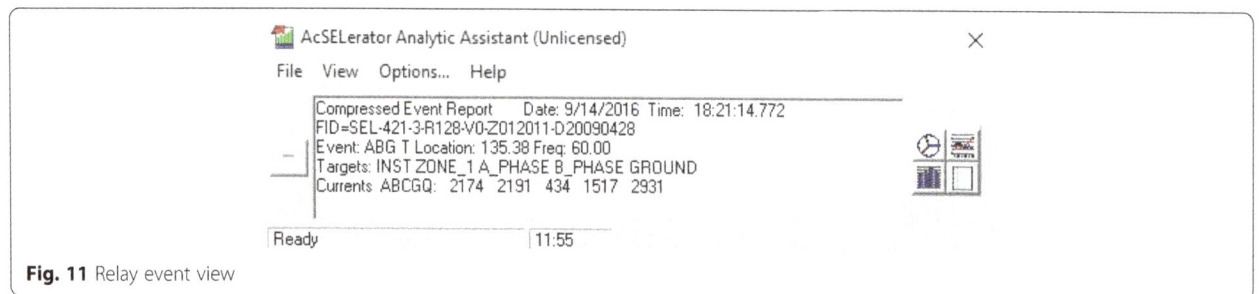

Fig. 11 Relay event view

Table 2 Comparison of accuracy with different fault location method

Fault Type	R_{Fault} Ω	Location km	Fault Location Error (%)		
			With two ends method	Relay Fault Location Function	With the proposed shceme
AB-G	0	125	2.6222	4.6133	−0.089

4.2 Fault near substations

This subsection discusses the faults near substations. In the other words, faults between substation A and point S and point T and substation B. In this condition, with failure of the measurement system in a substation, both ends methods cannot be applied. Thereafter, the measured values from the remote end can estimate the fault location with one-end algorithms. Whereas, in the proposed scheme, transient data of DFR installed in the line can be applied to obtain the desire objective.

To evaluate the proposed scheme, several faults with different locations, types, and resistances have been simulated. Testing of various fault location techniques are carried out by utilizing the hardware implementation in a laboratory environment. The result obtained from the preliminary analysis of the accuracy of different techniques are summarised in the Table 4.

Based on this experiment, it is observed that the proposed technique is more accurate than one-end fault location based measured values in substation.

Table 3 Comparison of accuracy of different fault between S and T

Fault Type	R_{Fault} Ω	Location km	Fault Location Error (%)		
			With two ends method	Relay Fault Location Function	With the proposed shceme
AB	0	100	3.4222	5.1733	0.2222
AB	0	125	2.7111	4.8311	0.0444
AB	50	100	3.3333	4.9333	0.6667
AB	50	125	2.6222	4.6133	−0.089
ABG	0	100	3.3333	4.9333	0.6667
ABG	0	125	2.6222	4.6133	−0.089
ABG	50	100	3.2444	139.2133	0
ABG	50	125	2.5333	143.2489	−0.356
BG	0	100	3.1111	6.6488	0.6667
BG	0	125	2.2667	7.1155	−0.622
BG	50	100	2.8444	8.8933	0.3556
BG	50	125	2.1333	10.6889	0.3556

Table 4 Comparison of accuracy of different fault between T and Substation B

Fault Type	R_{Fault} Ω	Location km	Fault Location Error (%)	
			One-end Function Location	With the proposed shceme
AB	0	175	4.4888	0.3111
AB	0	200	4.5066	−0.089
AB	50	175	9.9644	0.8
AB	50	200	12.44	0.2667
ABG	0	175	2.24	0.5778
ABG	0	200	4.4044	0.2222
ABG	50	175	191.87	0.8
ABG	50	200	228.4	−0.089
BG	0	175	8.2133	0.7111
BG	0	200	8.7911	0.2222
BG	50	175	17.1289	0.4
BG	50	200	23.3778	0.4

5 Conclusion

Recent development in electronic instrument transformer and communication technology provides new facility for power system functions. This paper has outlined a new fault location scheme for long transmission lines with the use of the event recorded by two DFRs which is installed in a transmission line. Preliminary analysis of testing the proposed fault locating scheme has shown satisfactory performance and a higher accuracy.

The proposed scheme can be utilized for other purposes such as protection functions. In future work, we intend to evaluate a similar scheme in protection applications.

Authors' contributions
SR: Hardware implementation, writer. MST: Technical editor, analysis of power system part, analysis of transient data. SM: Technical editor, network configurator. All authors read and approved the final manuscript.

Competing interests
The authors declare that they have no competing interests.

References
1. Saha, M. M., Izykowski, J. J., & Rosolowski, E. (2009). *Fault location on power networks*: Springer Science & Business Media.
2. Kezunovic, M. (2011). Smart fault location for smart grids. *IEEE Transactions on Smart Grid, 2,* 11–22.
3. Schweitzer, E. O., Guzman, A., Mynam, M. V., Skendzic, V., Kasztenny, B., & Marx, S. (2016). Protective relays with traveling wave technology revolutionize fault locating. *IEEE Power and Energy Magazine, 14,* 114–120.
4. Zimmerman, K., & Costello, D. (2005). Impedance-based fault location experience. In *58th annual conference for protective relay engineers, 2005* (pp. 211–226).
5. Kawady, T., & Stenzel, J. (2003). A practical fault location approach for double circuit transmission lines using single end data. *IEEE Transactions on Power Delivery, 18,* 1166–1173.

6. Izykowski, J., Rosolowski, E., & Saha, M. M. (2004). Locating faults in parallel transmission lines under availability of complete measurements at one end. *IEE Proceedings-Generation Transmission and Distribution, 151*, 268–273.

7. Izykowski, J., Rosolowski, E., Balcerek, P., Fulczyk, M., Saha, M.M., (2011). Fault location on double-circuit series-compensated lines using two-end unsynchronized measurements. *IEEE transactions on power delivery*, 26(4), pp.2072-2080.

8. Dutta, P., Esmaeilian, A., & Kezunovic, M. (2014). Transmission-line fault analysis using synchronized sampling. *Power Delivery IEEE Transactions on, 29*, 942–950.

9. de Pereira, C. E. M., & Zanetta, L. C. (2011). Fault location in multitapped transmission lines using unsynchronized data and superposition theorem. *Power Delivery IEEE Transactions on, 26*, 2081–2089.

10. Kang, N., Chen, J. and Liao, Y., (2015). A fault-location algorithm for series-compensated double-circuit transmission lines using the distributed parameter line model. *IEEE Transactions on Power Delivery*, 30(1), pp.360-367.

11. Tzu-Chiao, L., Pei-Yin, L., & Chih-Wen, L. (2014). An algorithm for locating faults in three-terminal multisection nonhomogeneous transmission lines using synchrophasor measurements. *Smart Grid, IEEE Transactions on, 5*, 38–50.

12. Dobakhshari, A.S. and Ranjbar, A.M., (2015). A novel method for fault location of transmission lines by wide-area voltage measurements considering measurement errors. *IEEE Transactions on Smart Grid*, 6(2), pp.874-884.

13. IEEE Draft Standard for Common Format for Transient Data Exchange (COMTRADE) for Power Systems, IEEE PC37.111/D4, January 2012 (IEC 60255-24 Ed.2), pp. 1-72, 2012.

14. Carson, J. R. (1926). Wave propagation in overhead wires with ground return. *Bell System Technical Journal, 5*, 539–554.

13

Development and research on integrated protection system based on redundant information analysis

Jinghan He, Lin Liu*, Wenli Li and Ming Zhang

Abstract

Facing the growth of energy demand and the worldwide interests in renewable energy sources, smart grid has been proposed in order to accommodate the needs of power grid development. In China, digital substations have been widely applied. Moreover, as the worldwide interest and development in smart grid have increased significantly, the digital information availability and communication capability of modern substation have been improved. Different with traditional substation protection configuration based on local information, Integrated Protection (IP) is used to denote the integration of several protective devices for multiple power equipment within the substation into one protective relay, obtaining all the real time information of the substation by communication network. The analysis of sharing information and cooperation among different protection functions helps to realize more reliable and sensitive fault detection. This paper describes several different integrated protection schemes and their respective advantages and disadvantages. The principles and functions applied in integrated protection systems are addressed to provide an overview of the current state of the technology. A new coordinated protection system based on inter-substation information is proposed and simulation study is discussed to verify the improved reliability and sensitivity.

Keywords: Integrated protection, Substation, Smart grid, Redundant information, IEC61850

Introduction

POWER systems are facing challenges due to the deregulation of the electricity market, environmental concerns and the impact of changing from a fossil fuel dominated system to one based on low carbon sources. The issues are further complicated by recent major blackouts, customers' expectation of high system reliability and the impact of power electronics equipment. Power systems need to be developed in the direction of high reliability, flexibility, intelligence and sustainability. All these challenges lead to high requirement on power system protection.

Traditional power systems usually have main and backup protections for every local plant. Main protection abides by the well-established and well-proven "dispersed and independent" principle. Backup protection issues tripping signals when the main protection or its associate circuit breaker fails to operate. These signals then trip the circuit breakers to isolate un-cleared faults from the network. They traditionally coordinate with the main protection using time delays and appropriate choice of settings [1]. The rapid development of power systems further complicates the setting and coordination of backup protection, and the resulting extended trip times risk operating safety and system stability.

The application of various intelligent secondary equipment and the construction of a reliable communication network are the fundamental factors for promoting the development of new protection systems. With the widely spread application of intelligent electronic devices (IEDs), information digitization and substation communications, the adoption of protection that uses comprehensive information sourced from the local region is required. Extensive studies have been carried out to improve protection performance, which generally involve enhanced utilization of digital and communications technologies.

* Correspondence: 11117363@bjtu.edu.cn
School of Electrical Engineering, Beijing Jiaotong University, Beijing 100044, China

The classification of these types of protection can be based on the information domain scope. Wide area protection is based on a wide area measurement system (WAMS) that implements a regional protection and control system [2–7]. This was developed from special integrity protection systems (SIPS) [8] and broadened the role of traditional relay protection from a point to a surface, so that the high standards of dependability and security inherent in modern protection systems can be maintained [9]. Wide area protection responds to various system disturbances through the analysis of wide area information and the evaluation of the system states. This type of protection involves the use of system-wide or regional information and the communication of selected local data to a remote location. It is designed to counteract the propagation of a large disturbance, and delivers better sensitivity and effectiveness as compared with traditional SIPS [10]. At present, because of communication delays and high requirement to Ethernet bandwidth, there are still many application issues of wide area protection for a large scale grid.

Protection schemes whose information domain is within a substation or within a local regional area network have also been proposed. The concept of integrated protection (IP) [11, 12] was proposed on the background of digital substation and smart grid. Integrated network protection (INP) is based on IP, and its information range is extended to several adjacent substations. All these studies have encouraged the protection and control system to utilize comprehensive information to achieve better performance and to satisfy the functional requirements of a modern grid.

Previous work summarized above indicates protection based on comprehensive information can adapt to the evolution of the power network and optimize its performance. China, U.S., Canada, Korea, and European Community (EC) countries have started research and development on integrated protection technologies and applications. For example, researchers in China applied integrated substation-area protection devices in 6 new generation intelligent substations in 2012 and the number has increased to 48 in 2014. The state grid corporation of China has announced that 65 % of the 110 kV substations will adopt integrated protection devices by 2020.

The rest of the paper is organized as follows. Section II describes several different kinds of integrated protection schemes and their respective advantages and disadvantages. Section III illustrates the protection principles and functions applied in integrated protection systems. A new coordinated protection system based on inter-substation information is proposed in Section IV, and finally, this paper is concluded in Section V.

Integrated protection schemes

Based on the optical transducer, data sampling with merging units and optic-fiber communication, the data collection part of protection device is separated. Different from conventional protection device, applying the intelligent breaker makes it possible that the protection device only keeps the data processing and communication part. With the rapid development and maturity of computer and electronic technology, high reliable integration and modularization of the protection device have become a development trend. Taking advantages of the compatibility with Common Information Model (CIM) of IEC 61850, communication between devices in substation automation systems becomes flexible. This technology has been applied in the protection scheme in intelligent substations and implemented in some latest power system protection products.

In the following, some of the different integrated protection schemes based on different communication infrastructures are presented, and their respective advantages and disadvantages are briefly explained.

Integrated substation protection scheme for rural substations

Rural substations are the vital nodes of rural smart grid and are commonly 35 kV substations in China, especially in the central and western regions. At the end of 2012, there are 16500 35 kV substations in State Grid and 1228 of them are new intelligent substations. On the background of smart grid and intelligent devices, research of integrated protection scheme for rural substations is driven by the development requirement.

A typical architecture of the integrated protection scheme of 35 kV substations consisting of three levels, is shown in Fig. 1. There is no process bus in the network and all the protection devices are connected to the merging units (MUs) and the intelligent units (IUs) in peer-to-peer mode. The station bus is in single-network structure considering the cost. Two integrated protection and control devices are in the bay level to receive the sample value information from the process level, and to conduct elaborate calculations and send decision signals to circuit breakers.

For new intelligent substations, two integrated protection central processing units (IPCPUs) are in the bay level and each one can protect all the primary devices in the substation. The adoption of dual IPCPUs is to improve the reliability, as one can work as the backup if the other fails. Because the substation area information is integrated in the IPCPU, it is easily to implement many protection functions, e.g. ground fault line detection, under frequency load shedding, voltage and reactive power automation control. GPS is used as the time server and the IRIG-B signals are used to synchronize all the data.

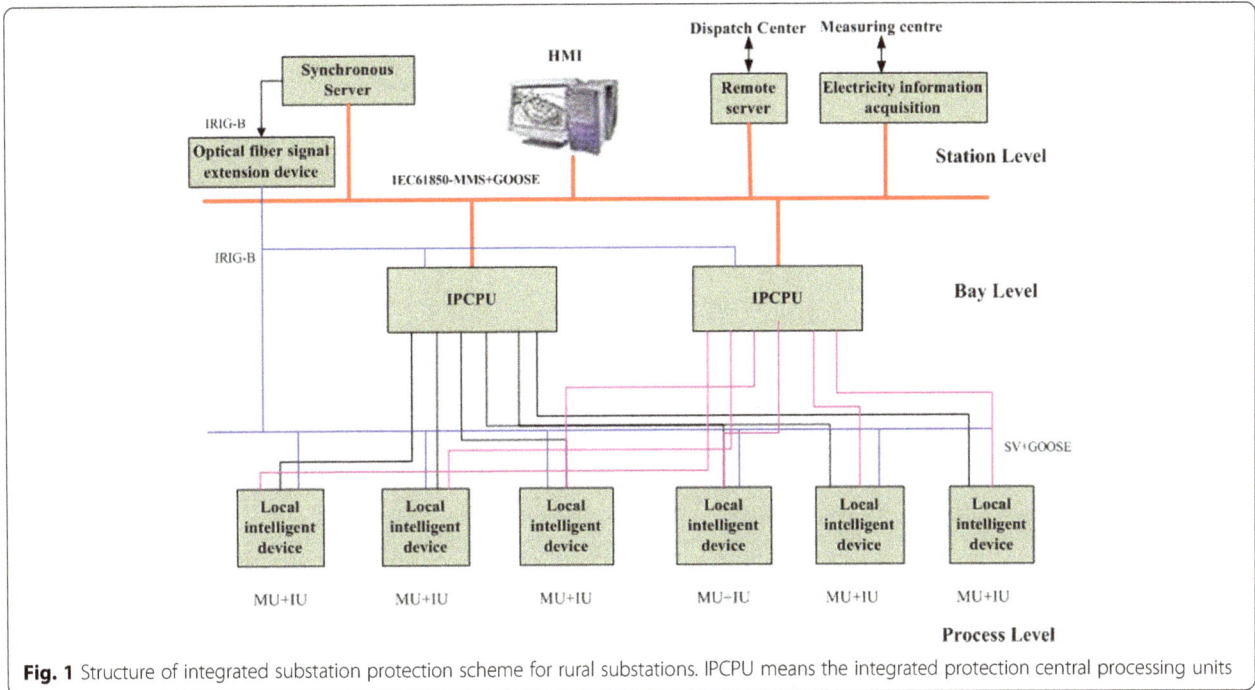

Fig. 1 Structure of integrated substation protection scheme for rural substations. IPCPU means the integrated protection central processing units

For refurbished substations, some protection functions can be set with MUs and IUs in the process level, and all these intelligent devices work together to implement distributed protection with no communication requirement. On the other hand, the IPCPU in bay level works as substation area protection including control and substation automation.

Some latest substation protection products based on this protection scheme have been applied in China, and some operation cases can be found in Table 1.

1) Advantages

The integrated substation protection is a cost-effective solution for rural substations, with simplified structure and reduced number of required equipment. The substation area information is integrated into the protection unit, and is easy to analyze comprehensive and redundant information to reach higher protection reliability and implement fault analysis. The peer-to-peer communication of the process bus is reliable and secure with low-latencies and sufficient bandwidth [13].

Table 1 Operation cases of latest substation protection products

No.	Substation Name	Voltage Level	Year
1	ShiMaChuan (New)	35 kV	2009
2	Weijin (Rebuit)	66 kV	2010
3	HaoTian (New)	35 kV	2011
4	DongTing (New)	35 kV	2011

The adoption of dual IPCPUs improves the system reliability.

2) Disadvantages

The optical-fiber interface has large numbers and the connection is complex, and thus this scheme is only suitable for small-scale substations. Furthermore, the scalability of this protection scheme is limited which does not meet the requirement of device networking.

Integrated substation protection scheme for high voltage substation

Different with rural substations, high voltage substations have much more bays and power equipment. Considering large numbers of information interface, different voltage-level sides have their own information sharing networks. Figure 2 shows the architecture of the integrated protection scheme of 110 kV substation. In the process level, the voltage and current signals acquired by the voltage transformers (VTs) and current transformers (CTs) are digitalized by the MUs and sent through the process bus to the bay level to achieve data sharing. The term "Integrated Substation Protection" (ISP) in the bay level, is used to denote the integration of several protective devices for multiple power equipment within the substation into one protective relay [1]. GPS signals are used to keep the synchronization of all the sample data. After the data processing and analysis of the sampling data from different devices in the substation, ISP send GOOSE messages including trip signals to the IU to realize the operation of the circuit breakers. Station level includes the Human Machine Interface (HMI)

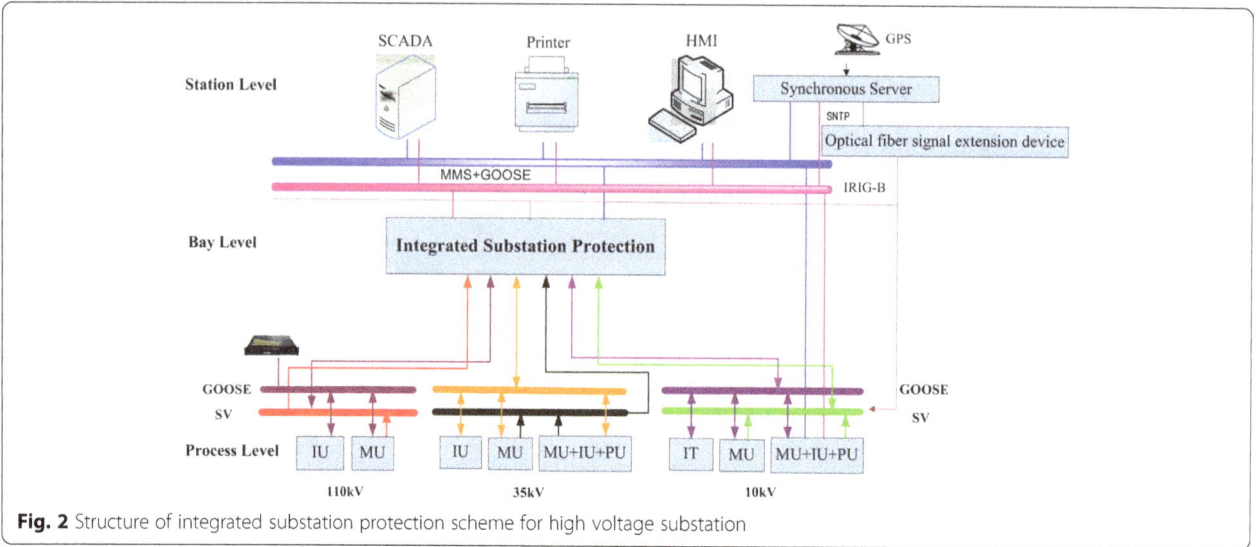

Fig. 2 Structure of integrated substation protection scheme for high voltage substation

and SCADA system. Through the station bus, the status data of various components are available to operators for monitoring and operation purposes. The manual control signals flowing in the opposite direction can also be issued through the network to perform some control functions by the protection units (PUs). Considering the communication reliability, both the process bus and station bus are in dual-network structure.

One protection product based on this protection scheme has been applied in LongChang 110 kV traction substation in 2011.

1) Advantages

   This protection scheme has good scalability and the information sharing network structure can meet the requirement of smart grid development. This communication network can also save optical-fiber interfaces and reduce infrastructure costs in project implementations. On the other hand, the integrated substation protection is based on substation-domain information and has better reliability.

2) Disadvantages

   The data sampling relies on external clock (GPS) but the transmission time delay of switch network is not a fixed value, and thus, the SV synchronization has a high risk. For GOOSE network, its transmission time delay is not fixed and the tripping time of the circuit breakers may be effected. Nowadays, the cost of fiber interface switch is high and thus, some idle interfaces should be considered as backup in project construction. The application cost should be given more attention.

Hierarchical regional area protection

The hierarchical regional area protection is a protection system involving the information of the whole power

grid to achieve reliable and adaptive fault detection and clearance. It is a combination of protection information, functions and coordinated strategy. There are three protection subsystems: local protection subsystem, substation area protection subsystem and wide area protection subsystem (as shown in Fig. 3).

Local protection subsystem (LPS): protection devices are installed for every local item of plant. Main protection abides by the well-established and well-proven "dispersed and independent" principle, and reliable and fast fault clearance is achieved based on local data analysis.

Substation area protection subsystem (SAPS): this kind of protection is different to the ISP, because its integrated information is the status data of the local protection devices and circuit breakers. SAPS can also communicate with adjacent substations through the synchronous digital hierarchy (SDH) [14], to receive status of the opposite terminal circuit breaker. Through centralized analysis, useful decisions can be given based on redundant information, in order to improve the system operation reliability. The SAPS also works as the sub-substation's wide area protection subsystem.

Wide area protection subsystem (WAPS): in this subsystem, protection and control decisions are based on wide area PMU information, and it responds to various system disturbances through the analysis and evaluation of the system states.

Coordinated strategies: the WAPS collects the status information of the SAPS and control signals are issued though the SAPS to relevant equipment. The SAPS collects the data of LPS and can send control signals immediately without cooperating with the LPS.

1) Advantages

   Taking advantages of broadening information domain, extensive information and coordinated

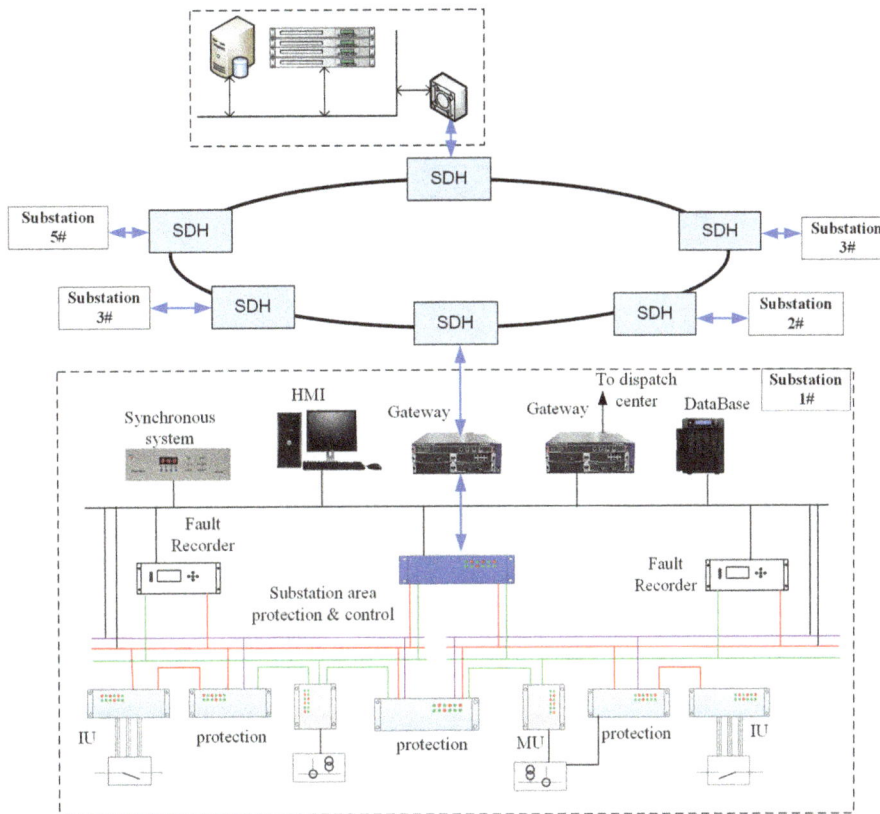

Fig. 3 Structure of hierarchical regional area protection. SDH means synchronous digital hierarchy

strategies, the pressure of the traditional local protection an be relieved. In addition, based on the regional information, exact faulted section can be located and the protection sensitivity and reliability is significantly improved.

2) Disadvantages

The SAPS and WAPS are both dependent on the communication network. The protection system is likely to lose their functions when the communication network fails. The integrated information of SAPS is the status data, thus it is easy to implement data interaction. However, problems like data missing and high error rate challenge the communication reliability, and the fault tolerance needs to be considered further. On the other hand, the interaction of analog data must pay attention to the synchronization and time delay of the switch. For the WAPS, the PMU configuration is another problem, and a good solution is to keep the balance of enough data and economic cost.

Protection scheme based on multi-agent systems

Multi-agent technology is a powerful new technique for use in distributed protection systems due to its autonomous, cooperative, and proactive nature. There are already lots of researches on the application of multi-agent to protection systems [15, 16]. For different types of integrated protection schemes presented above, protection functions can be designed in a modularizing mode and implemented based on the multi-agent technology. A protection scheme based on multi-agent systems is analyzed below, as an example.

In the protection scheme, integrated protection functions are divided into different agents according to different missions. Three levels are defined for agent system (as shown in Fig. 4): execution level, coordination level, and organizing level. In the execution level, there are measuring agent (M Agent), state diagnose agent (SD Agent) and tripping agent (T Agent). Coordination level includes the basic protection functions and others based on information interaction: searching agent, fault detection agent, fault line selection agent. The top level is the organizing level which works as the agent controller (named as recombination agent).

The agents in the coordination level can not only work independently, but also coordinate with other agents to realize data sharing, backup function and adaptive setting, especially when the network has disturbances.

The agents in the execution level operate based on the communication network, whereas its interaction

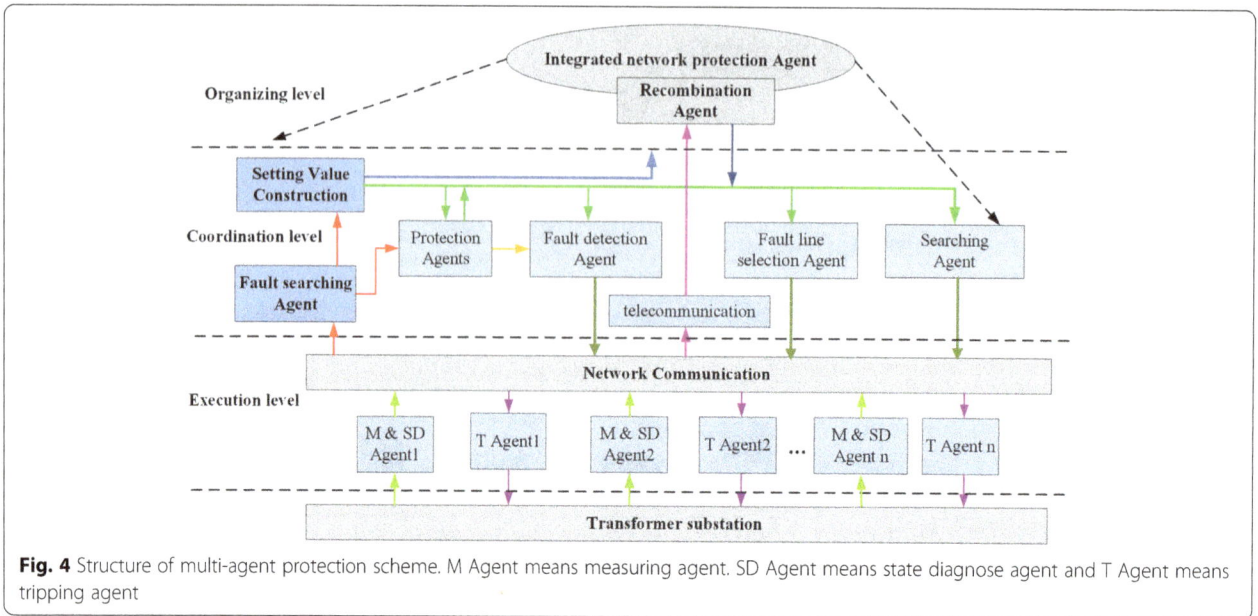

Fig. 4 Structure of multi-agent protection scheme. M Agent means measuring agent. SD Agent means state diagnose agent and T Agent means tripping agent

with the coordination level is solved by software logics. The top level controls all the agents of lower levels by recombination agents.

1) Advantages

The application of multi-agent technology improves the use of redundant information, and the protection functions and configurations become adaptive to the change of power system resulting in higher protection adaptability and robustness.

2) Disadvantages

The design of the agent structure is complex, including hardware and software. When the scale of the power system is large or the interaction of adjacent substations is considered, optimal agent model needs to be further studied.

Principles and functions

In the integrated protection schemes, traditional protective principles can work in the centralized relay to protect all the apparatuses within the substation. Based on the further analysis of redundant information, some novel principles are proposed and some solved issues and automation functions are presented below.

Traditional protective principles

The integrated protection scheme is mainly based on the well-established overcurrent (OC) protection technique. Reference [17] presents a new integrated protection scheme based on the overcurrent protection principle. Reference [11] presents an approach to improve the overcurrent protection by combining the adaptive current and voltage instantaneous protections.

With the rapid development of optical fiber communication technology, current differential protection, owing to its simplicity of principle, higher sensitivity and inherent ability of phase selection, has been extensively used [18]. Current differential relaying works on the detection of the unbalance in current flow into and out of a definite protected zone and it is especially attractive to subtransmission systems since the dynamic changes of power system impedance and power direction are solved in a simple and elegant manner. References [19] and [20] proposed a multi-zone fault clearing protection with a current differential scheme of a fixed protection zone, providing integrated primary and backup protection functions. In the integrated protection scheme [21], an integrated current differential relay, interfaced to the CTs on all of the output lines connected to the substation bus and neighboring substations at the remote ends of the lines, is responsible for the protection of all the transmission lines associated with the substation.

CT saturation caused by external faults may result in a differential current increase or a restraint current decrease leading to protection maloperation. The directional information principle is a good solution to solve this problem. A protection method based on polarity comparison is introduced in [22]. In [23], a directional comparison bus protection is presented.

An integrated protection scheme based on distance protection technique [24] is installed within a substation, and the centralized distance relay is implemented with multiple distance settings to cover all the protected sections. By utilizing the ratio between the sum of the two terminal voltages of the line and that of the two terminal currents, which is defined as integrated impedance, a novel transmission line pilot protection principle is presented in [25]. The research in [26] extends the work of Rockefeller [27] and proposes an integrated, hierarchical protection system based on distance relays where settings are adapted to ensure optimized performance under widely varying power system operating conditions.

Novel protection principles

A new protection scheme for high-speed protection of transmission lines, namely, Integrated Positional Protection scheme, is proposed in [28]. The relay which is based on the fault generated high frequency transient current signals incorporates two novel protection principles of positional protection techniques with and without the synchronization of Global Positioning System (GPS) respectively. The integrated protection technique based on transient current polarity comparison is presented in [29]. The polarity of the transient current component which comes from the faulted direction on the transmission line is different from the one which comes from the other emanating lines. Protection using this principle can protect not only all lines associated with a bus, but also the emanating parallel lines. As shown in Fig. 5, when the fault occurs at F_1, the polarity of the transient current component I_{NM} is opposite to that of the I_{NLL1} and I_{NLL2}, while the polarities of I_{NLL1} and I_{NLL2} are the same. When the fault occurs at F_2, the polarity of the transient current component I_{NLL1} is opposite to that of the I_{NM} and I_{NLL2}, while the polarities of I_{NM} and I_{NLL2} are the same. The polarity calculation is based on modulus maximum principle of wavelet transform.

A scheme of Substation-area Differential Backup Protection oriented to intelligent substations is proposed in [30]. Through defining four types of differential zones, fault section can be detected precisely by ordinal searching. The bus protection comprehensive algorithm [31] of information integration and information decentralization is presented based on the integrated protection. Peaks-cutting-criterion and interpolation fitting compensation methods are also proposed, considering the influence of electronic transformer on bus differential protection.

Moreover, a bus protection method based on current direction comparison is proposed for the integrated protection. For a typical bus structure shown in Fig. 6 (a), the current directions of $\dot{I}_1$ and $\dot{I}_2$ are the same feeding in to the bus internal fault, whereas the current directions of $\dot{I}_1$ and $\dot{I}_2$ are opposite when it is a bus external fault as shown in Fig. 6 (b).

Reference [32] uses the ratio of the bus voltage to the sum of the branch currents to detect internal faults and external faults, and this ratio uses the integrated information and is defined as the integrated impedance. Some papers have proposed power differential protective principles, which analyze the integrated comprehensive information to reflect the power unbalances caused by system faults.

Some issues can be solved by integrated information

The term IP is used to denote the integration of protective devices for multiple power apparatuses within a substation into one protective relay. Regional and comprehensive information is the most significant advantage of IP. Thus, additional functions and criterions can be proposed through analyzing the redundant information. A novel scheme based on the sine degree is proposed for distinguishing the transformer magnetizing inrush current and the power system fault currents [33]. Similarly, more effective CT saturation criterions and VT breaking criterions are proposed based on integrated information analysis.

Other substation automation functions

Maintaining the operation of a continuous process is generally the highest priority when part of the energy supply system is lost. Noncritical loads may be shedded when the utility supply fails. In that case, the process must be powered by whatever available on-site generation. Taking advantages of integrated regional information, an optimal comprehensive cost model for under frequency load shedding (UFLS) is proposed in [34], and other optimal load shedding strategies are presented in [35, 36].

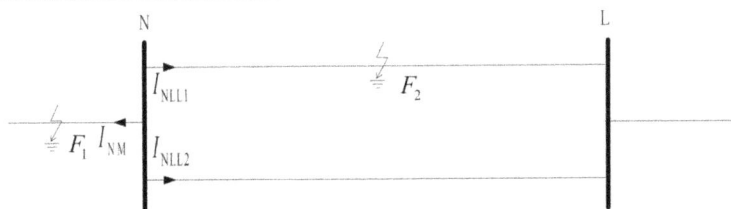

Fig. 5 Internal and external faults of parallel lines. F1 is an external fault of the parallel lines and F2 is an internal fault. Three currents (I_{NM}, I_{NLL1} and I_{NLL2}) are analyzed to demonstrate the polarity protection principle

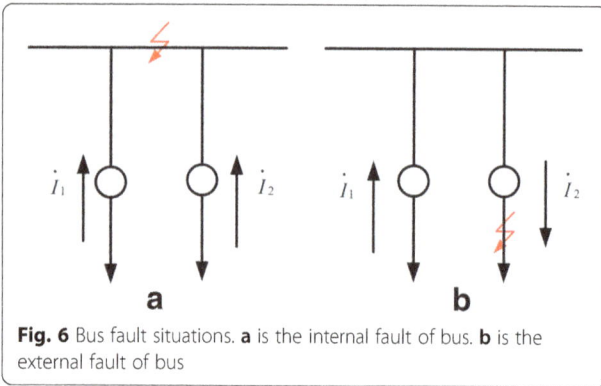

Fig. 6 Bus fault situations. **a** is the internal fault of bus. **b** is the external fault of bus

A new coordinated protection system

Most researches described above focus on the analysis of data, which depends on integrated information acquisition. Protection considering coordination of different units and functions needs further study. A novel solution utilizes the information acquisition mode of a digital substation and the communication network of a smart grid is proposed in this section.

Referred as a coordinated substation protection (CSP), it implements the corresponding protection configuration principles and concentrates on the coordination and cooperation among different intelligent electronic devices (IEDs) within a substation. It utilizes the communication network and the data acquisition capability available within a digital substation. Compared with the existing integrated protection approach, IEDs with distributed processing have more advantages in terms of functional support and information utilization. The local information coordination of IEDs improves the reliability of the fault identification process, and the CSP has stronger independence and higher reliability compared

with the traditional backup protection scheme. The detailed functional realization, communication network and the operating principles are discussed below.

Functional realization

CSP is a new application concept which utilizes the standardized communication protocol and the secure information network available in a digital substation. The Process Level accomplishes synchronous sample value acquisition, action command execution and breaker status uploading. The Bay Level consists of a number of IEDs which work in a harmonious and coordinated manner to provide reliable fault detection. The coordination of IEDs includes information sharing and coordinated backup tripping. The Station Level integrates useful data and events for substation management and provides communication with the control center or a SDH network.

Figure 7 shows the structure of a substation based on coordination and cooperation between different protection IEDs. In terms of functionality, an IED integrates the function module (FM), local coordination module (LCM) and remote coordination module (RCM). FM is responsible for the protection algorithm, which processes the samples received from the process bus, performs the protection calculations and sends the action commands to the breaker. At the same time, FM can communicate with other IEDs to share information and deliver coordinated tripping of the breakers in a coordinated manner. LCM is the module which monitors the requirements of the local IED. It establishes the appropriate tasks and sends the commands. RCM receives remote commands from other IEDs and analyses their requirements, or receives remote information and use

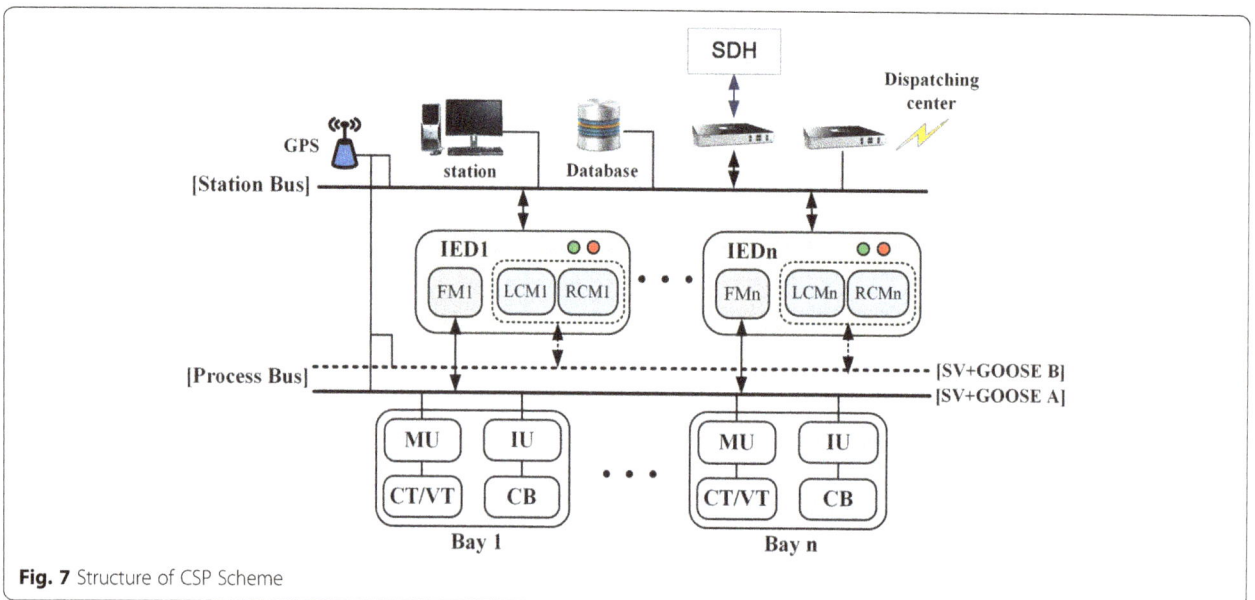

Fig. 7 Structure of CSP Scheme

this for local processing. The protection uses an owner-less structure and provides redundancy in the case of IED failures. The IED is based on a modulus design principle and is easy to recombine or extend functionality via the standard module interface.

CSP is substation-domain protection system, and its protection range is elements within the substation and the outgoing-lines. Coordinated IEDs implement local high speed protection decisions, whilst other IEDs exchange information with adjacent substations to allow the protection of outgoing-lines. If a primary equipment (CT, VT or CB) is in an abnormal working state, the IEDs work in a coordinated mode, using data sharing and coordinated tripping to improve the reliability of the local substation protection.

Communication network

A reliable communication network is one of the key elements in a CSP scheme. The communications network includes two parts: the substation based process level communication network and communications between adjacent substations.

The process level communication network is shown in Fig. 7. Based on the IEC61850-9-2LE standard, the communications between the bay level and the process level are via the process bus and a high performance local area network (LAN) [5] which carries sampled values (SV) and GOOSE messages. Considering the large quantities of information and the network load capacity, a dual process network is established within a substation. The FM performs protection calculations and communicates with the primary equipment through the LAN "SV + GOOSE A". The LCM and RCM are responsible for coordination, and communicate with their corresponding modules in the other IEDs through LAN "SV + GOOSE B". To synchronize the different data sources, LAN communication should

have an accurate and reliable GPS synchronous time-marker.

One substation can exchange the information of outgoing-lines with the adjacent substations, to implement cooperation between the different substations. Network communications with adjacent substations are based on SDH technology and operate in a self-healing ring mode. The substations connect with all the other IEDs in the ring, allowing flexible data acquisition to be achieved. As shown in Fig. 8, substations S_2 and S_3 upload data to the SDH network using an SDH device, and substation S_1 can download the necessary data from the SDH network for protection decisions in its substation domain.

CSP for the substations implements rapid fault judgment and the coordination working mode, and consequently it improves protection reliability. The local protection function which analyzes the local information, works as the main protection and the coordination working mode is based on coordinated sharing information to detect the fault when the main protection fails to clear the fault. Independent operation and tripping without time delays helps to promote the application.

Methods

Current differential protection is not affected by bi-directional power flow, and can identify faults correctly and rapidly, without the need for VTs to provide directionality. The differential principle can be easily applied within a substation and involves the cooperation of multiple IEDs by sharing information and using functional coordination to implement an adaptive differential protection scheme. The basic principle is to use the nearest correct data when the local data is erroneous. In addition, if the local protection or its breaker has not operated correctly, the protection trip boundary will be broadened to ensure the protection can clear the fault.

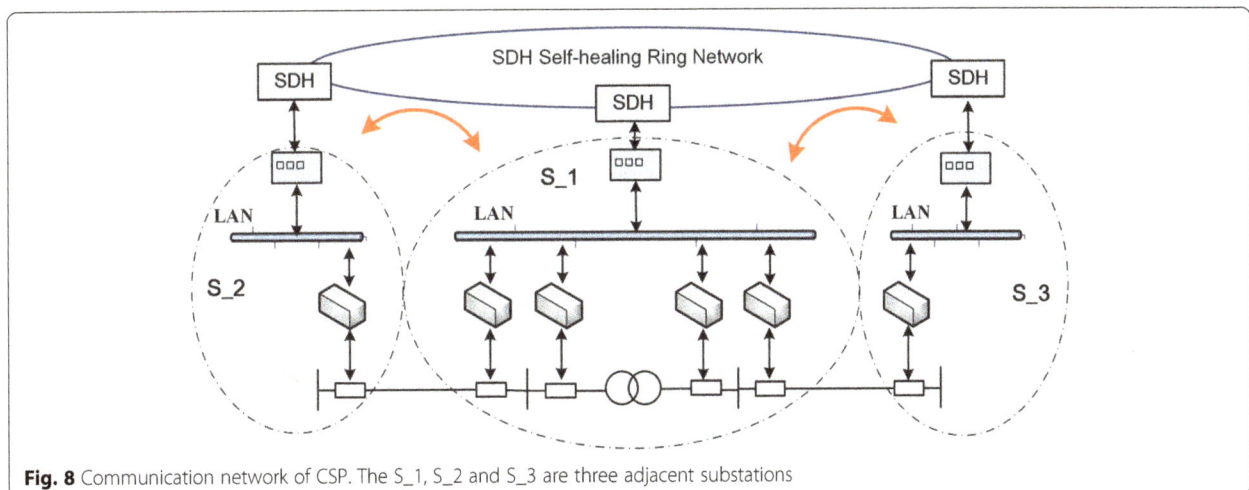

Fig. 8 Communication network of CSP. The S_1, S_2 and S_3 are three adjacent substations

Using the CSP concept, an IEC61850 based system for a typical 110 kV-35 kV-10 kV substation is designed as shown in Fig. 9. On the basis of local protection area of single instrument and approaches of information sharing, multi-level extended protection regions are established. M1 to M19 denote measurement devices at different locations. Several different protection regions are described in terms of protection scale, data included and function level in Table 2.

The protection system implements substation domain protection based on the differential principle. As an example, the fault location in the system as shown in Fig. 9 is considered to analyze the coordination working mode. The current positive direction is assumed to be from a bus to a line. Current pairs $(\dot{I}_{14}, \dot{I}_{16}, \dot{I}_{17})$, $(\dot{I}_3, \dot{I}_4, \dot{I}_{16})$ and $(\dot{I}_4, \dot{I}_{19})$ are the currents at the terminals of transformer 1, 35 kV bus 1 and line 4–19, respectively. The differential current dif_1 of transformer 1 is $|\dot{I}_{14} + \dot{I}_{16} + \dot{I}_{17}|$. If the local data $\dot{I}_{16}$ is not available, it can be replaced by $-\dot{I}_4$ and $-\dot{I}_3$ through information coordination with IED (35 kV bus 1). The information sharing is achieved through the process bus LAN "SV + GOOSE B" and the new differential current dif_2 is $|\dot{I}_{14} + \dot{I}_{17} - (\dot{I}_4 + \dot{I}_3)|$. The minus operator is used because

$(\dot{I}_4 + \dot{I}_3)$ are defined with an opposite direction to $\dot{I}_{16}$. Similarly, if $\dot{I}_4$ is not available, it can be replaced by $-\dot{I}_{19}$ through coordination with IED (line) and the differential current dif_3 is $|\dot{I}_{14} + \dot{I}_{17} - \dot{I}_3 + \dot{I}_{19})|$.

Results

A single phase-to-ground fault was applied at the fault location F and the three differential currents (dif_1, dif_2, dif_3) were calculated as discussed above and are shown in Fig. 10. As seen, the differential currents increase significantly immediate after fault occurrence and the trajectory intersects the operation threshold at 0.503 s. It is obvious that, differential protection using information sharing can judge the fault effectively. The partial enlarged figure shows the differential currents dif_2 and dif_3 are higher than dif_1. The reason is that information coordination brings greater measurement errors and distributed capacitance current, both of which contribute to the differential current. In a real application, this error can be large when one considers the communication error of the IEDs and the data conversion error. Therefore, the protection setting should take this error into consideration.

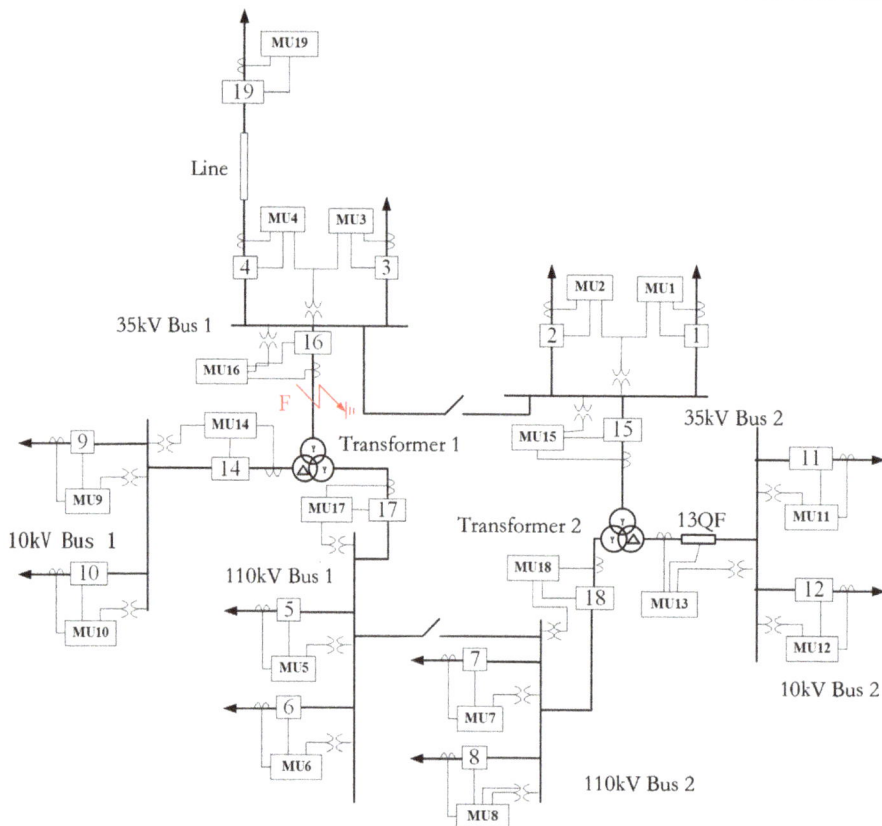

Fig. 9 A typical 110 kV-35 kV-10 kV substation system diagram. F means a fault location

Table 2 Division of functional regions

Functional region	Protection Scale	Data	Function Level
1	Transformer 1	M14, M16, M17	I
2	35 kV Bus 1	M3, M4, M16	I
3	Transformer 1 & 35 kV Bus 1	M3, M4, M14, M17	II
4	Transformer T & 35 kV Bus 1 & Line 4–19	M3, M14, M17, M19	III

Figure 11 depicts the variations of instantaneous current, RMS voltage, active power and reactive power of the MU 17. The circuit breakers are tripped at 0.59 s and it can be seen in Fig. 11 that the current becomes zero at the instant fault section clearance whereas the voltage, active/reactive powers become zero a short time after the fault isolation. The total fault clearing time takes into consideration of the time delay components which are sensor delay (5 ms), communication delay (5μs which is negligible), integrated protection unit delay (dependent on the protection principle) and circuit breaker delay (90 ms). The time of IED coordination is not considered in this paper. However, the time cost is related to the communication delay and coordination strategy. Actually, the data networks of smart substations have short communication time cost, thus the application of simplified information coordination should be given more attention.

On the other hand, when the circuit breaker fails to clear the fault, adjacent circuit breakers could work in a coordinated mode to reduce the outage area. For example, when a fault occurs at the line 4–19 and circuit breakers 4 & 19 fail to clear the fault section, protection IED of the 35 kV Bus 1 would send tripping commands to circuit breaker 16 to clear the fault instead.

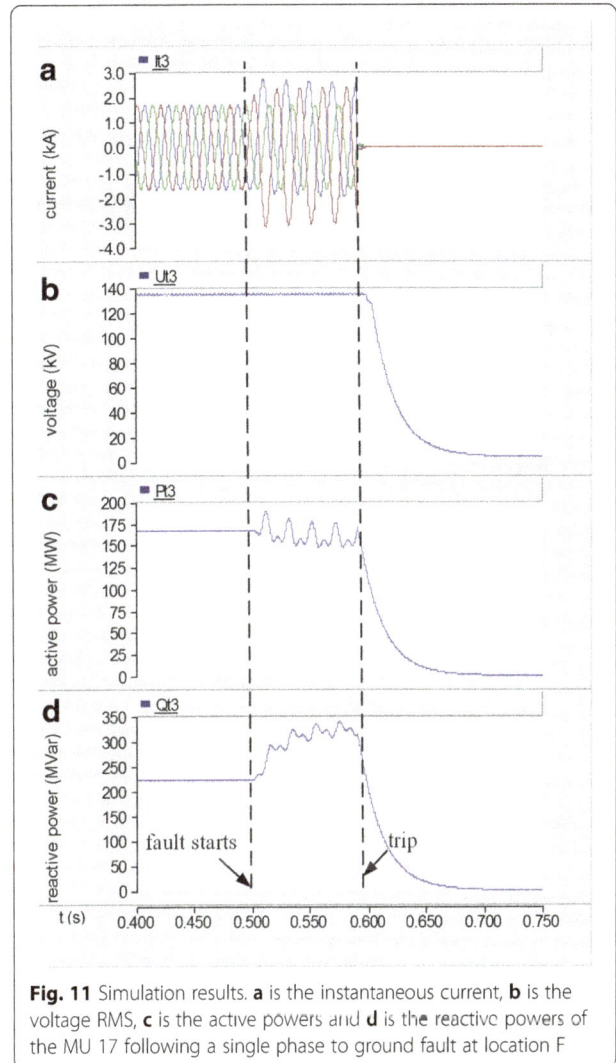

Fig. 11 Simulation results. **a** is the instantaneous current, **b** is the voltage RMS, **c** is the active powers and **d** is the reactive powers of the MU 17 following a single phase to ground fault at location F

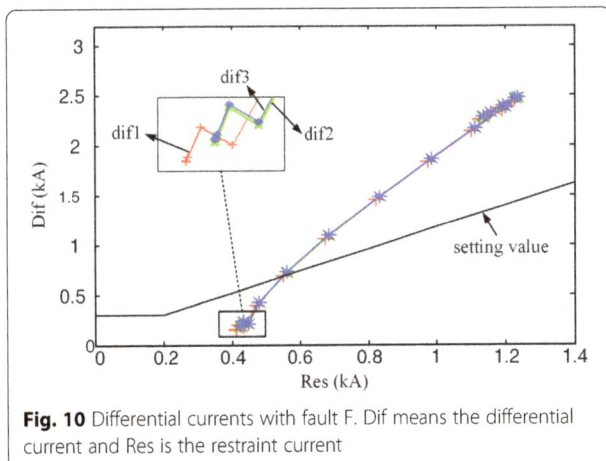

Fig. 10 Differential currents with fault F. Dif means the differential current and Res is the restraint current

Discussion

The application costs of the proposed CSP system are related to the costs of the communications infrastructure, hardware architecture and software applications. Actually, using third-party public-switched data networks in active distribution network is suitable from the economic standpoint and the communication time cost may be short. Applying the differential princple to the proposed protection system, there is no additional cost for VTs. However, the cost of protection IEDs should be given more attention.

Conclusion

In recent years, the progresses in microprocessor, communication and transducer technologies have provided new means for the design and development of new generation of power system protection scheme. The term "Integrated Protection" is used to denote the integration of several protective devices for multiple power equipment within the substation into one protective relay, obtaining all the real time information of the substation by communication network. The business driver behind functional integration is cost reduction by reduced hardware, wiring and installation times. It is also expected that it will lead to a reduction in life-time operational costs.

In this paper, several different kinds of integrated protection schemes and their respective advantages and disadvantages are discussed. Protection principles and functions applied in integrated protection systems are also presented. A new coordinated protection system based on inter-substation information is proposed and simulation results illustrate and verify the feasibility of the scheme.

Acknowledgment
This work is supported by the National Natural Science Foundation of China under Grant 51277009, and the Application Technology Research and Engineering Demonstration Program of National Energy Administration under Grant NY20150302.

Authors' contributions
JH contributed to the study design and analysis and drafted the manuscript; LL was involved in data acquisition, analysis and revision of the manuscript; WL worked on aspects of the study relating to inter-substation information protection system; MZ was involved in data acquisition and revision of the manuscript. All authors have read and approved the final manuscript.

Competing interests
The authors declare that they have no competing interests.

References
1. Blackburn, JL. (1987). *Protective relaying principles and applications*. New York, Marcel Dekker.
2. Ackeman, T, Anderson, M, & Seder, L. (2001). Distributed Generation: a Definition. *Electric Power System Research, 57*, 195–204.
3. Jiang, ZH, Li, FX, Qiao, W, Sun, HB, Wang, JH, Xia, Y, Xu, Z, Zhang, P. (2009). A Vision of Smart Transmission Grids. In *PES '09 IEEE Power & Energy Society General Meeting Calgary, AB, Canada.* (pp. 6–10). Piscataway: IEEE.
4. Chen, L, Zhang, KJ, Xia, YJ, Hu, G. (2012). Study on the substation area backup protection in smart substation. In *IEEE PES Asia-Pacific Power and Energy Engineering Conference (APPEEC) Shanghai, China.* (pp. 1–4). Piscataway: IEEE.
5. Li, ZX, Yin, XG, Zhang, Z, & He, ZQ. (2013). Wide-area protection fault identification algorithm based on multi-information fusion. *IEEE Transactions on Power Delivery, 28*, 1348–1355.
6. Serizawa, Y, Tanaka, T, Fujikawa, F, Sugiura, H, Shioyama, T, Kimura, Y. (2012). Use case study on a decentralized modular device network for wide-area monitoring, protection and control. In *2012 IEEE Power And Energy Society General Meeting, San Diego, CA, USA.* (pp. 1–8). Piscataway: IEEE.
7. Su, S, Li, KK, & Chan, WL. (2010). Adaptive agent-based wide-area current differential protection system. *IEEE Transactions on Industry Applications, 46*, 2111–2117.
8. Abdulhadi, I, Coffele, F, Dysko, A, Booth, C, Burt, G. (2011). Adaptive protection architecture for the smart grid. In *2011 2nd Innovative Smart Grid Technologies (ISGT Europe) Anaheim, Manchester, UK.* (pp. 1–8). Piscataway: IEEE.
9. Adamiak, MG, Apostolov, AP, Begovic, MM, Henville, CF, Martin, KE, Mechel, GL, Phadke, AG, & Thorp, JS. (2006). Wide Area Protection-Technology and Infrastructures. *IEEE Transactions on Power Delivery, 21*, 601–609.
10. Ingelsson, B, Sweden, SK, Lindstrom, PO, Karlsson, D, & Runvik, G. (1997). Wide-area protection against voltage collapse. *Computer Applications in Power IEEE, 10*, 30–35.
11. Li, J, He, JH, Zhang, H, & Xie, FX. (2011). Research on adaptive protection based on integrated protection. *Advanced Power System Automation and Protection (APAP), 2*, 848–852.
12. Zhang, H, He, JH. (2008). Design of a real-time substation communication system for integrated protection. In *Transmission and Distribution Conference and Exposition: Latin America, Bogota, Colombia.* (pp. 1–5). Piscataway: IEEE.
13. Gungor, VC, Sahin, D, Kocak, T, Ergut, S, Buccella, C, Cecati, C, & Hancke, GP. (2011). Smart Grid Technologies: Communication Technologies and Standards. *IEEE Transactions on Industrial Informatics, 7*, 293–297.
14. Lin, XN, Li, ZT, & Wu, KC. (2009). Principles and implementations of hierarchical region defensive systems of power grid. *IEEE Transactions on Power Delivery, 24*, 30–37.
15. Giovanini, R, Hopkinson, K, Coury, DV, & Thorp, J. (2006). A primary and backup cooperative protection system based on wide area agents. *IEEE Transactions on Power Delivery, 21*, 1222–1230.
16. Serizawa, Y, Imamura, H, & Kiuchi, M. (2001). Performance evaluation of IP-based relay communications for wide-area protection employing external time synchronization. *Proceedings of the IEEE Power Eng. Soc. Summer Meet, 2*, 909–914.
17. Bo, ZQ, He, JH, Dong, XZ, Caunce, B.RJ (2006). Overcurrent Relay based Integrated Protection Scheme for Distribution Systems. A. In *International Conference on Universities Power System Technology, Chongqing, China.* (pp. 1–6). Piscataway: IEEE.
18. Xu, ZY, Du, ZQ, Ran, L, et al. (2007). A current differential relay for a 1000 kV UHV transmission line. *IEEE Transactions on Power Delivery, 22*, 1392–1399.
19. Serizawa, Y, Myoujin, M, Kitamura, K, Sugaya, N, Hori, M, Takeuchi, A, et al. (1998). Wide-area current differential backup protection employing broadband communications and time transfer systems. *IEEE Transactions on Power Delivery, 13*, 1046–1052.
20. Kangvansaichol, K, & Crossley, PA. (2003). Multi-zone current differential protection for transmission networks. *Proc. IEEE Power Engineering Soc. Transmission Distribution Conf. Expo, 1*, 359–364.
21. Wang, HG, Du, DX, Bo, ZQ, Dong, X. (2006). An Integrated Current Differential Protection Scheme. In *2006 International Conference on Power System Technology, Chongqing, China.* (pp 1–6). Piscataway: IEEE.
22. Song, S, & Zou, G. (2015). A Novel Busbar Protection Method Based on Polarity Comparison of Superimposed Current. *IEEE Transactions on Power Delivery, 30*(4), 1914–1922.
23. Zadeh, MRD, Sidhu, TS, & Klimek, A. (2011). Implementation and testing of directional comparison bus protection based on IEC61850 process bus. *IEEE Transactions on Power Delivery, 26*(3), 1530–1537.
24. Bo, ZQ, He, JH, Dong, XZ, Caunce, BRJ. (2006). An integrated protection scheme based on a distance approach. In the *41st International Universities Power Engineering Conference, Newcastle upon Tyne, UK.* (pp. 800–803) Piscataway: IEEE.
25. Suonan, J, Deng, X, & Liu, K. (2011). Transmission line pilot protection principle based on integrated impedance. *IET Generation Transmission and Distribution, 5*, 1003–1010.
26. Stedall, B, Moore, P, Johns, A, Goody, J, & Burt, M. (1996). An investigation into the use of adaptive setting techniques for improved distance back-up protection. *IEEE Transactions on Power Delivery, 11*(2), 757–762.
27. Rockefeller, GD, Wagner, CL, Linders, JR, Hicks, KL, & Rizy, DT. (1988). Adaptive Transmission Relaying Concepts for Improved Performance. *IEEE Transactions on Power Delivery, 3*(4), 1446–1456.
28. Bo, Z.Q, Han, M, Klimek, A, Zhang, BH, He, JH, Dong, XZ. (2009). A Centralized Protection Scheme Based on Combined Positional Protection Techniques. In *2009 IEEE Power & Energy Society General Meeting, Calgary, AB, Canada.* (pp. 1–6). Piscataway: IEEE.
29. Jinghan He, Biao Zhang, Z. Q. Bo. (2007). Parallel line integrated protection based on transient current polarity comparison. In *2007 Klimek Universities Power Engineering Conference, Brighton, UK.* (pp. 597–601). Piscataway: IEEE.
30. Gao, H, Liu, Y, Zou, G, Cui, D, Liu, M, Li, X. (2014). Principle and Implementation of Substation-Area Backup Protection for Digital Substation. In the *12th IET International Conference on Developments in Power System Protection, Copenhagen, Denmark.* (pp. 1–5) Piscataway: IEEE.
31. Wu, CY, He, JH, Zhang, H, & Zhi, ZQ. (2012). Design of bus protection for integrated protection. *Power System Protection and Control, 40*, 125–130.
32. Suonan, J-I, Deng, X-y, Alimu, J, et al. (2010). A novel principle of integrated impedance based bus-bar protection. *Power System Protection and Control, 38*, 1–7.
33. Han, JH, Li, JZ, Yao, B, & Ou, ZJ. (2007). New Approach of Transformer Inrush Detected Based on the Sine Degree Principle of Current Waveforms. *Proceedings of the CSEE, 27*, 54–58.

34. Han, JH, Bo, DD, Wang, XJ, & Ye, HD. (2013). An Optimal Algorithm of Comprehensive Cost for Under Frequency Load Shedding. *Power System Technology, 37*, 3461–3465.
35. Bo, DD, Han, JH, Wang, XJ, & Ye, HD. (2014). Adaptive UFLS scheme based on grey correlation analysis. *Power System Protection and Control, 42*, 20–25.
36. Han, JH, Chen, ZL, Ye, HD, Liu, L, & Xu, YK. (2015). Optimization Scheme for Under-Voltage Load Shedding Wide-Area Configuration and Substation-Area Setting. *Power System Technology, 39*, 2333–2339.

14

A diagnostic method for distribution networks based on power supply safety standards

Junhui Huang[1], Shaoyun Ge[2], Jun Han[1], Hu Li[1], Xiaomin Zhou[2], Hong Liu[2*], Bo Wang[2] and Zhengfang Chen[3]

Abstract

In order to overcome the shortages of diagnostic method for distribution networks considering the reliability assessment, this paper proposed a method based on power supply safety standards. It profoundly analyzed the security standard of supply for urban power networks, and established quantitative indicators of load groups based on different fault conditions. Then a method suitable for diagnostic evaluation of urban distribution networks in China was given. In the method, "N-1" calibration analysis of the distribution network was conducted. Then the results are compared with quantitative indicators of load groups on different conditions deriving the diagnostic conclusions and the standard revision is discussed. The feasibility and accuracy of the method is finally verified in the case study.

Keywords: Distribution network, Diagnostic process, N-1 calibration analysis, Power supply security

Introduction

With the rapid development of social economy and continuous improvement of people's living standards, the terminal users of urban power networks require more secure and reliable supply [1]. Therefore, the major problem and technical difficulty the urban power networks are facing in China are how to invest moderately to meet the security and reliability of supply which can be accepted by the users [2]. To solve the technical difficulty, the primary task is finding an appropriate method to diagnose power grid structure and investigate the realistic condition of grid operation. At present, reliability assessment is usually used.

However, as the main method of security diagnosis, reliability assessment has many shortages. It mainly focuses on three aspects, which are the frequency, duration and range of interruptions. Reliability indexes can reflect supply condition at every point of the system [3]. At present, the frequency of interruptions mainly relies on statistics of people, which cannot realistically reflect the situation of the grid due to great personal factors.

Although the duration and range of interruptions can be obtained by the network, quantitative indicators are used to determine whether they comply with safety standards. Therefore, the criterion of supply security needs to be studied by utilizing the data sources and operational experience and basing on the practical situation of urban power networks, in order to provide the theoretical fundamentals and solutions of improving security of urban power networks in China.

To obtain the method of security diagnosis, this paper firstly analyses the basic concept on the safety standards of power supply and obtains the quantitative indicators by confirming the relevant regulations and evaluation index of "supply security". Then, a set of diagnostic evaluation method, which is suitable for Chinese urban distribution networks, is constructed based on these indexes and "N-1" calibration results. The security of the grid in different conditions are analyzed and the solutions of improving the security of networks are proposed. In this way, the diagnosis for the distribution networks changes from qualitative indicators to quantitative ones, which is a guide to the development of the distribution networks and improvement of operation and maintenance in different regions.

* Correspondence: liuhong@tju.edu.cn
[2]Key Laboratory of Smart Grid of Ministry of Education, Tianjin University, Tianjin 300072, China
Full list of author information is available at the end of the article

Methods

Security criterion analysis of supply

Learning from the advanced technologies and ideas of othercountries and considering the background in China, the National Energy Administration issued the power supply security standard, DT/T 256–2012, for urban networks. The reliability level of supply not only depends on the quality of components and the level of operation and maintenance but also the damage of the load after components' outage [4]. The security criterion of supply and safety standards should be set to meet the requirements of security diagnosis [5].

Power supply security standards for urban areas

The regulations are comparatively specific in the power supply security standards for urban areas in China, which chooses common load groups in distribution networks. The security level suitable for distribution networks is indicated in Table 1.

Supply security criterion analysis for urban areas

According to the network structure in China, the capacity of the common equipment is shown in Table 2 [6].

In this table, the loading rate of transformer is 50 % ~ 70 %. The capacity of the line is calculated by $S=\sqrt{3}UI$, where U is the voltage of the line, and I is the carrying capacity of the line. Considering lines can not be operated with full loading, 40 % ~ 50 % of the capacity was picked as the transmission capacity of the line.

The load groups are divided into 6 levels A to Fin terms of their sizes, based on the power supply security standards for urban networks in China [7, 8]. The higher level the load is, the shorter the restoring time is after the circuit outage, and the higher the extent of recovery is.

This paper focuses on the security diagnosis of distribution networks where the levels A to C of the load groups are selected for analysis.

A. The load group ranges between 0 and 2 MW. The typical capacity of a 10/0.4 kV transformer is from 0.015 to 1MVA. Two neighboring section switches divide several (usually less than 5) 10/0.4 kV transformers into a unit. The load of every unit is from 0 to 2MV and does not need to be supplied by the other circuits.

B. The load group varies from 2 to 12 MW. The transmission capacity of a 10 kV line is 3–4MV. The common capacity of a 35/10 kV transformer is 31.5MVA, 20MVA, 16MVA and 5MVA. The typical capacity of a 35/10 kV substation can be 5*2MVA, 16*2MVA, whose load is between 5 and 10 MW.

C. The load group ranges from 12 to 180 MW. The transmission capacity of a 35 kV line is 12–15MV. The common capacity of a 110/10 kV transformer is 63MVA, 50MVA, 40MVA, 31.5MVA and 20MVA. The typical capacity components of a 35/10 kV substation are 31.5*2MVA, 40*2MVA, 50*2MVA, 63*2MVA and 63*4MVA, whose load is between 30 and 170 MW.

The quantification of the restoration capacity is different at the different levels of the load groups [9]. The specific analysis can be stated as follows:

(1) Load groups-2 MW. 2 MW is the maximum load of a segment on a medium voltage 10 kV line. Load group-2 MW represents a load of MV line without the outage segment.

(2) Load groups-12 MW. 12 MW is a load of a 35 kV line. Load group-12 MW represents a total load of a 110/10 kV substation minus the load of a 35 kV line.

(3) Load groups-60 MW. 60 MW is a load of a 110 kV line. Load groups-60 MW represents a total load of a 220/110 kV substation minus the load of a 110 kV line.

(4) 2/3 of load groups. By analyzing the situation of China, the typical capacity, in summary, is 2/3 of the annual peak load (the typical load is the level under which the load in 80 % of evaluation time should be). So the load group of C does not lose the load in the case of the "N-1" planned outages while losing a part of the load in the case of the "N-1" failure outage. Network reconfiguration by the remote operation can regain 2/3 of load groups. After 3 h, all load groups recover.

Table 1 Security standards of supply for urban China (common load groups in distribution networks)

Supply level	Scope of load groups (MW)	"N-1" outage	"N-1-1" outage	comment
A	≤2	After maintenance: restore load groups	No requirement	
B	2 ~ 12	(1) In 3 h: load restored = load groups-2 MW (2) After maintenance: restore load groups	No requirement	
C	12 ~ 180	(1) In 15 min: load restored ≥ min (load groups-12 MW, 2/3 load groups). (2) In 3 h: restore load groups	No requirement	*[a]

*[a]: User group is generally supplied by two (or more than two) normally-closed circuits or one circuit but can be switch to other circuits by artificial or automatic switch. The load group is the maximum load of the use group
The table above shows that the "N-1" outage is the main point to be considered in distribution networks

Table 2 Capacity of common equipment in China

Equipment	Common capacity of transformer/Limiting capacity of line (MVA/MW)	Load of transformer/transmission capacity of line (MW/MW)	Load level
10/0.4 kV Transformer	1, 0.8, 0.5, 0.4, 0.315, 0.2, 0.05	0.5	A
10 kV Line	8	3 ~ 4	B
35/10 kV Transformer	20, 16, 5	5 ~ 10	B
110/10 kV Transformer	63, 50, 40, 31.5, 20	30 ~ 170	C
35 kV Line	30	12 ~ 15	C
110 kV Line	100	40 ~ 50	C

Security diagnostic process
Security diagnostic process of supply
After analyzing the power supply security standards for urban areas in China and establishing the basis of quantitative indicators, the security diagnostic process can be established, which is shown in Fig. 1.

In this figure, based on the quantitative outcomes of the power supply security standards for urban areas in China, a "N-1" verification module was firstly used for grid analysis, and then deal with the results together with the comparison module to derive conclusions and solutions. The whole assessment process is specifically introduced as follows.

"N-1" Verification module
The inputted network structure is mainly verified with the "N-1" contingency by this module. The components that cannot pass the "N-1" verification are marked, and the load transfer schemes of components are given and

the sizes of the loads are recorded if passing the verification [10]. Based on the "N-1" calibration, topology analysis, connection mode analysis, power flow calculation and transfer scheme analysis are used to verify the grids, completely and comprehensively [11, 12].

The basic data processing module consists of three parts: topology analysis, connection mode analysis and power flow calculation, which is shown in Fig. 2.

The raw data of the system is combining through, and the result is converted to be used by power flow calculation mentioned in "N-1" verification. The connection between main transformers and lines, and the information of switches' state are obtained from the topology analysis to prepare data for a set of switching operation choices in the calculation of "N-1" verification. The output power of the main transformers and the power flow through components can be obtained by the power flow calculation to prepare data for the maximum transfer load of the main transformers and

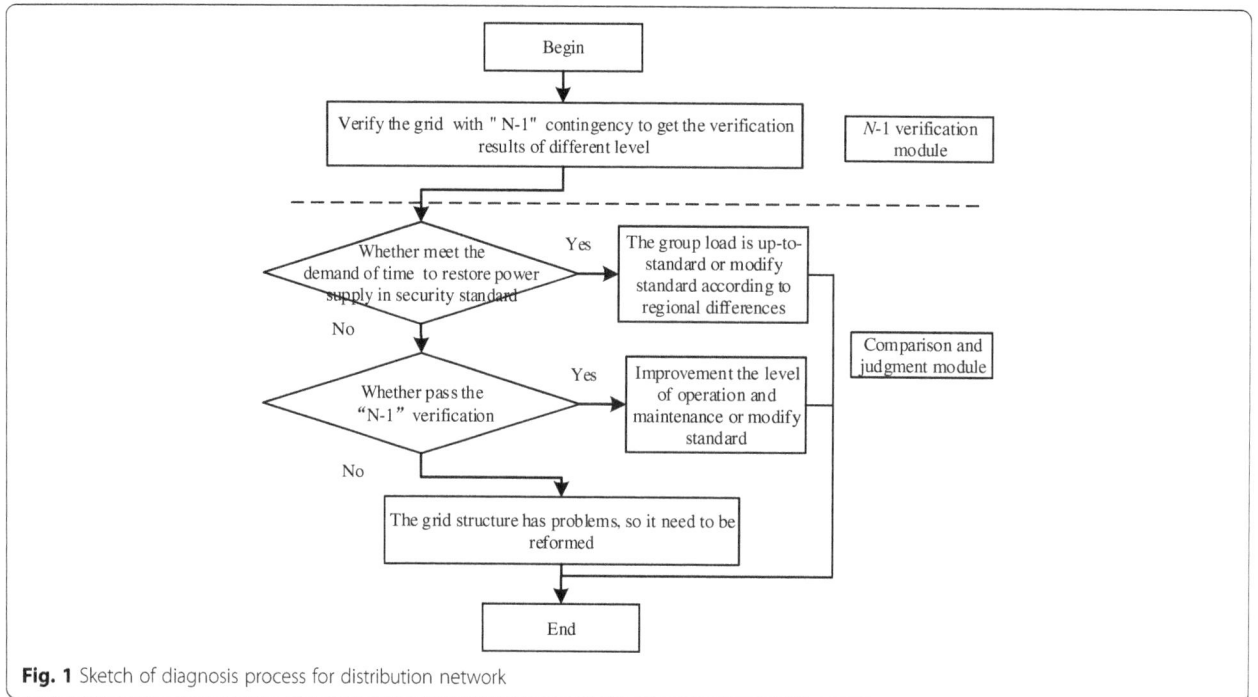

Fig. 1 Sketch of diagnosis process for distribution network

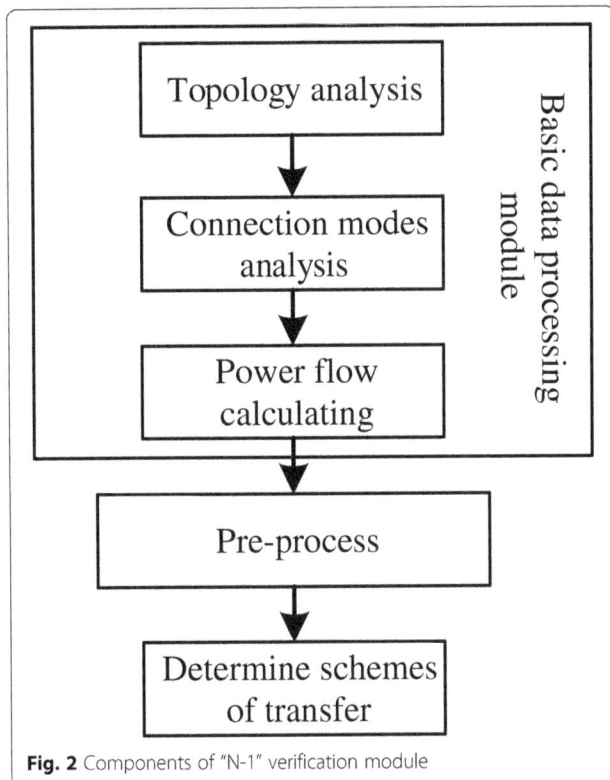

Fig. 2 Components of "N-1" verification module

the lines. Furthermore, connection mode analysis and power flow calculation are combined to prepare for "N-1" verification.

The pre-processing has two functions: eliminating the data which cannot obviously pass the verification and put them into the result table; and selecting the data which can obviously pass the verification and put them into the result table. The rest data is put into the table of components that need to be verified.

The data that cannot obviously pass the verification includes the single radial line obtained through the topology and connection mode analysis, which has no way to be transferred. If there's a failure at outlet section of the substation bus, it will lead a wide outage in the area. So the single radial line can be directly determined.

The data that can obviously pass the verification is the situation where the main transformer with the maximum capacity in the substation has a failure and another main transformer can supply all the loads in the substation. So this situation can be directly determined.

The pre-processing effectively reduces the number of components meant to be verified, which can simplify the process and improve the speed of verification.

Determining schemes of transfer are the core parts of the "N-1" verification module. Its function is to

verify the components and record the size of the transferred load.

Comparison and judgment module
After analyzing the grid with the "N-1" verification module, the Verification Result Table can be derived, which records whether the components pass the verification, the way to transfer and the size of the transferred load. The comparison is conducted from 3 aspects: failure of the line, failure of the main transformer and the outage of the substation [13].

Furthermore, according to the size of transferred loads and the length of transferring time, comparing with the power supply security standards for urban areas in China, the analyzing results are obtained and the solutions of improving the security of networks can be proposed.

By comparing the Verification Result Table with power supply security standards for urban areas in China, following a differential and hierarchical way to check one by one, the comparison result has a high reliability. Based on this, the network structure is further analyzed.

The load groups of different levels all meet the demand in power supply security standards for urban areas in China. This kind of loads meets the details of the standard from two aspects of transferring time and the size of transferred loads. Because different regions possess differing development levels, if the practice time is usually shorter than the standard, the standard can be modified to drive the development the network structure and the level of operation and maintenance.

There are load groups of some levels do not meet the power supply security standards for urban areas in China. This kind of loads can not meet the details of standard from two aspects that load can't be transferred or transferred load don't meet load capacity required [14]. The Result Table of "N-1" verification is checked to get the information whether the components pass the verification.

(1) If the component can pass the verification, the reason of being substandard is that the time of manual reconstructing is too long when the load is transferred, so the level of operation and maintenance should be improved [15, 16]. At the same time, the practical situation of the region should be considered. If they can't generally meet the standard, the standard needs to be modified to fit the region.

(2) If only a part of the component cannot pass the verification, the reason of being substandard is structure of power network, so the network should be reformed [17]. How to reform the network is a difficult issue which is not discussed in this paper.

Results and discussion

Combined with the diagnostic method for the distribution networks based on power supply safety standards, a case study is conducted to verify the correctness of the proposed method.

Diagnosis of region-A

According to the diagnostic method and process above, the region-A is analysed as a case to verify the feasibility of this method. And the practical case of the structural parameters and the analysis results are displayed as follows.

There are three substations in the region-A, whose diagram is shown in Fig. 3. Station A1 is a 110 kV substation located in the northwest, and the power supplies a part of the industrial applications at the north of Road and a few residential applications. Currently, there are 26 lines in the coverage of this regional power supply. Station A2 is a 35 kV substation in the southwest, which mainly supplies for the western residential applications. Correspondingly, there are 16 lines. And the 35 KV station A3 located in the mid-east like A2, mainly supplies for the western residential applications, but it has 14 lines. The A2 and A3 cannot be interconnected because of the natural barrier.

The time standards of the operation and maintenance level are: 10 h for breakdown maintenance; 5 h for manual reconstruction; 20 min for remote control reconstruction and 60 s for automatic switching.

"N-1" Calibration results of region-A

According to the above method, region-A is verified with "N-1" calibration, which includes the main transformers and lines.

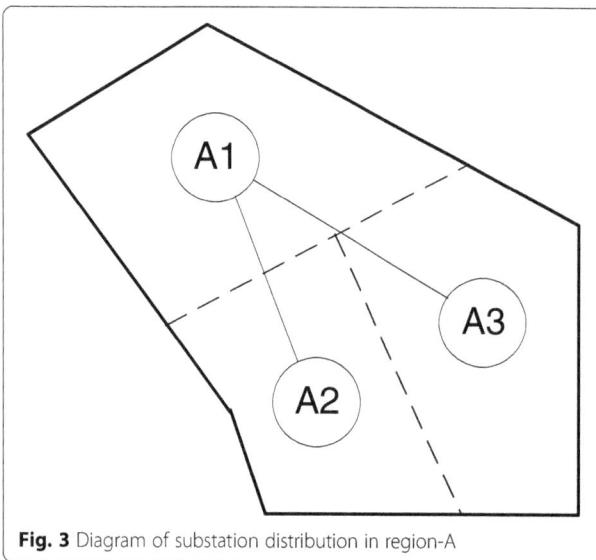

Fig. 3 Diagram of substation distribution in region-A

According to the diagnostic process, the "N-1" calibration results of the lines and main transformers are obtained, as indicated in Tables 3 and 4.

Comparative analysis results of region-A

When there is a fault at the line, it is divided into two parts, which are the faulty section and non-faulty section. For the faulty section, the power is restored after maintenance tasks, so it is necessary to ensure that whether maintenance time could meet the safety standard requirements. For the non-faulty section, the transferring time is checked because its load can be transferred by the link line. The following analysis assumes that the automation level of region-A is relatively low, and the transferring is based on the manual switching operation.

According to the analysis results above, the detailed descriptions are stated as follows.

(1) Line 5, 6, 7, 8 in Station A1, Line 39, 40, 41, 42 in Station A2, and Line 51, 52, 53, 54, 55, 56 in Station A3. These single-radiation lines cannot pass the "N-1" calibration. When there are faults, the loads cannot be transferred, so the power supply for the non-faulty section comes after maintenance. The load class of some lines (Line 5, 6, 8, 39, 40, 41, 42, 51, 52, 53, 55 and 56) can be expressed as A, and the maintenance time of them is 10 h, which cannot meet the safety standard requirements. Besides, carrying the large load, Line 7 in Station A1 and Line 54 in Station A3 can be expressed as B and cannot transfer all of load required when faults happen, so they must be upgraded to transfer load and their transferring time should meet the safety standard requirements.

(2) Line 10 in Station A1 and Line 45 in Station A3. These lines with single tie cannot satisfy the "N-1" calibration owing to the lack of transferring capacity. For the non-faulty section, either 5 h for transferring or 10 h for maintaining to restore the power supply cannot meet the safety standard requirements.

(3) Remain lines. They have ties and could pass "N-1" calibration. But similarly, 5 h for transferring to restore the power supply would not meet the safety standard requirements.

According to the analysis above, the operation and maintenance level of this region cannot satisfy the safety standard requirements; especially, there are two lines (Line 7, 54) which need to be improved, so the operation and maintenance level of this region should be upgraded.

When there are faults in the main transformers, internal transferring of station should be considered firstly. It can be completed automatically, and the time is about 60 s. If the loading rates of main transformers are high,

Table 3 "N-1" calibration results of 10 kV lines in region-A

Substation	Line name	Connection type	Line load (kW)	Translational load (kW)	Whether pass	Substation	Line name	Connection type	Line load (kW)	Translational load (kW)	Whether pass
A1	Line 1	Single-tie	1524	2476	Yes	A2	Line 29	Multi-tie	1120	2880	Yes
A1	Line 2	Multi- tie	1653	1347	Yes	A2	Line 30	Multi-tie	1333	2267	Yes
A1	Line 3	Multi-tie	2387	1613	Yes	A2	Line 31	Multi-tie	1519	2481	Yes
A1	Line 4	Single-tie	1559	2441	Yes	A2	Line 32	Multi-tie	1021	1979	Yes
A1	Line 5	Single-radiation	1346	–	No	A2	Line 33	Single-tie	1105	2895	Yes
A1	Line 6	Single-radiation	1847	–	No	A2	Line 34	Single-tie	1231	2469	Yes
A1	Line 7	Single-radiation	2044	–	No	A2	Line 35	Single-tie	1152	2648	Yes
A1	Line 8	Single-radiation	1523	–	No	A2	Line 36	Single-tie	1226	2774	Yes
A1	Line 9	Multi-tie	2102	1860	Yes	A2	Line 37	Multi-tie	1237	2763	Yes
A1	Line 10	Single-tie	1879	2121	No	A2	Line 38	Single-tie	1032	2868	Yes
A1	Line 11	Multi-tie	1865	2135	Yes	A2	Line 39	Single-radiation	1255	–	No
A1	Line 12	Multi-tie	1623	2377	Yes	A2	Line 40	Single-radiation	1065	–	No
A1	Line 13	Multi-tie	1116	2884	Yes	A2	Line 41	Single-radiation	1126	–	No
A1	Line 14	Multi-tie	1356	2644	Yes	A2	Line 42	Single-radiation	1176	–	No
A1	Line 15	Multi-tie	1645	2355	Yes	A3	Line 43	Multi-tie	1744	2256	Yes
A1	Line 16	Multi-tie	2496	1504	Yes	A3	Line 44	Multi-tie	1206	2794	Yes
A1	Line 17	Multi-tie	1212	2788	Yes	A3	Line 45	Multi-tie	2343	1857	No
A1	Line 18	Single-tie	1742	2258	Yes	A3	Line 46	Multi-tie	1542	2258	Yes
A1	Line 19	Single-tie	1435	2565	Yes	A3	Line 47	Single-tie	1615	2385	Yes
A1	Line 20	Single-tie	1318	2682	Yes	A3	Line 48	Multi-tie	1375	2625	Yes
A1	Line 21	Multi-tie	1764	2236	Yes	A3	Line 49	Single-tie	1464	2536	Yes
A1	Line 22	Single-tie	2317	1683	Yes	A3	Line 50	Single-tie	1717	2283	Yes
A1	Line 23	Single-tie	1435	2565	Yes	A3	Line 51	Single-radiation	1032	–	No
A1	Line 24	Single-tie	1767	1206	Yes	A3	Line 52	Single-radiation	1390	–	No
A1	Line 25	Single-tie	1232	2768	Yes	A3	Line 53	Single-radiation	1258	–	No
A1	Line 26	Single-tie	2315	1685	Yes	A3	Line 54	Single-radiation	1015	–	No
A2	Line 27	Single-tie	1235	2565	Yes	A3	Line 55	Single-radiation	1265	–	No
A2	Line 28	Single-tie	1167	1206	Yes	A3	Line 56	Single-radiation	1034	–	No

internal transferring of station would not supply the power to all lines, so the excess lines would be cut. And then connections between stations should be taking into account. If there is a sufficient connection capacity between stations, the load will be transferred by other stations. Accordingly, whether the transferring time reaches the safety standard requirements can be confirmed. The rest part, which cannot be transferred, gets power supply after maintenance. The results after failure of transformers are shown as Table 5.

According to the power supply safety standards, it requests 8 h for breakdown maintenance, 3 h form anual reconstruction to restore the power supply and 60 s for automatic switching.

Table 4 "N-1" calibration results of main transformers in region-A

Region	Substation	Load (MW)	Main transformer capacity (MVA)	Total capacity of 10KV Part (MVA)	Translational load of link line (MVA)	Translational Load of main trans (MVA)	Whether pass	Loss load
A	A1	41.3	2×40	80	5.7	40	No	1.3
	A2	23	2×20	40	5.7	20	No	3
	A3	21	2×20	40	0	20	No	1

Table 5 Failure analyses results of transformers in region-A

Region	Substation name	Load (MW)	Main transformer Capacity (MVA)	Total capacity (MVA)	Translational load of main trans (MW)	Recovery time	Translational load of lines (MW)	Recovery time	Loss load (MW)	Recovery time
A	A1	41.3	2 × 40	80	40	60 s	1.3	5	0	–
	A2	19	2 × 20	40	19	60 s	3	5	0	–
	A3	20	2 × 20	40	20	60 s	0	–	1	10

According to the analysis results above, the detailed descriptions are shown as follows.

(1) Substation A1: Assuming that each main transformer could carry 130 % of the load when faults happen in another transformer, then the total load of the faulty transformer would be transferred through automatic internal transferring of station in 60 s, which reaches the safety standard requirements. Assuming that each main transformer could carry 100 % of the load, and the automatic switching could make a part of loads transferred, the remaining part would be transferred through the connection between stations, therefore there will be 1.3 MW load transferred in 5 h, which cannot reach the safety standard in view of the overlong time.

(2) Substation A2: Assuming that each main transformer could carry 130 % of the load when faults happen in another transformer, then the total load of the faulty transformer would be transferred through automatic internal transferring of station in 60 s, which reaches the safety standard requirements. Assuming that each one could carry 100 % of the load, and the automatic switching could make a part of loads transferred, the remain part would be transferred through the connection between stations, therefore there will be a 3 MW load transferred in 5 h, which cannot reach the safety standard requirements in view of the overlong time.

(3) Substation A3: Assuming that each main transformer could carry 130 % of the load when faults happen in another transformer, then the total load of the faulty transformer would be transferred through automatic transferring in station in 60 s, which reaches the safety standard requirements. Assuming that each one could carry 100 % of the load, and the automatic switching could make a part of loads transferred, different from the above, the remaining part would not be transferred due to the lack of connection between stations; therefore the 1 MW load would get power supply only after the 10 h maintenance, which cannot reach the safety standard requirements.

Summary of analysis results

A2 and A3 cannot be interconnected because of the natural barrier, so the connection condition of the whole region is poor. Carrying a large load, defined as Class B, Line 7 in Station A1 and Line 54 in Station A3 cannot transfer loads to reach the standard when there is a failure. Obviously, it is necessary to upgrade these lines. Besides, a lot of single-radiation lines urge the region to improve the structure of the network. The low operation and maintenance level also makes region A cannot reach the power supply safety standard. However, it is more appropriate to revise this standard to avoid the negative impact of the strict rules on regional development. Therefore, in order to adapt to the actual situation and promote the level of operation and maintenance, 9 h for breakdown maintenance and 4 h form anual reconstruction are set up.

Conclusions

As the power supply safety standards need to be improved, the diagnostic method using the reliability index cannot offer the quantitative analysis of the power grid structure. The diagnostic method for distribution networks based on power supply safety standards has the following characteristics:

(1) The power supply safety standard has profound theoretical basis, a practical background and strict logic. It is a complete quantitative criterion, which is simple, clear and easy to use. According to the actual situation of Chinese urban power grids, it can fully reflect the structure performance level of Chinese power grid currently, based on practical experience in many years and the results of the survey.

(2) The diagnostic method has a distinct and strong structure. The standards considering the actual situation are not only distinguished from different levels of loads, but also from the grid structure level.

(3) In view of the different levels of structure, it is significant to make a quantitative index distinguishing the difference, so this standard has a great guiding significance for the diagnosis of the distribution networks.

To summarize, this standard can be used to give appropriate solutions of improving the security of networks and revise the safety standard on the basis of the regional characteristics.

Authors' contributions
JH and SG proposed the diagnostic process of the distribution network. JH, HL, XZ, and HL carried out the theoretical studies and drafted the manuscript. BW and ZC performed the case study. All authors read and approved the final manuscript.

About the Authors'
Junhui Huang (1965-), male, received bachelor degrees from Nanjing Engineering College (Southeast University) in 1987. He was deputy chief engineer of the Jiangsu Electric Power designs institute from 1987 to 2006, deputy director of the Jiangsu Electric Power Company Power Grid Planning Research Center from 2006 to 2011, and is director of the Jiangsu Electric Power Company Economic Research Institute Power Grid Planning Assessment Center now. His major is the planning and management of power network, Email: xshhjh@126.com.
ShaoyunGe (1964-), male, received master degrees from School of electric automation engineering in Tianjin University and acquired his Ph.D. from Hong Kong University. He is now a Professor of Tianjin University. His major is urban power system programming and automation of distribution system. Moreover he has published more than 20 papers in IEEE. Email: syge@tju.edu.cn.
Jun Han (1985-), male, received bachelor degrees from Hunan University in 2008 and acquired master and Ph.D. degrees from School of electric automation engineering in Tianjin University in 2013 and is an engineer in the Jiangsu Electric Power Company Economic Research Institute. His major is the planning of distribution network and the application of the theory, Email: hjchallenge@126.com.
Hu Li (1979-), male, received bachelor and master degrees from Xi'an Jiaotong University in 2000 and 2003, respectively. He worked at the Jiangsu Electric Power designs institute from 2003 to 2006, the Jiangsu Electric Power Company Power Grid Planning Research Center from 2006 to 2011, and work at the Jiangsu Electric Power Company Economic Research Institute now. His major is the planning and designing of distribution network. Email: tigerli110@163.com.
Xiaomin Zhou (1992-), female, received bachelor degrees at School of electric automation engineering in Tianjin University and now is studying for Master at the same University, Her major is the planning of distribution network, Email: zhouxiaomin@tju.edu.cn
Hong Liu (1979-), male, received master and Ph.D. degrees from School of electric automation engineering in Tianjin University in 2005 and 2009, respectively. Major in the planning and evaluation of distribution network and Smart Grid, Email: liuhong@tju.edu.cn.
Bo Wang (1993-), female, now is studying for Master at School of electric automation engineering in Tianjin University, Her major is the planning of distribution network, Email: Wangbo@163.com.
Zhengfang Chen (1963-), male, Bachelor anda Engineerin the Suzhou Electric Power Company, Major in the planning of distribution network, Email: wlys2000@163.com.

Competing interests
The authors declare that they have no competing interests.

Author details
[1]State Grid Jiangsu Economic Research Institute, Nanjing, Jiangsu province, 210008, China. [2]Key Laboratory of Smart Grid of Ministry of Education, Tianjin University, Tianjin 300072, China. [3]State grid Suzhou Power Supply Company, Suzhou, Jiangsu province, 215000, China.

References
1. Ming D, Jing Z, & Shenghu L. (2004). Research on the Model for Distribution Network Reliability Evaluation Based on Sequential Monte-Carol Simulation [J]. *Power System Technology, 28*(3), 38–42 (in Chinese).
2. Qunhui G, An L & Ji W. (2003). A Practicable Algorithm to Forecast and Evaluate Reliability of Power Supply [J]. *Power System Technology, 27*(12), 76–79. (in Chinese).
3. Xiu-ren L, Zhen R, & Wenying H. (2005). Unascertained Mathematical Approach to Management of Uncertainty in Reliability Evaluation of Distribution Systems [J]. *Automation of Electric Power Systems, 29*(17), 28–33 (in Chinese).
4. Yu H, & Jidong L. (2001). Cost-benefit Analysis and Evaluation of Power Network Planning [J]. *Power System Technology, 25*(7), 32–35 (in Chinese).
5. Qian Z, Yang X, & Wang X. (2008). Discussion on Security of Power Supply Engineering Recommendation ER P2/6 of UK [J]. *Power System Technology, 32*(18), 96–102 (in Chinese).
6. Zhichao Z, Yan Z, Wang W, & Mao J. (2004). Quantitative Evaluation of Substation Supply Reliability [J]. *Automation of Electric Power Systems, 28*(9), 66–69 (in Chinese).
7. Wenyuan L. (2002). Incorporating Aging Failures in Power System Reliability Evaluation[J]. *IEEE Trans. On Power Systems, 17*, 3.
8. Mingtian F, & Zuping Z. (2008). *Researchon the Development Strategy of distributed networks in China[M]*. Beijing: China Electric Power Press (in Chinese).
9. Wenyuan Li & Billinton R. (2003). Common Cause Outage Models in Power System Reliability Evaluation[J]. *IEEE Trans. On Power Systems, 18*, 2.
10. Pengxiang B, Jian L, Chunxin L, & Wenyuan Z. (2002). A Refined Genetic Algorithm for Power Distribution Network Reconfiguration [J]. *Automation of Electric Power Systems, 26*(2), 57–61 (in Chinese).
11. Liu C-C, & Lee SJ. (1988). An expert system operational aid for restoration and loss reduction of distribution systems [J]. *IEEE Transactions on Power Systems, 3*(2), 619–626.
12. Nagata T, Sasaki H, & Yokoyama R. (1995). Power system restoration by joint usage of expert system and mathematical programming approach [J]. *IEEE Transactions on Power Systems, 10*(3), 1473–1478.
13. Guo Y. (2003). *Power System Reliability Analysis [M]* (p. 1). Beijing: Tsinghua University Press. in Chinese.
14. Lee Willis H. (1997). *Power Distribution Planning Reference Book [M]*. NewYork: MarcelDekke.INC.
15. Fan Mingtian. (2007). Comparison the Standards of Power Supply Security for Urban Power Network at Home and Abroad [J]. *Distribution & Utilization, 24*(5), 5–8. in Chinese.
16. Zhu H, Cheng H, & Zhang Yan HZ. (1999). A Review of Electric Power Network Flexible Planning [J]. *Automation of Electric Power Systems, 23*(17), 38–4. 1 (in Chinese).
17. JianNing WZ, Wei N, & Yang X. (2007). Constructing Nanjin Urban Power Network with Advanced World Idea of Power Network Planning [J]. *Distribution & Utilization, 24*(5), 1–4 (in Chinese).

15

Studies on the active SISFCL and its impact on the distance protection of the EHV transmission line

Chao Li[1], Bin Li[2*], Fengrui Guo[3], Jianzhao Geng[1], Xiuchang Zhang[1] and Tim Coombs[1]

Abstract

The active saturated iron-core superconductive fault current limiter (SISFCL) is a good choice to decrease fault current. This paper introduced the principles and impedance characteristic of the active SISFCL. Then, it shows the current-limiting effects of the SISFCL. Besides, the impact of the active SISFCL on the distance protection of the EHV transmission line is evaluated. Based on that, the coordination scheme of the distance protections is proposed. A 500 kV double-circuit transmission system with SISFCLs is simulated by Electro-Magnetic Transients Program including DC (EMTDC). Simulation tests demonstrate the correctness and validity of theoretical analyses.

Keywords: Superconductive Fault current limiter (SFCL), Saturated iron core, Distance protection, Current-limiting effects

Introduction

The growth of the load and increasing demand for the stability of the power system lead to the combination of large capacity generating station and interconnection of electric power transmission and distribution systems. Consequently, the fault current is rising so greatly that the interrupting capacity of the existing electric equipment such as breakers could not withstand it [1].

One way to solve the problem is to replace the current equipment with the ones of larger capacity, which will definitely cost huge. And another method is to limit the fault current, which seems more promising. Traditional fault current limiting measures such as adopting the series reactor or the high short-circuit impendence transformer, can be equivalent to integrating an inductance or resistor with constant value to the grid. However, the method has to be at the expense of decreasing the stability of the grid and increasing the network loss. Under the circumstance, the superconductive fault current limiter (SFCL) can realize the limitation of the fault current effectively overcoming above shortcomings, because of the characteristic that it can present low impedance in the normal operation and high impedance the in the fault time. And the

characteristic has broadened its application prospects in the power grid [2].

Currently the most important types of SFCL contain resistive type, bridge type, dc biased iron core type, shielded iron core type and fault current controller type [3]. And those superconductive fault current limiters can mainly be classified into two types according to different fault current limiting characteristics, that is, the resistive and inductive type. For the application in the EHV transmission line, the resistive type has to deal with several difficult problems such as high voltages at cryogenic temperatures to ensure a reliable electrical insulation, recovery time after quench to meet the requirements of the circuit breaker reclosing system. Owing to the advantages of requiring no quench of superconducting coils during operation and a smaller cryogenic system needed, the application of the dc biased saturated iron-core SFCL (SISFCL) in the EHV transmission line seems more promising. It was firstly proposed by B. P. Raju in 1980's [4]. Lots of research and experiments were carried out to develop these kinds of devices and practical prototypes were installed in distribution and transmission power grids. Among them, one type called the active SISFCL is developing very fast in China [5–7]. And a 500 kV active SISFCL is being manufactured and schemed to integrate to the EHV transmission line in China South Grid.

* Correspondence: binli@tju.edu.cn
[2]Key Laboratory of Smart Grid of Ministry of Education, Tianjin University, Tianjin 300072, China
Full list of author information is available at the end of the article

Except for the device SFCL itself, its interactions with the power grid also attract the concerns of the grid company. Therefore, it's indeed necessary to investigate the impact of the SFCL on the coordination between the SFCL and the protective relay which is able to detect fault and abnormal operation conditions with high sensitivity and reliability [8, 9]. However, the topic of the coordination between the active SISFCL and the protection was not discussed directly with depth studies. And we will discuss the topic in the following. Besides, we introduce the working principle of the saturated iron core superconductive fault current limiter and its current-limiting effects in the electric power transmission line. There is no doubt that the application of superconducting fault current limiters in the power grid will be promising if their negative impacts on the operation of the power grid could be significantly reduced.

Working principle and current limiting effects of an active SISFCL

Working principle and impedance characteristic of the active SISFCL

As shown in Fig. 1, the active SICSFCL is composed of two iron cores, ac coils, and a superconductor dc circuit with high-speed switch control. The ac coils wounded around the iron cores are integrated to the power grid and the inductance of the ac coils displays the impedance characteristic of the active SISFCL, which is low impedance during normal power transmission and high impedance for fault current limiting. The dc magnetization circuit with high-speed switch control has the following functions: driving the two iron cores into deep saturation during normal condition, stopping the dc current supply at once a fault current occurs and providing the dc magnetizing again current as soon as the fault is cleared. It is

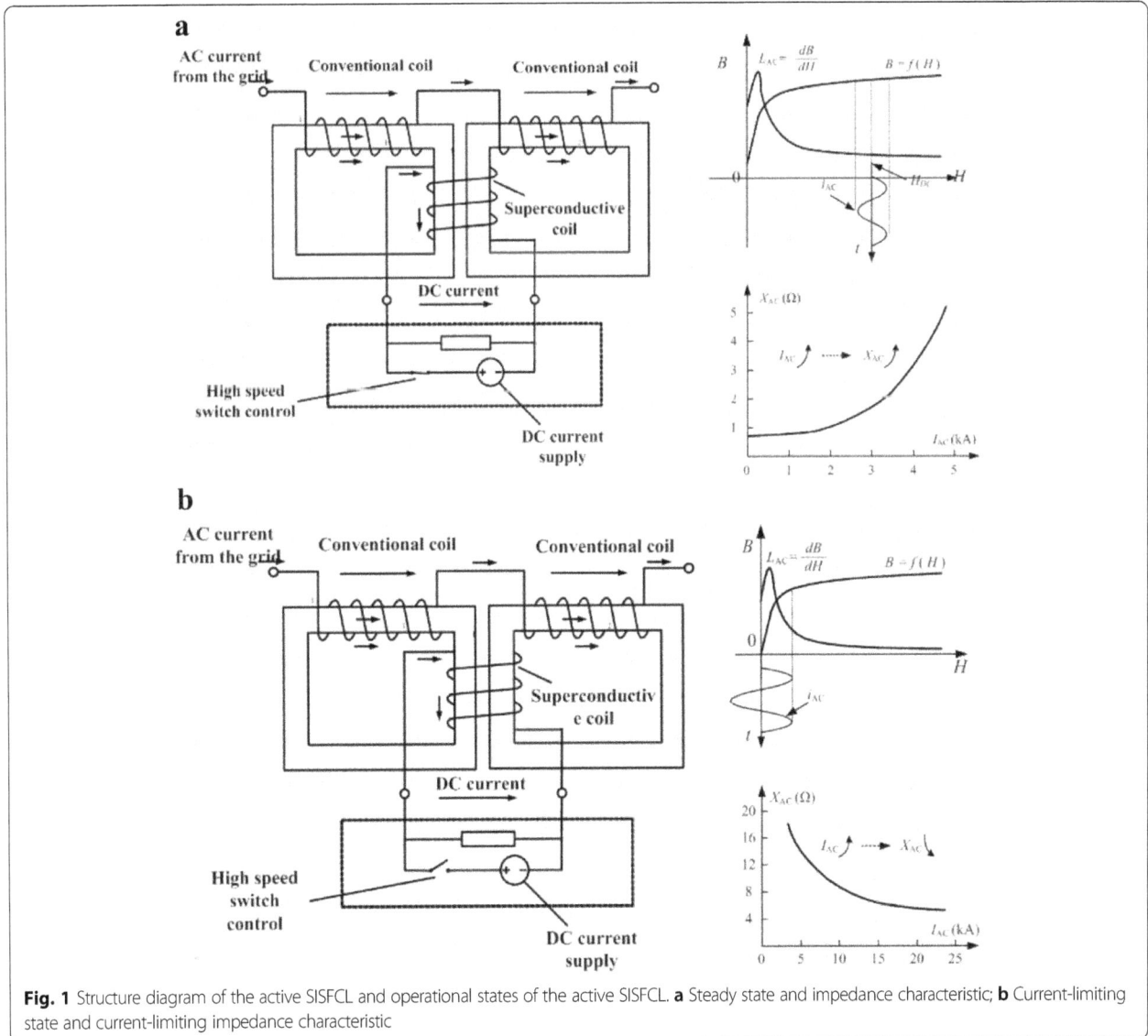

Fig. 1 Structure diagram of the active SISFCL and operational states of the active SISFCL. **a** Steady state and impedance characteristic; **b** Current-limiting state and current-limiting impedance characteristic

Table 1 Current of the three-phase fault and single-phase fault

	Current of three-phase fault(kA)		Current of single-phase fault(kA)		Numbers of substation exceeding 63kA
	Max	Average	Max	Average	
2010	62	45.7	61.8	41.6	0
2015	70.9	47.9	72.9	43	13
2020	78.7	52.7	72.5	48	17

63kA is the interrupting capacity of the breakers in the 500 kV substation

the superconductor dc circuit with high-speed switch control that help the active SISFCL overcome the disadvantages of the conventional SISFCL such as high cost, large size and induced overvoltage in the dc circuit. Consequently, two operational states of the active SISFCL are invited: the steady sate when the high speed switch is on and the current-limiting state when the high speed switch is off.

The two operational states with different impedance characteristics are depicted in Fig. 1a and b.

(1) Steady state: during the normal operation condition, both of the two iron cores are driven into saturation at points H_{DC} by the dc biased current in the dc superconductive magnetization circuit. The normal current in the ac coils is low enough to keep the cores fully saturated. Thus, the magnetic permeability is so small that the active SISFCL shows low impedance, as shown in Fig. 1a. Besides, the impedance of the active SISFCL is nonlinear positive correlation with the current flowing through the active SISFCL.

(2) Current-limiting State: the high-speed switch turned off dc current supply at the onset of the current limiting stage, decoupling the dc coils and ac coils in less than ten milliseconds. The de-saturating iron cores came to the points where the magnetic permeability of the iron core is large enough to make the SISFCL show a high impedance value,

as shown in Fig. 1b. In addition, the impedance of the active SISFCL decreases with the increase of the fault current.

As for the previous 35 kV and 220 kV active SISFCLs, the six iron-core frames of the three phases are conjugated in a hexagonal structure, sharing only one dc superconductor bias coil. Different from the previous three-phase-in-one structure of the 35 kV and the 220 kV SISFCLs, the 500 kV device is comprised of three independent single-phase SISFCLs, which makes it possible that the DC circuits of each single-phase SISFCL operate on its own.

Due to nonlinear characteristic of the fundamental magnetization curve, the impedance of the active SISFCL is not constant but relates to the current flowing through the SISFCL and its operation state.

Current-limiting effect of the active SISFCL
With the interconnection of the 500 kV power grid, the Guangdong Province Grid in China is achieving full closed loop operation in 2015. Hence, the fault current in 2010 and the fault current forecasted to be in 2015 and 2020 in shown in Table 1 [10].

Electromagnetic Transient Programming including DC (EMTDC) software has been used to investigate the current-limiting effects of the active SISFCL. The simulation model has been built according to the real data from 500 kV transmission system in China, as shown in Fig. 2.

Assuming a single-phase fault and a three-phase fault occurred in the outlet of the substation m respectively, the fault current of the transmission line with and without the active SISFCL is shown in Fig. 3. For the single phase fault, the RMS value of the fault current is as large as 49kA without installing the active SISFCL in the substation. However, the RMS value of the fault current decreased to 17kA in the presence of the active SISFCL. Similarly, the current of the three-phase fault can go down to 32.45% with the active SISFCL.

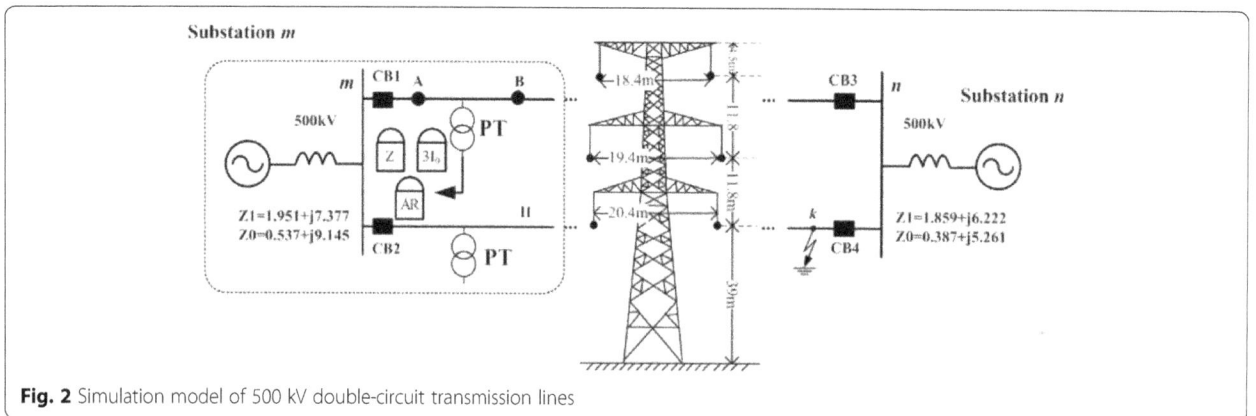

Fig. 2 Simulation model of 500 kV double-circuit transmission lines

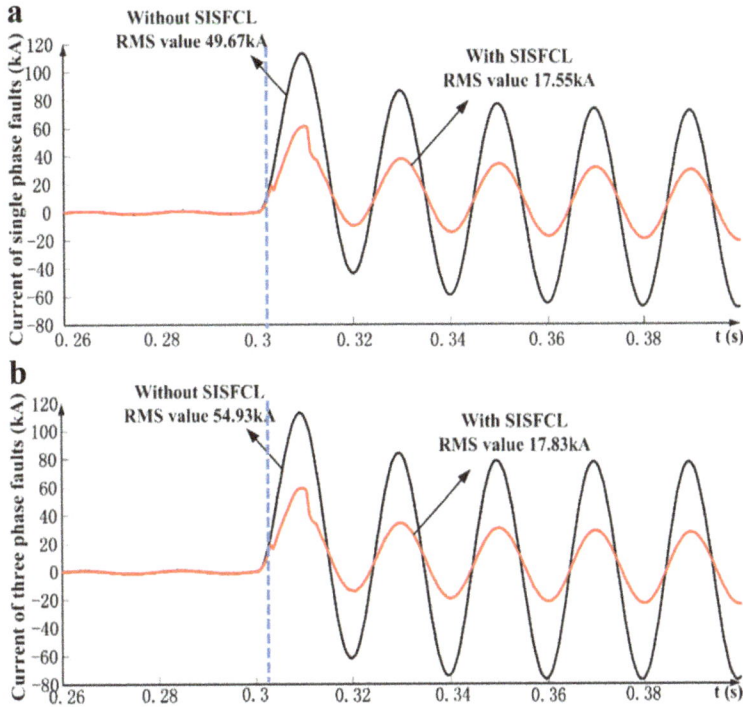

Fig. 3 Current-limiting effect of the active SISFCL: **a** for single phase fault **b** for three phase fault

Coordination of the active SISFCL and the distance protection
Impact of the active SISFCL on the distance protection for the EHV transmission line
The objective is to determine how the location of the SISIFCL impacts the distance protection, which is a type of protection widely applied in the EHV transmission line.

For the convenience of daily maintenance and repairs, it's a good option to install the SISFCL in the transformer substation (see Fig. 2). Hence we have two choices to locate the PTs for distance protection, namely point A and point B. If the SISFCL is installed at side A, it means the distance protection of the local side will not contain the SISFCL and the distance protection of the location will not be influenced by the SISFCL but the remote side will be. If installed at side B, the SISFCL will be included by the distance protection of local side and its impact on the distance protection of local and remote side should be analyzed in detail.

Distance relays are usually used to protect high voltage transmission lines. They can respond to the impedance of the distance from the fault location to where the relay is installed. In order to avoid maloperation and misoperation of the distance relays, the three-zone distance protection is adopted.

It's a fundamental principle of distance relay that the voltage and current used to energize the appropriate relay are such that the relay will measure the positive sequence impedance to the fault.

Principles for coordination of the active SISFCL and the distance protection
In order to ensure the sensitivity for the fault occurring at the end of the transmission line, we must be definite about the impedance of the SISFCL. However, the impedance of the active SISFCL has two characteristics: a) the impedance is variable. b) the impedance is a nonlinear function of the current flowing through the SISFCL. Therefore, the two impedance characteristics make the impact of the active SISFCL on the distance protection coordination more complex than that of other SFCLs.

According to above analysis, the impedance of the SISFCL is a function of the current and the current determine whether the active SISFCL stays in steady state or current-limiting state. Besides, there is great difference in value of the current under the circumstance of different fault points, different fault types, different system operation modes and so on.

According to the principle of the distance protection, the second zone of the distance protection should contain the impedance of the whole protected line and the largest impedance of the active SISFCL and its sensitivity for the fault shall be guaranteed. But whether the second

Table 2 The impedance of different fault points on transmission line MN

Fault position (%)	15	45	60	75	end
Impedance (Ω)	8.85	9.2	9.6	9.9	10.8

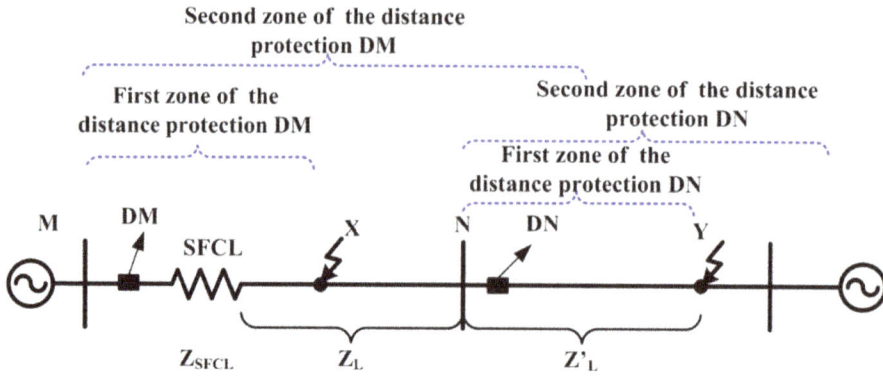
Fig. 4 Configuration of the distance protection

zone will overreach the first zone of the next distance protection shall also be verified. The conventional scheme for verifying overreach is suitable for the distance relay, which includes the active SISFCL, owing to the special impedance characteristics which is introduced above. In the following, a new scheme for verifying overreach will be discussed in detail.

On the base of Fig. 1, the impedance of the active SISFCL is negative correlation with the current flowing through. That's to say, the smaller the current is, the larger the impedance of the SISFCL will be. Hence, for all the faults X happened on the transmission line MN, the largest impedance of the active SISFCL will appear when the fault occurred at the Point N. The simulation by PSCAD/EMTDC verified above analysis, as shown in Table 2.

In Fig. 4, the largest impedance of the SISFCL ($Z_{SFCL.max}$) appears when a phase-to-phase fault happens at the end of the transmission line under the circumstance of the system minimum operating mode. The second zone of the distance protection DM should be expressed as:

$$Z_{II.DM} = K_{sen}(Z_L + Z_{SFCL.N.\,max}) \quad (1)$$

When a fault happens at point Y of the next neighbor transmission line, the smallest impedance of the active SISFCL should be used to verify whether the distance protection DM will overreach. And the its smallest impedance appears in the situation of the largest fault current, namely the current of a three-phase fault in the system maximum operating mode. Hence the measure impedance of the local distance relay can be expressed:

$$Z_m = Z_L + Z_{SFCL.Y.\,min} + Z_L' \quad (2)$$

If $Z_m > Z_{II}$, it means there is no overreaching for DM.
If $Z_m < Z_{II}$, it means there is overreaching for DM.

Hence, the current flowing through the active SISFCL during the fault at point Y is smaller than that when the fault happens in the Z_L. Consequently, the impedance of

the active of SISFCL will become larger and help to avoid the overreach of the distance protection of DM. The characteristic of the SISFCL is very helpful for the coordination of the distance protection for a long transmission line with a much shorter neighbor transmission line.

Conclusions

This paper demonstrates the active SISFCL is a good solution to the problem of increasing fault current, using the real data from South China Grid. The results have shown that the application prospects of the active SISFCL is promising. According to its operational principle in the current-limiting state, the impedance of the active SISFCL decreases with the increase of the fault current nonlinearly. This characteristic can help to avoid the overreach of the distance protection. Besides, a setting scheme for the distance protection in the EHV transmission has been proposed.

Authors' contributions
Manuscript is approved by all authors for publication.

Competing interests
The authors declare that they have no competing interests.

Author details
[1]EPEC Superconductivity Group, University of Cambridge, Cambridge CB3 0FA, UK. [2]Key Laboratory of Smart Grid of Ministry of Education, Tianjin University, Tianjin 300072, China. [3]State Grid Tianjin Electric Power Company, Tianjin 300010, China.

References
1. Zhang, Y., & Dougal, R. A. (2012). *State of the art of Fault Current Limiters and their applications in smart grid*. San Diego: IEEE General Meeting of Power and Energy Society.
2. Ye, L., Lin, L. Z., & Juengst, K.-P. (2002). Application studies of superconducting fault current limiters in electric power systems. *IEEE Transactions on Applied Superconductivity, 12*(1), 900–903.
3. Noe, M., & Steurer, M. (2007). High-temperature fault current limiters: concepts, applications and development status. *Superconductor Science and Technology, 20*(3), 15–27.

4. Raju, B. P., Parton, K. C., & Bartram, T. C. (1982). A current limiting device using superconducting d.c. bias applications and prospects. *IEEE Transactions on Power Apparatus and Systems, 101*, 3173–3177.

5. Xin, Y., Gong, W. Z., & Cao, Z. J. (2007). Development of saturated iron core HTS fault current limiters. *IEEE Transactions on Applied Superconductivity, 17*(2), 1760–1763.

6. Hong, H., Cao, Z., Zhang, J., et al. (2009). DC magnetization system for 35 kV/90 MVA super-conducting saturated iron-core fault current limiter. *IEEE Transactions on Applied Superconductivity, 19*(3), 1851–1854.

7. Xin, Y., Gong, W. Z., Niu, X. Y., et al. (2009). Manufacturing and test of a 35 kV/90 MVA saturated iron-core type superconductive fault current limiter for live-grid operation. *IEEE Transactions on Applied Superconductivity, 19*(3), 1934–1937.

8. Li, B., Li, C., & Guo, F. (2014). Overcurrent Protection Coordination in a Power Distribution Network With the Active Superconductive Fault Current Limiter. *IEEE Transactions on Applied Superconductivity, 24*(5), 5602004.

9. Jin-Seok, K., Sung-Hun, L., & Jae-chul, K. (2012). Study on application method of superconducting fault current limiter for protection coordination of protective devices in a power distribution system". *IEEE Transactions on Applied Superconductivity, 22*(3), 5601504.

10. Wei, L., et al. (2013). *Prospects for application of superconducting fault current limiter in Chinese power system* (Proceedings of 2013 IEEE International Conference on Applied Superconductivity and Electromagnetic Devices Beijing, China, pp. 513–516).

16

Cost reduction of a hybrid energy storage system considering correlation between wind and PV power

Lin Feng[1*], Jingning Zhang[1], Guojie Li[1] and Bangling Zhang[2]

Abstract

A hybrid energy storage system (HESS) plays an important role in balancing the cost with the performance in terms of stabilizing the fluctuant power of wind farms and photovoltaic (PV) stations. To further bring down the cost and actually implement the dispatchability of wind/PV plants, there is a need to penetrate into the major factors that contribute to the cost of the any HESS. This paper first discusses hybrid energy storage systems, as well as chemical properties in different medium, deeming the ramp rate as one of the determinants that must be observed in the cost calculation. Then, a mathematical tool, Copula, is explained in details for the purpose of unscrambling the dependences between the power of wind and PV plants. To lower the cost, the basic rule for allocation of buffered power is also put forward, with the minimum energy capacities of the battery ESS(BESS) and the supercapacitor ESS(SC-ESS) simultaneously determined by integration. And the paper introduces the probability method to analyze how power and energy is compensated in certain confidence level. After that, two definitions of coefficients are set up, separately describing energy storage status and wind curtailment level. Finally, the paper gives a numerical example stemmed from real data acquired in wind farms and PV stations in Belgium. The conclusion presents that the cost of a hybrid energy storage system is greatly affected by ramp-rate and dependence between the power of wind farms and photovoltaic stations, in which dependence can easily be determined by Copulas.

Keywords: Probability analysis, Copula, Correlation, Capacity optimization, Hybrid ESS

Introduction

Owing to the fact of the fast development of renewable energy and the increased concern of the environmental and sustainable impact of fossil fuels, wind farms and photovoltaic plants have been widely increasingly built around the world over past decades [1]. Wind farms and photovoltaic plants are currently not confined in the conventional off-grid generation systems. Those on-grid variable power sources, however, exert an adverse impact on the operation and control of conventional power grid [2]. The power of wind farms and PV plants is normally much more dependent upon the on-site landform, topography and climate. As a result, they inherently contain the congenital defects, including intermittency, fluctuation and undispatchablility. Hence, energy storage

systems which nowadays have become less expensive are introduced in such systems to ease the instability tendency of grids, with the ability of compensating for intermitted and fluctuant outputs of wind/PV plants. As the cost gradually falls, energy storage systems have been put into use in grids on a larger scale. Single type energy storage systems cannot meet the demands in real applications, considering the power and energy requirement at different time scale. As a result, hybrid energy storage systems turn out to be the feasible choice. The most common hybrid energy storage system is composed of batteries, such as advanced lead-acid or lithium-ion batteries, and supercapacitors [3].

Refs. [4, 5] present a fundamental frame to analyze the cost in a hybrid energy storage system, pointing out the ramp rate is the principal factor that has effect on the cost. Copula functions are discussed in [6, 7]. The space-time complementarities between wind farms and PV stations are emphasized in [8].

---

* Correspondence: fenglin@sjtu.edu.cn
[1]School of Electrical Information and Electronic Engineering, Shanghai Jiao Tong University, Shanghai, China
Full list of author information is available at the end of the article

Ref. [9] cautiously explored the wind curtailment phenomenon in nations partly driven by wind power generation, such as America, Spain and Denmark. A new wind-ESS combined control method for surpassing Wind curtailment are in detail discussed in [10].

To further bring down the cost and actually implement the dispatchability of wind/PV plants, hybrid energy storage systems will be increasingly pervasive in the foreseeable future. Therefore, there is a need to penetrate into the major factors that contribute to the cost of the any HESS.

In this study, based on the cost of HESS by considering the ramp rate mentioned in [4], Copulas functions are proposed to analyze the dependence between wind and PV power. Then the cost of HESS is analyzed considering the ramp rate as well as the dependence between wind and PV power. Simulation studies are carried out to verify the above analysis.

Hybrid energy storage system

A hybrid energy storage system often owns the merits of individual single energy storage system. A battery energy storage system, such as advanced lead-acid batteries, has the advantages in pricing and large-scale using. It, however, poorly performs in the situation where power soars dramatically, or on the contrary, drops instantly. So, it is supposed to accumulate massive less fluctuant energy. With the more costly price, a supercapacitor energy storage system is not equipped with possibility of extensive using, but it does well in the moments when power moves fast.

The Fig. 1 depicts the diagram of a common fundamental generation system with a HESS, in which power equipment are omitted. The arrows point out the direction of power in an instant.

In Fig. 1, P_v represents the output power of photovoltaic stations. P_w represents the power of wind farms. P is the aggregated power of P_v and Pw. Buffered power is represented by P_h. P_{sc} is the power of the supercapacitor energy storage system. P_b is the power of the battery energy storage system. P_d represents the dispatched power into the grid.

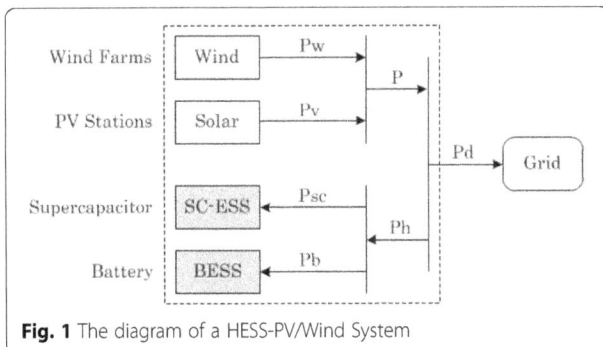

Fig. 1 The diagram of a HESS-PV/Wind System

Ramp rate

The ramp rate represents the change rate of power. More specifically, ramp rates, including ramp-up rates and rate-down rates, can be approached in two respects. First, they cannot be separated from wind farms or PV stations. For not only protecting wind turbines, photovoltaic cells and girds, but also ensuring that the power fluctuation of wind farms and PV plants can be regulated by other conventional power sources, like fossil-fueled plants, there is a limit to maximum change rate of active power in the wind farms and photovoltaic stations. Tables 1 and 2 bring the standards in China.

Second, ramp rates also go for the energy storage system. Owing to the chemical properties and specifications by battery manufacturers, a battery energy storage system has the boundaries of maximum ramp rate as well. There is no limit to ramp rates in the supercapacitor energy storage system, which undoubtedly assumes the superiority of supercapacitor energy storage systems.

Method

By Copulas, the dependences, or correlations between random variables can be set up. On one hand, marginal distributions can easily be obtained with joint probability distribution. On the other hand, it is not easy to get the joint distribution when marginal distributions are known. The emergence and improvement of Copulas functions, to some extent, resolves the problem [6].

Definition

In 1959, Sklar put forward the thinking of Copula function, which divides N-dimension joint probability distribution into N marginal distribution functions and one copula function that describes the correlation degree between variables. Nelsen gave the strict definition of the Copula function in 1999. Copula, denoted as $C(u_1, u_2, ..., u_N)$, is a connection function that links joint probability distribution $F(X_1, X_2, ..., X_N)$ of random variables $X_1, X_2, ..., X_N$ with their individual marginal distribution, $F_{X1}(x_1), ..., F_{XN}(x_N)$, which can be explicitly formulated as follows.

Table 1 The limitation of ramp-rates in wind farms

Installed capacity	Maximum variation in 10 min(MW)	Maximum variation in 1 min(MW)
<30	20	6
30–150	capacity/1.5	capacity/5
>150	100	30

Technical rule for wind power plant connected to Power Grid in 2009

Table 2 The limitation of ramp-rates in PV stations

Project scale	Maximum variation in 10 min(MW)	Maximum variation in 1 min(MW)
small	capacity	0.2
medium	capacity	capacity/5
large	capacity/3	capacity/10

Technical rule for photovoltaic power station connected to Power Grid in 2011

$$F(X_1, X_2, ..., X_N) = C\left[F_{X_1}(x_1), F_{X_2}(x_2)...F_{X_N}(x_N)\right] \tag{1}$$

Common copulas

Supposing Pearson coefficient is, a two-dimensional Gaussian Copula function is as follows,

$$C(u, v; \rho) = \int_{-\infty}^{\Phi^{-1}(u)} \int_{-\infty}^{\Phi^{-1}(v)} \frac{1}{2\pi\sqrt{1-\rho^2}} \exp$$

$$\left[-\frac{s^2 - 2\rho st + t^2}{2(1-\rho^2)}\right] dsdt \tag{2}$$

A two-dimensional t-Copula function with free degree k is as follows,

$$C(u, v; \rho, k) = \int_{-\infty}^{t_k^{-1}(u)} \int_{-\infty}^{t_k^{-1}(v)} \frac{1}{2\pi\sqrt{1-\rho^2}}$$

$$\left[1 + \frac{s^2 - 2\rho st + t^2}{k(1-\rho^2)}\right]^{-(k+2)/2} dsdt \tag{3}$$

Archimedean copulas are defined as,

$$C(u_1, u_2, ..., u_N) = \begin{cases} \phi^{-1}[\phi(u_1), \phi(u_2), ..., \phi(u_N)], \\ 0 \end{cases}$$

$$\sum_{i=1}^{N} \phi(u_1) \leq \phi(0)$$

$$\textit{other} \tag{4}$$

The measurement of correlation

There are several measurements of correlation, or dependency, between random variables. With regard to different usage, massive of them can be adopted, such as Pearson coefficient ρ, Kendall rank correlation coefficient τ, Spearman rank correlation coefficient ρ and tail dependence λ.

If marginal distributions of continual random variable (X, Y) are $F(x)$ and $G(y)$, the relationships between Copula function $C(u, v)$ and Kendall rank correlation

coefficient τ, Spearman rank correlation coefficient ρ, tail dependence λ are as follows, respectively.

$$\tau = 4 \int_0^1 \int_0^1 C(u, v) dC(u, v) - 1 \tag{5}$$

$$\rho_s = 12 \int_0^1 \int_0^1 C(u, v) dudv - 3 \tag{6}$$

$$\lambda^{up} = \lim_{u \to 1^-} \frac{1 - 2u + C(u, v)}{1 - u} \tag{7}$$

$$\lambda^{lo} = \lim_{u \to 0^+} \frac{C(u, v)}{u} \tag{8}$$

Results

Dispatchability

Generally, a generation plant bids against each other. It is common for bidders in several power trading markets to forecast its power (nowadays, more and more forecast data are purchased from third-party service providers) and to propose generation schedule N hours before the market begins every trading day, which can be exemplified with Nordic power exchange market. The mechanism is called N-hour rule. To simulate the operation of real world, a forecast example is given below, with the hypothesis that dispatched power is derived from hourly average of forecasting power. A forecast algorithm using historical power data can be found in Fig. 2.

The allocation of P_b and P_{sc}

P_h is the aggregated power of P_b and P_{sc}. P_h is the power that buffers from/into hybrid energy storage system. Y is the ramp rate. P_b and P_{sc} are confirmed in (9) and (10) [4].

Supposing

$$\begin{aligned} P_{B,r}(t_{i-1}) + Y \cdot \Delta t \geq P_{H,r}(t_i) \\ P_{B,r}(t_i) = P_{H,r}(t_i), P_{SC,r}(t_i) = 0 \end{aligned} \tag{9}$$

Otherwise,

$$\begin{aligned} P_{B,r}(t_i) = P_{B,r}(t_{i-1}) + Y \cdot \Delta t \\ P_{SC,r}(t_i) = P_{H,r}(t_i) - P_{B,r}(t_i) \end{aligned} \tag{10}$$

The determination of energy capacity

After integrating P_b and P_{sc} with time, E_b and E_{sc} can be achieved. E_b and E_{sc} display the energy variation, or the energy level, of the energy storage system [4]. In Fig. 3, the zero-interface represents the assumptive initial energy level. The energy storage system is charged when the curve goes up with system being discharged as the curve comes down. Thus, the minimum energy capacity of the BESS or the SC-ESS is based on its own moving

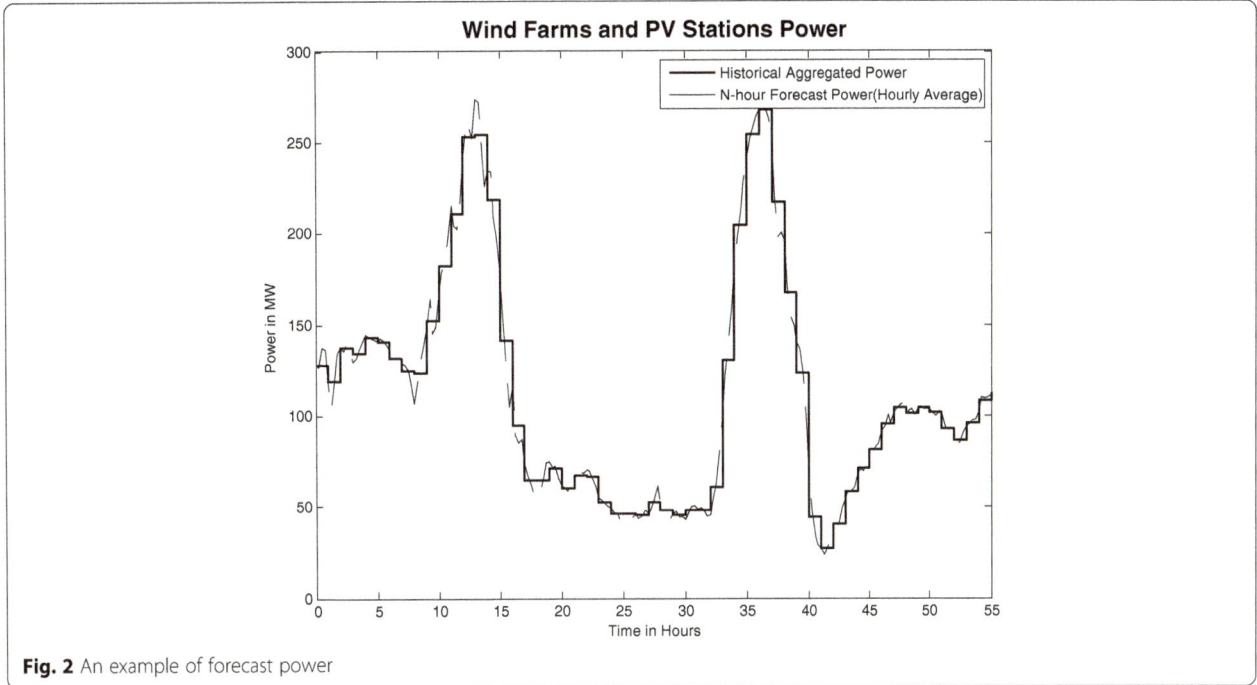

Fig. 2 An example of forecast power

range, respectively. This implies that the best energy capacity is no less than these least value.

Cost function

Solving (9) and (10), every ramp rate Y corresponds to a set of $P_b(t)$ and $P_{sc}(t)$, from which $E_b(t)$ and $E_{sc}(t)$ can be gained. The fit value of P_b is the maximum of $P_b(t)$; the same for P_{sc}. Note that the fit values of E_b and E_{sc}

depend primarily on their difference between upper and lower bounds in Fig. 3.

Then, the basic cost function is as follows [4].

$$f = k_1 \times E_b + k_2 \times E_{SC} + k_3 \times P_b + k_3 \times P_{SC} \quad (11)$$

where k_1, k_2, k_3 and k_4 are obtained in [11].

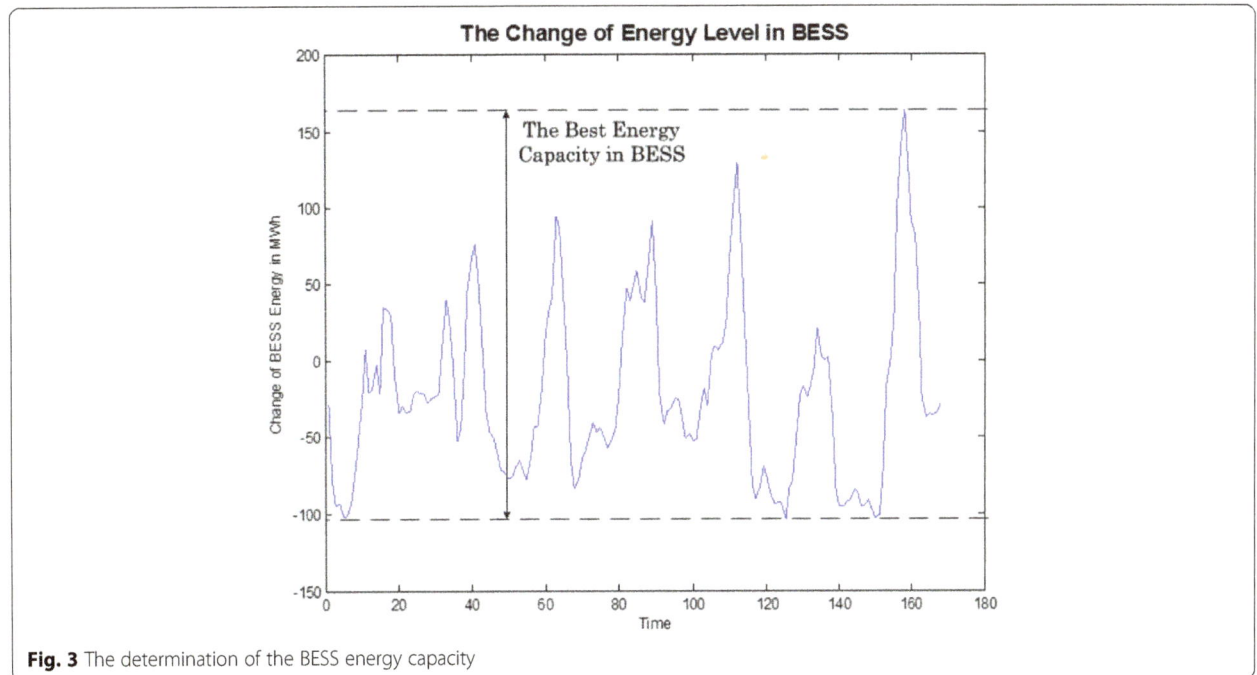

Fig. 3 The determination of the BESS energy capacity

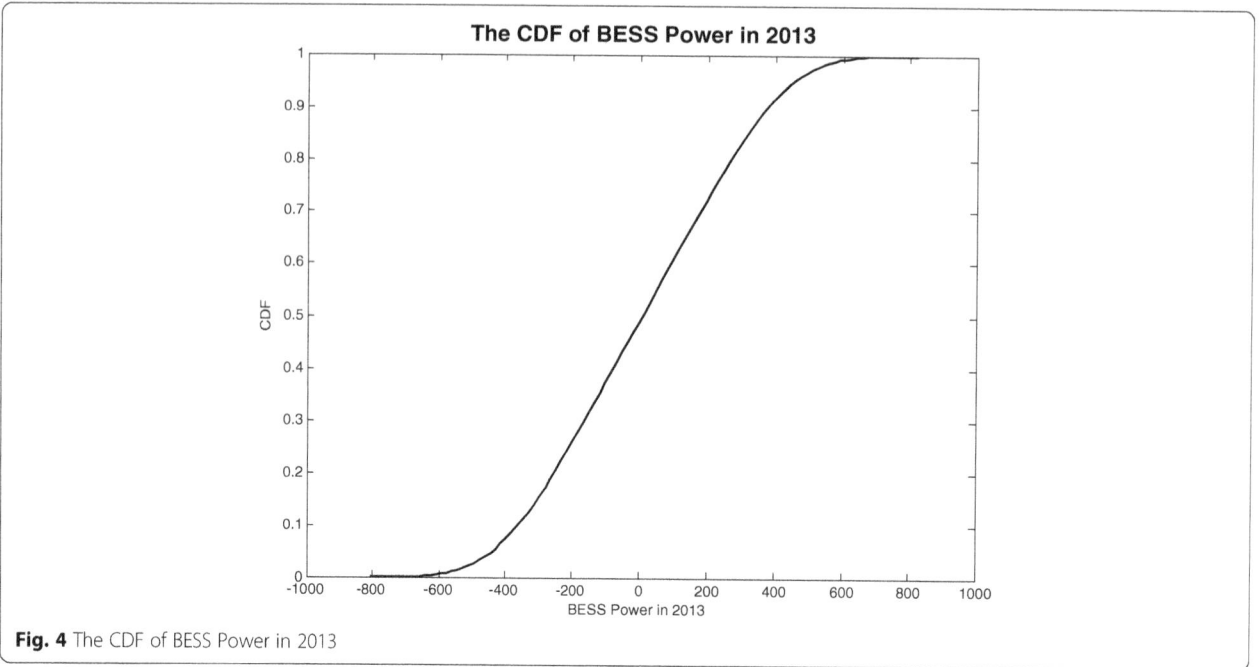

Fig. 4 The CDF of BESS Power in 2013

Statistical observation

To minimize the cost f, the proper ramp rate Y should be selected. Meanwhile, $P_b(t)$, $P_{sc}(t)$, $E_b(t)$ and $E_{sc}(t)$ are successively determined. It is natural that the Cumulative Distribution Function (CDF) of these capacities is obtained as shown in Fig. 4 to depict the complete spectacle in statistical manners [4].

Definitions of coefficients

To make the problem more intuitive, here we separately define two types of coefficients, wind curtailment coefficient and energy storage coefficient. This approach will undoubtedly construct the connection between energy storage status and wind curtailment condition. The similar solar power abandonment is ignored for parallel approach.

The key to the problem is that we make a probabilistic modification to wind power output. In other words, P_w is replaced by $[1-\xi(t_i)]\cdot P'_w$; note that P'_w is the original 100 % of wind power.

Now we choose $\xi(t_i)$ to describe the level of wind curtailment and the definition of $\xi(t_i)$ is as follows,

$$\xi(t_i) = \begin{cases} 0, & zero\,wind\,curtailment \\ 0\sim1, & partly\,wind\,curtailment \\ 1, & complete\,wind\,curtailment \end{cases}$$

Then, the wind curtailment coefficient is denoted by,

$$C = \frac{\sum_{i}^{i+k} \xi(t_i)}{T/\Delta t} \times 100\%$$

Finally, the energy storage compensation coefficient is represented,

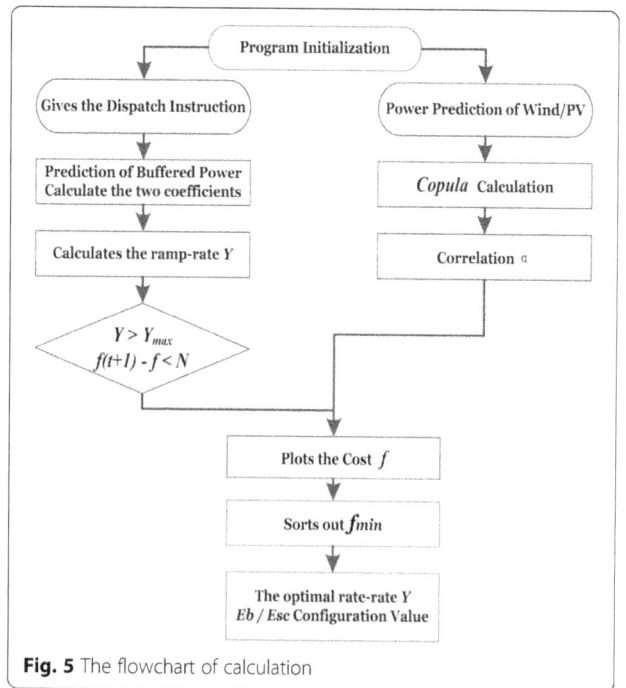

Fig. 5 The flowchart of calculation

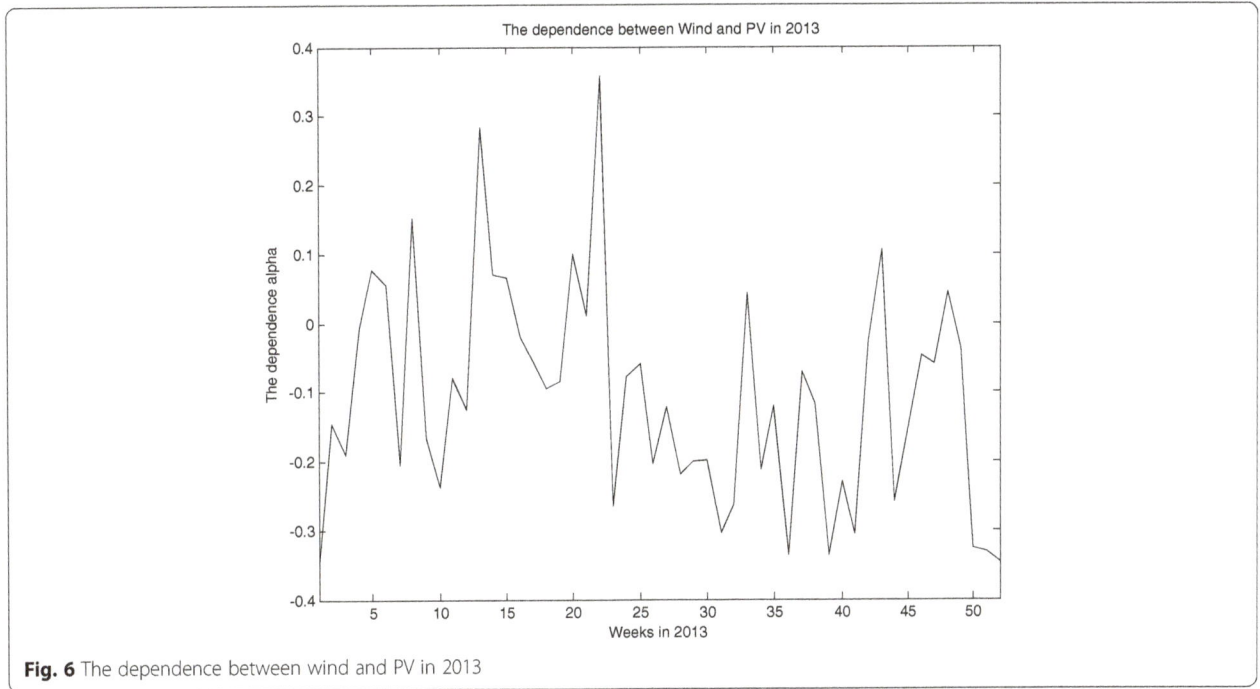

Fig. 6 The dependence between wind and PV in 2013

$$\lambda = \frac{\sum\limits_{i}^{i+k}|[1-\xi(t_i)]P_w + P_v - P_{H,r}|\cdot\Delta t}{\sum\limits_{i}^{i+k}P_h\cdot\Delta t}$$

As a result, we obtain that

- $\lambda > 1$, the cost of hybrid energy storage system goes up
- $\lambda = 1$, the cost of hybrid energy storage system remains unchanged
- $\lambda < 1$, the cost of hybrid energy storage system goes down

Procedure to determine the optimum BESS ramp-rates
The MATLAB program is developed following the methods mentioned above. The flowchart is shown in Fig. 5.

Discussion
Here, the paper exhibits a numerical example. All the data are based on real historical scene. The installed capacities of wind farms and PV stations are 143.45 MW and 431.17 MW, respectively, which are located at Belgium (http://www.elia.be).

The dependence using copulas
The paper adopts Copulas to readily calculate the dependences between wind farms and photovoltaic stations, which are shown in Fig. 6.

From the above dependences, three typical sets of data, representing three different common dependences, are picked up. The copula PDF of one of the three sets of data is given in Fig. 7.

The buffered power
The paper utilizes the method in [12] to forecast the dispatchable power P_d into the grid in next N hours. Note that the following procedures are based merely on one set of data. Then, P_h is obtained after P deducting P_d. An example of P_h is shown in Fig. 8.

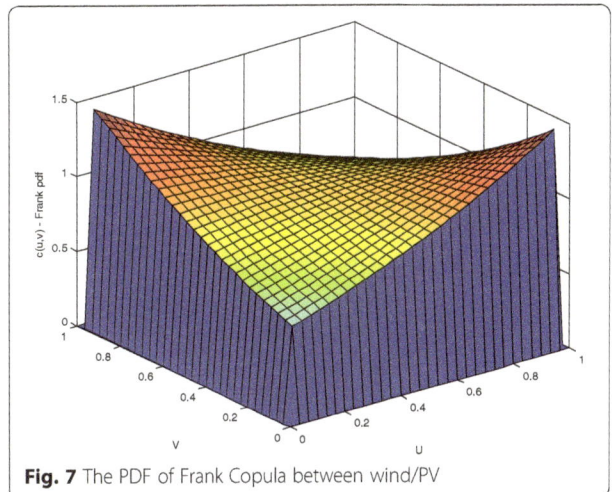

Fig. 7 The PDF of Frank Copula between wind/PV

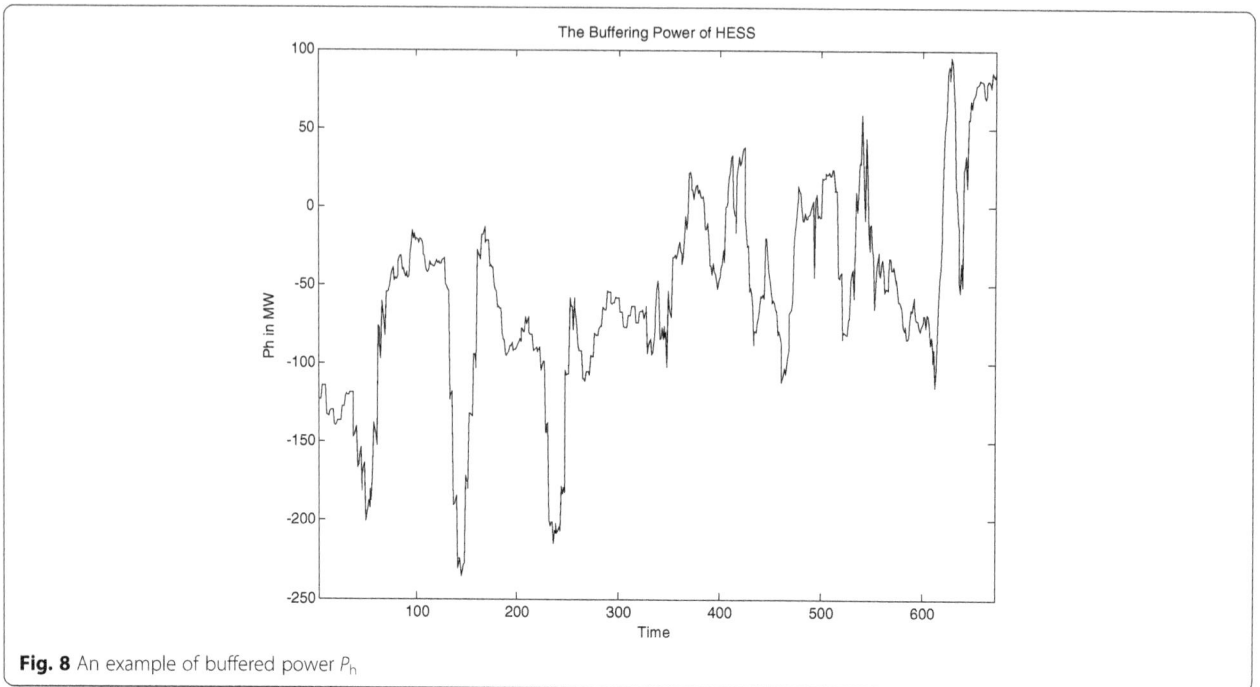

Fig. 8 An example of buffered power P_h

Power and energy capacities

Once ramp rate Y is decided, a set of corresponding data is determined as well, including P_b, P_{sc}, E_b and E_{sc}, where energy level E_b and E_{sc} are from the integral of output powers. Here, the designed MATLAB Program endeavors to traverse Y from 0 through 10. Then, a great number of sets of $\{P_b, P_{sc}, E_b, E_{sc}\}$ are acquired by the method mentioned in Subsection Cost function. The

allocation of P_b and P_{sc} is revealed in Fig. 9. It is not difficult to find out that majority of P_h is supported by the BESS, while the minority is assumed by SC-ESS in this case.

The essence of cost function

As seen in (11), the cost of the HESS can be obtained by traversing ramp rate Y. By the input of three sets of data,

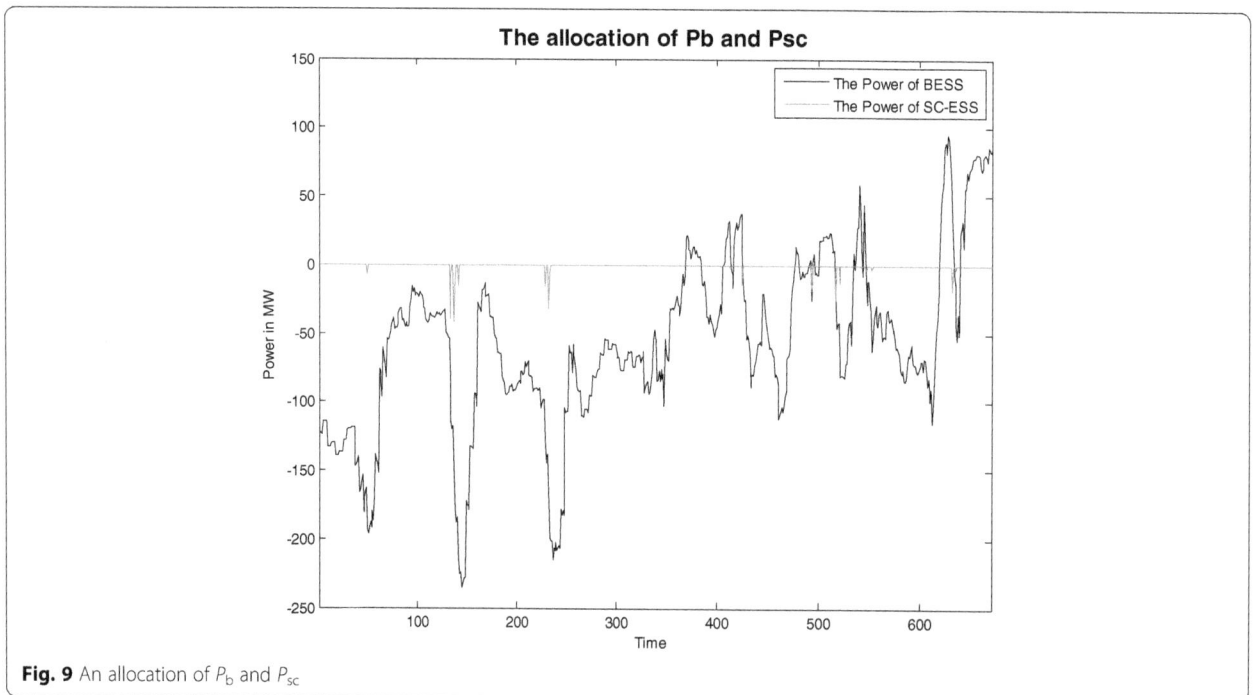

Fig. 9 An allocation of P_b and P_{sc}

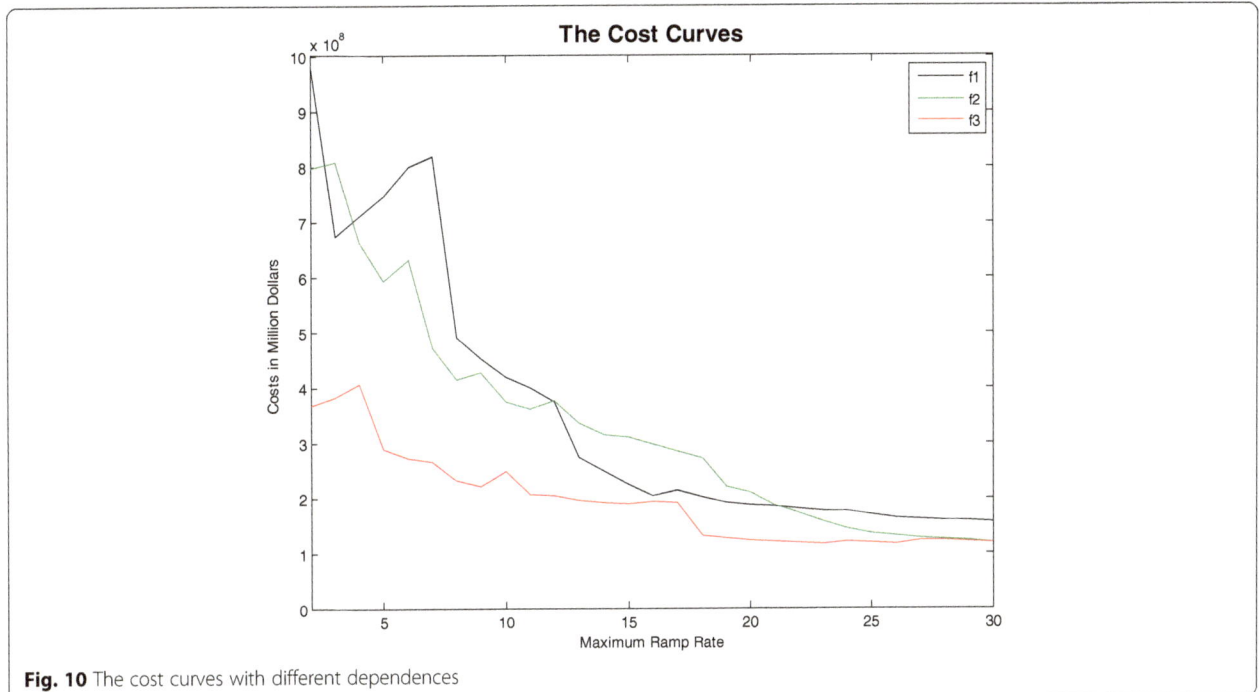

Fig. 10 The cost curves with different dependences

the Fig. 10 intuitively demonstrates the relation between cost and dependence.

Hence, the essence of the HESS cost is as follows.

$$f = f(Y, \alpha, \lambda; t) \qquad (12)$$

In (12), f_1, f_2 and f_3 correspond the dependence -0.1, -0.2 and -0.3, respectively. As the dependence α between wind

farms and PV stations goes up, the cost f diminishes. On determining the best ramp rate Y, the minimum energy capacities of the BESS and the SC-ESS are also confirmed. The cost curves with different lambda are showed in Fig. 11.

Conclusion

The paper concluded that the cost of a hybrid energy storage system is greatly affected by ramp-rate and

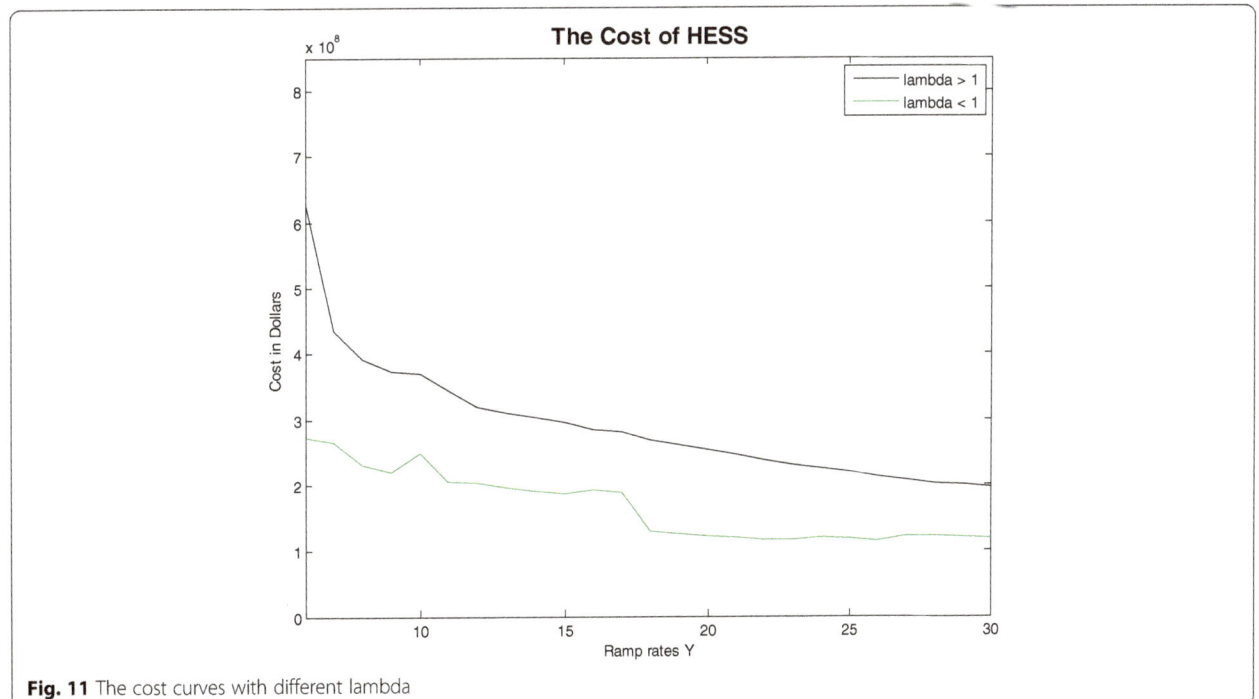

Fig. 11 The cost curves with different lambda

dependence between the power of wind farms and photovoltaic stations, in which dependence can easily be determined by Copulas. A numerical example shows as the dependence between wind farms and PV stations goes up, the cost decreases. Moreover, the cost is also influenced by energy storage compensation coefficient.

Acknowledgements
This work was supported by Shanghai Science and Technology Committee (13231204002) and National Key Technology R&D Program of China (2015BAA01B02).

Authors' contribution
LF and JNZ carried out the theoretic studies, calculated the numerical example and drafted the manuscript; GJL and BLZ participated in the theoretic studies. All authors read and approved the final manuscript.

Competing interests
The authors declare that they have no competing interests.

About the Authors
L. Feng She is now a lecturer in the Dept. of Electrical Engineering, Shanghai Jiao Tong Univ., Shanghai, China. Her research interests are the control and the integration of renewable energy, and Microgrid.
J. N. Zhang He is currently pursuing the Master Degree in SEIEE from Shanghai Jiao Tong University. His research interests are capacity optimization and energy storage systems.
G. J. Li He is now a professor in the Dept. of Electrical Engineering, Shanghai Jiao Tong Univ., Shanghai, China. His current research interests include power system analysis and control, wind and PV power control and integration, and Microgrid.
B. L. Zhang He is currently working for Shanghai Power & Energy Storage Battery System Engineering Tech. Co. Ltd.. His research interests are the electric power system design and the integration of new energy resources.

Author details
[1]School of Electrical Information and Electronic Engineering, Shanghai Jiao Tong University, Shanghai, China. [2]Shanghai Power & Energy Storage Battery System Engineering Tech. Co. Ltd., Shanghai, China.

References
1. Ashwin Kumar, A. (2010). A study on renewable energy resources in India. In *International Conference on Environmental Engineering and Applications (ICEEA)* (pp. 49–53).
2. MacGill, I. F. (2012). *Impacts and best practices of large-scale wind power integration into electricity markets -Some Australian perspectives* (Power and Energy Society General Meeting, pp. 1–6).
3. Dai, H. (2010). *A Study on Lead Acid Battery and Ultracapacitor Hybrid Energy Storage System for Hybrid City Bus* (Optoelectronics and Image Processing (ICOIP), 2010 International Conference on. IEEE, Vol. 1, pp. 154–159).
4. Wee, K. W., Choi, S. S., & Vilathgamuwa, D. M. (2011). *Design of a renewable—hybrid energy storage power scheme for short-term power dispatch* (Electric Utility Deregulation and Restructuring and Power Technologies (DRPT), 2011 4th International Conference on. IEEE, pp. 1511–1516).
5. Yao, D. L., Choi, S. S., & Tseng, K. J. (2011). Design of short-term dispatch strategy to maximize income of a wind power-energy storage generating station. In *Innovative Smart Grid Technologies Asia (ISGT)* (pp. 1–8).
6. Papaefthymiou, G. (2009). Using Copulas for Modeling Stochastic Dependence in Power System Uncertainty Analysis. *IEEE Transactions on Power Systems, 24*(1), 40–49.
7. Brunel, N. J.-B. (2010). Modeling and Unsupervised Classification of Multivariate Hidden Markov Chains With Copulas. *IEEE Transactions on Automatic Control, 55*(2), 338–349.
8. Zhang, N., Kang, C., Xu, Q., Jiang, C., Chen, Z., & Liu, J. (2013). Modelling and Simulating the Spatio-Temporal Correlations of Clustered Wind Power Using Copula. *Journal of Electrical Engineering and Technology, 8*(6), 1615–1625.
9. Wang, Q.-k. (2012). Update and Empirical Analysis of Domestic and Foreign Wind Energy Curtailment. *East China Electric Power, 40*(3), 378–381.
10. Zhang, N., Kang, C., Xu, Q., Jiang, C., Chen, Z., & Liu, J. (2013). Wind-ESS Combined Control for Suppressing Ramp Rates of Wind Power. *Automation of Electric Power Systems, 37*(13), 17–23.
11. Chen, H., Cong, T. N., Yang, W., Tan, C., Li, Y., & Ding, Y. (2009). Progressin Energy Storage System: a Critical Review. *Progress in Natural Science, 19*(3), 291–312.
12. Mahoney, W. P., Parks, K., Wiener, G., Liu, Y., Myers, W. L., & Sun, J. (2012). A wind power forecasting system to optimize grid integration. *IEEE Transactions on Sustainable Energy, 3*(4), 670–682.

17

Research on the protection coordination of permanent magnet synchronous generator based wind farms with low voltage ride through capability

Renfeng Tao[*], Fengting Li, Weiwei Chen, Yanfang Fan, Chenguang Liang and Yang Li

Abstract

To coordinate the protection of PMSG (permanent magnet synchronous generator), collector circuits and outgoing lines, a comprehensive and improved protection method of PMSG based wind farms with LVRT (low voltage ride through) capability is proposed. The proposed method includes adding a short time delay to the collector network current protection zone I and a directional protective relaying to the collector network protection, installing grounding transformers and zero sequence current protection, and generator low-voltage protection action improvement. A LVRT scheme consisting of variable resistance dumping circuit, grid side dynamic reactive power control and reactive power compensation control is proposed. The fault characteristics of PMSG based wind farms are analyzed, and a PMSG based wind farm in Dabancheng, Xinjiang, is used as an example to analyze typical wind farm protection configuration, the setting values considering LVRT requirements, and the coordination problems. Finally, an improved wind farm protection coordination methodology is proposed and its validity is verified by simulation.

Keywords: LVRT, Protection Coordination, PMSG, Wind Farm Fault Characteristics

1 Introduction

As a clean and renewable energy, wind power is developing rapidly in China. In 2015, the new wind power installed capacity has reached 30,753 MW, an increase of 32.6% compared to 2014, and the total installed capacity has reached 145,362 MW, ranked first in the world.

Because of wind power's stochastic and variable operation nature, the stable operation of grid integrated with large scale of wind power and the protection configuration of wind farms are facing enormous challenges. Therefore, for maintaining the stability and safe operation of the power systems, wind turbines require to have the low voltage ride through capability [1–5]. Nevertheless, in the event of grid side fault or voltage dip, wind farms often simply trip to protect the wind turbines [6, 7]. To solve the problem, many advanced LVRT technologies have been adopted by wind turbine manufacturers based on

GB19963(2011), including reducing input power of the wind turbine side converter by controlling the pitch angle or limiting wind turbine electric-magnetic power [8], eliminating the unbalanced power on the DC bus by installing power dumping circuit or energy storage devices [9], and enhancing grid-side converter output power by using var compensation devices, adding auxiliary converter, or adopting reactive power control strategy [10]. Though those measures can improve the LVRT capacity the wind turbine and wind farm fault characteristics can be changed [11], which potentially lead to misoperation and malfunction of their protection [12–14].

PMSG is an important type of wind power generator in China [15], and its market share has significantly increased in the last few years. For the protection coordination of PMSG, collector circuits and outgoing lines, this paper proposes a comprehensive and improved protection method of PMSG based wind farms with LVRT capability. The impact of various LVRT techniques on wind farm fault characteristics is considered. The rest of the paper is

* Correspondence: taorenfeng@126.com
College of Electrical Engineering of Xinjiang University, Xinjiang, China

organized as follows. Section 2 analyzes and summarizes wind farm fault features and the research status on protection configuration. Section 3 proposes the LVRT scheme consisting of variable resistance dumping circuit, grid side dynamic reactive power control and reactive power compensation control, and analyzes the PMSG wind farm fault characteristics. In Section 4, a PMSG based wind farm in Dabancheng, Xinjiang, China is used as an example to analyze the typical wind farm protection configuration, its setting values considering LVRT requirements and its existing issues on the coordination among its protections. In Section 5, an improved wind farm protection coordination methodology is proposed and its validity is verified by simulation. Finally, conclusion and perspectives for future work are given in Section 6.

2 Research status of wind farm protection
2.1 Research status of wind farm fault characteristics
Existing researches have mainly focused on the criterion of fault current with little emphasis on fault characteristics of wind turbines (especially PMSG) with LVRT capability. Short circuit currents of different types of wind turbines show significant differences due to the different control strategies in [16–19], which can have a great impact on the protection setting. According to the flux balance equations and frequency-domain analysis, mathematical expressions of the short circuit current for doubly-fed induct generator (DFIG) are derived without considering the effect of the crowbar [20, 21]. However, the short circuit current characteristics will be different when crowbar circuit is used in DFIG wind turbines [22, 23]. The PMSG short circuit current does not normally exceed the rated current of the grid side converter [24]. Control process of the full power converter that quickly isolates PMSG faults is analyzed in [25], and the short circuit current flowing into the grid is shown to be controllable. Therefore, the control strategy of LVRT of PMSG can affect the short circuit current characteristics, and the protection setting and configuration.

2.2 Research status of wind farm protection configuration
At present, most of the researches focus on the protection of the collector networks and outgoing lines of wind farms with little on protection and coordination of wind farms with LVRT capability. It has shown that wind farm integration and its fault current attenuation characteristics may reduce the reliability of current protection in power grids [26, 27]. Thus, the setting of wind farm collector network instantaneous protection should not only consider its short circuit current characteristics, but also require installing directional element if necessary [28, 29]. An adaptive protection method based on short circuit current characteristics is presented in [30, 31], which counters for different wind power output, short circuit current level

and wind farm types. However, the presented researches have little consideration on LVRT control strategy for wind farm protection configuration and coordination.

3 PMSG wind farm fault ride through feature based on LVRT control strategy
3.1 LVRT methodology applied in PMSG
At present, there are three LVRT control methods used in PMSG, including reducing the generator side converter input power, eliminating the unbalanced power at the DC bus, and increasing the output power of the grid side converter. Although these methods can enhance wind farm LVRT capability, they all have noticeable shortcomings. For example, pitch angle control has slow response time and limiting the electromagnetic power of wind turbines leads to slow system recovery, discharging circuits have difficulty in heat dissipation and occupy large volumes, energy storage devices have high cost and complex control, reactive power compensation has slow response and limited capacity, and auxiliary circuits require large volumes and have high cost.

3.2 An improved and comprehensive LVRT control strategy
Based on the structure characteristics of PMSG and without additional devices, a comprehensive LVRT scheme consisting variable resistance dumping circuit, reactive power control on the grid side and reactive power compensation is proposed in this paper, as shown in Fig. 1.

The control strategy of the variable resistance dumping circuit is

$$\alpha = \frac{U_{dc}^2}{R_{d\max}(P_g-P_s)} \tag{1}$$

where α is the resistance accommodation factor of the dumping circuit, whose value range is (0, 1]. U_{dc} is the converter DC bus voltage. P_g and P_s are the generator side input power and the grid side output power, respectively. $R_{d\max}$ is the maximum value of the dumping resistor, which can be calculated as:

$$R_{d\max} = \frac{U_{dc\max}^2}{\Delta P_{\min}} \tag{2}$$

where $U_{dc\max}$ is the maximum DC bus voltage of the converter, and $\Delta P_{\min}$ is the power difference between the generator and grid side when the dumping circuit is initiated to operate.

The dynamic reactive power control regulation is shown in Fig. 2, which changes the converter power factor to control its reactive power output based on different voltage dips. The power factor function can be obtained based on the wind turbine operation data.

The reactive power compensation device is formed of thyristor switched capacitors (TSC) in parallel with

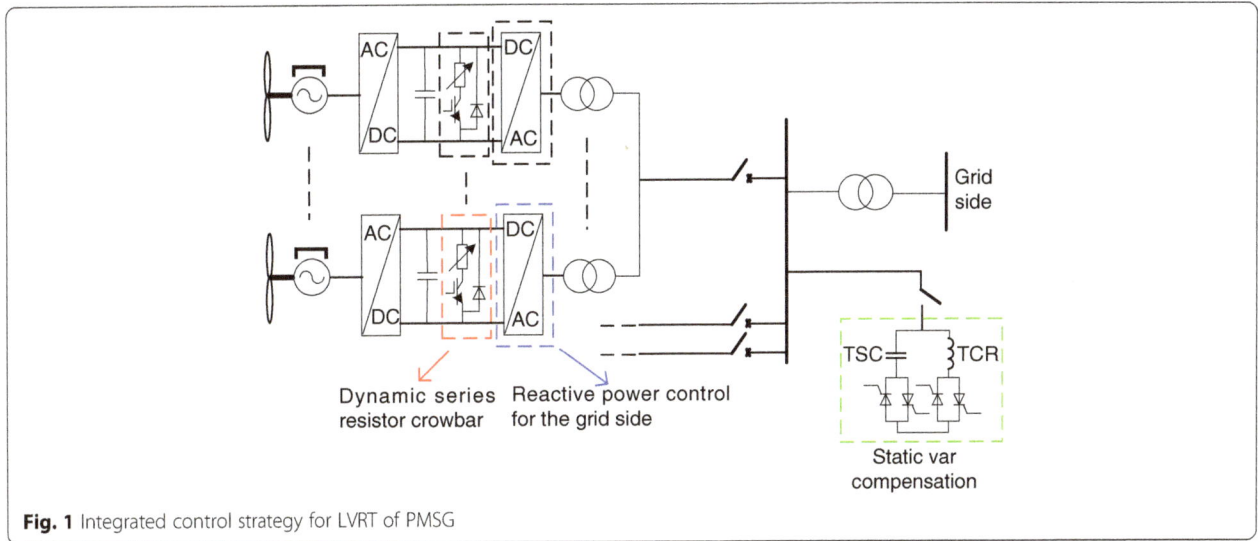

Fig. 1 Integrated control strategy for LVRT of PMSG

thyristor controlled reactors (TCR). The numbers of TSC and TCR installed can be modified according to the required reactive power compensation at certain voltage dip value and requirement to ensure grid voltage stability.

3.3 Fault ride through characteristics of wind turbines with LVRT capability

The wind farm considered consists of four 1.5 MW PMSGs and the reactive power compensation capacity is

1.5Mvar. When a three-phase fault with a duration of 0.1 s occurs on the high voltage side of the main wind farm transformer at $t = 2.0$ s, the output active/reactive power, the terminal voltage and current of the wind turbines are shown in Fig. 3. The simulation results show that: 1) The proposed LVRT method not only increases the wind farm LVRT capability but also improves the wind turbine stability after its LVRT; 2) The fault current has been significantly increased, up to 1.9 p.u., which can potentially adversely affect the protection operation and coordination as the current magnitude is usually used as the action criterion. Similar conclusions can be drawn for single phase-to-ground, two phase-to-ground and phase-to-phase faults.

4 Protection configuration and coordination analysis of PMSG based wind farm

4.1 Wind farm typical protection configuration and its coordination analysis

Figure 4 shows a typical protection configuration of a PMSG based wind farm in China, which references to the wind farm located in Dabancheng, Xinjiang, China. It has 33 LVRT enabled 1.5 MW PMSG wind turbines. All the wind turbines are connected to the grid through 3 collection lines with 35/110 kV step-up transformers. Its equivalent impedance is shown in Fig. 5.

In order to analyze the protection coordination of the wind farm, the fault location is shown in Fig. 6. K1 to K9 represent wind farm internal faults and K10 represents the wind farm outgoing line fault.

4.2 Wind turbine protection configurations based on the LVRT requirement

Wind turbine protection mainly includes voltage protection, frequency protection, current protection zone I and III, and phase unbalance protection. The protection

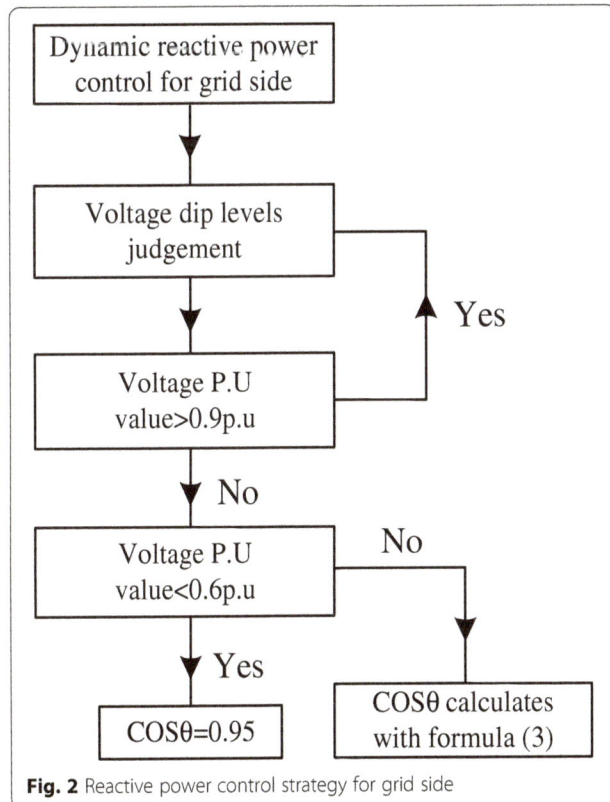

Fig. 2 Reactive power control strategy for grid side

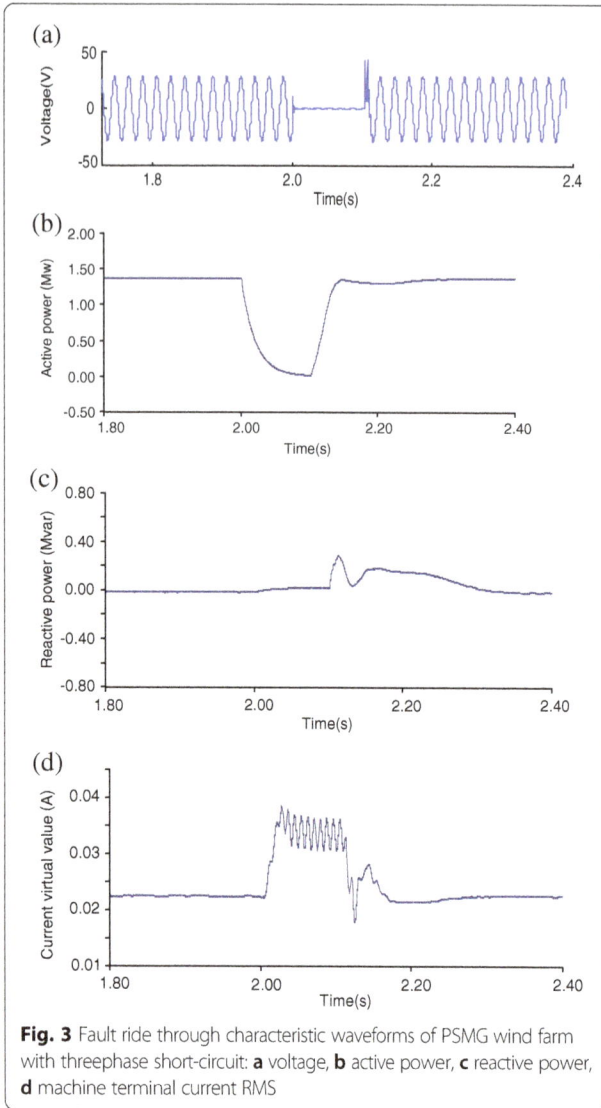

Fig. 3 Fault ride through characteristic waveforms of PSMG wind farm with threephase short-circuit: **a** voltage, **b** active power, **c** reactive power, **d** machine terminal current RMS

configuration before and after the improvement are compared in Table 1.

4.3 Protection and coordination analysis of power collection line

The protection of wind farm collection network mainly adopts three-section current protection, and its setting values are shown in Table 2.

At 1.5 s, a three-phase short circuit fault with a duration of 1.2 s occurs at K1. The RMS currents of the 3 collection lines are shown in Fig. 7. The result shows that: ① When the fault occurs, the wind turbine protection instantaneously opens the breaker at the low voltage side to cut G6 off; ② Because the terminal voltage near wind turbine G6 is lower than 0.2p.u., G14, G15, G16, G17 and G21 are also tripped off instantaneously; ③ The current flowing through protection 1 is larger than its zone I setting value, so the fault is cut off. ④The fuse cannot act instantaneously. ⑤Although the current flowing into K1 from the non-fault collector network is less than its zone I setting value, it is larger than that of zone III. When the fault line current protection zone I fails to act, the protection zone III of the non-fault collector may act to trip all its connected wind turbines.

The simulation results show that, when the fault occurred at K1, it is designed to be cut off simultaneously by the breaker at the low voltage side of G6's transformer and the fuse at its high voltage side whereas the protection 1 is only considered as the backup protection of the fuse. However, as the fuse and protection 1 is not coordinated, a large numbers of wind turbines near G6 are falsely tripped off. At the same time, the current protection zone III of the non-faulted collectors may act and trip all the wind turbines, which is not in accordance to the stable

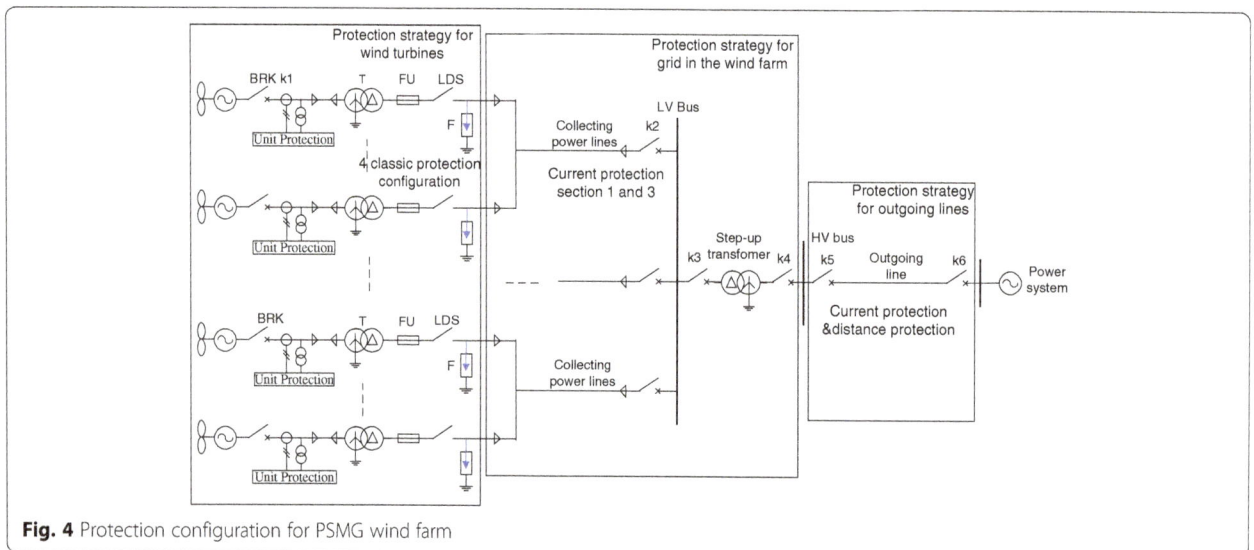

Fig. 4 Protection configuration for PSMG wind farm

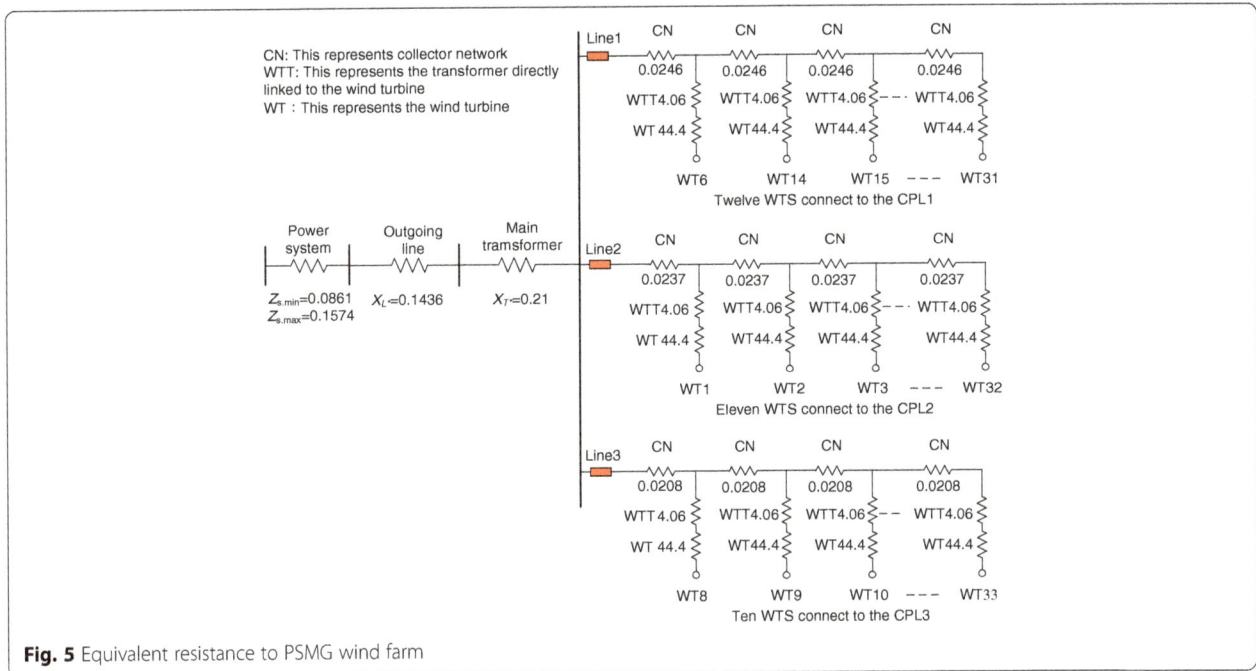

Fig. 5 Equivalent resistance to PSMG wind farm

operation of the wind farm. The problems caused by a three-phase fault at K3 and K5 are similar.

Figure 8 shows the RMS current of the collector network when a three-phase fault occurs at K2 at 1.505 s. The analysis shows that when the fault occurs, the protection zone I of collector line 1 can remove the fault instantaneously. However, there still exist two problems: ①When the fault occurs at K2, the reverse current provided by the non-fault collector network is less than the setting value of the current protection zone I, but greater than that of the zone III. If the

protection zone I of the fault collector line fails to operate, the protection zone III of the non-fault collector lines may act to trip off all the connected wind turbines. ②When the fault is close to the bus of the collector network, the generator terminal voltage difference between the fault collector and the non-fault collector will be small, and low voltage protection of all the wind turbines will act immediately, leading to the disconnection of the whole wind farm. Problems caused by the fault at K4 and K6 are similar to the case of K2.

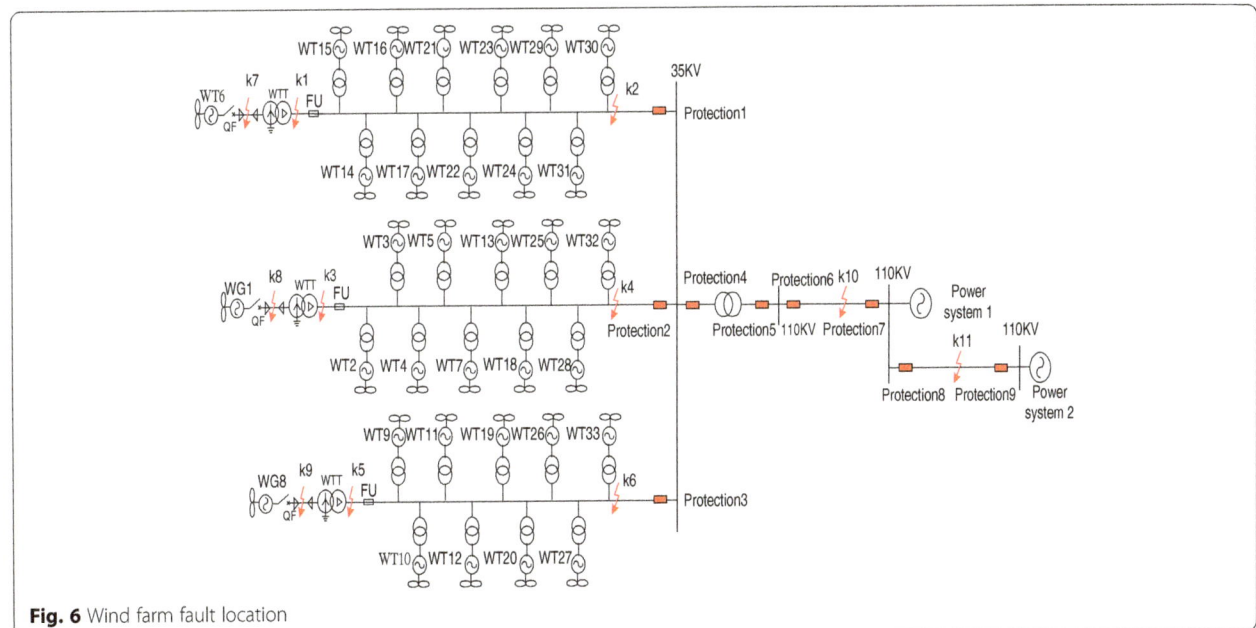

Fig. 6 Wind farm fault location

Table 1 Comparison between the conventional protection and improved protection

Protection		Without LVRT		LVRT	
		Settings	Time(s)	Settings(U_N)	Time(s)
Voltage violation protection	Upper limit protection	$1.15U_N$	0 s	$1.15U_N$	0
		$1.10U_N$	0.1	$1.10U_N$	2
	Lower limit protection	$0.9U_N$	0.1	$0.9U_N$	2.2
				$0.5U_N$	1.3
				$0.2U_N$	0.66
Frequency protection	Upper limit protection	50.5Hz	0.1	52.5Hz	2
	Lower limit protection	49.5Hz	0.1	47.5Hz	2
Over current protection	Primary section	Phase to phase fault	0	NO	
	Zone III	$2I_N$	1.5	NO	
Voltage unbalance protection	$0.1U_N$		0.3	$0.1U_N$	3
Current unbalance protection	$0.15I_N$		0.1	$0.15I_N$	1

Table 5 shows the RMS current when a phase-to-ground fault occurs at K1 ~ K6. The analysis shows that when the fault happens, the currents flowing through protection 1 ~ 3 are less than the setting values of their current protection of zone I, II and III. This will likely lead to over voltage and potentially cause the fault to become a phase-to-phase one resulting in the expansion of the accident. Thus, there requires some coordination in the collector network protection (Table 3).

Studies show that the wind farm protection coordination functions well when three-phase fault occurs at K7, K8, and K9, which are all at the low voltage sides of the wind turbine transformers.

4.4 Protection and its coordination analysis of outgoing line

Based on Figs. 4 and 5, the protection of the wind farm outgoing lines includes the zero-sequence current protection and the distance protection. Their setting values are listed in Table 4.

Table 5 shows the maximum and minimum voltages of the wind turbines when a three-phase fault occurs at K10 of the outgoing line. The fault starts at 1.5 s and is cleared at 1.7 s. The results show that: ① After the fault occurs, the outgoing line's main protection will act within 0.1 s; ② All the currents flowing through the

collector protections (the protection 1 to 3) are smaller than their setting values. There is no bad coordination. ③ All the PMSG terminal voltages are larger than 0.2pu, and thus the turbines will not trip. When a single phase-ground fault occurs at the same place, similar conclusions can be drawn.

However, it needs to be noted that the generator minimum voltage is very close to 0.2p.u. If the fault location is near the PCC (point of common coupling), the voltage will be lower than 0.2p.u. and the generator protection may act and potentially lead to the trip of the whole wind farm, which is not in accordance to wind farm operation and successful reclosure of the outgoing line. It will be beneficial for the stability of wind power integrated grid when the generator can last 0.3 s to 0.6 s at very low voltage. Meanwhile, because of the wind farm

Table 2 Three-section current protection setting of the collector network

Collector network		1	2	3
Protection I	Current(A)	1672	1800	1895
	Time(s)	0	0	0
Protection II	Current(A)	1117	1168	1252
	Time(s)	0.4	0.4	0.4
Protection III	Current(A)	440	400	370
	Time(s)	0.7	0.7	0.7

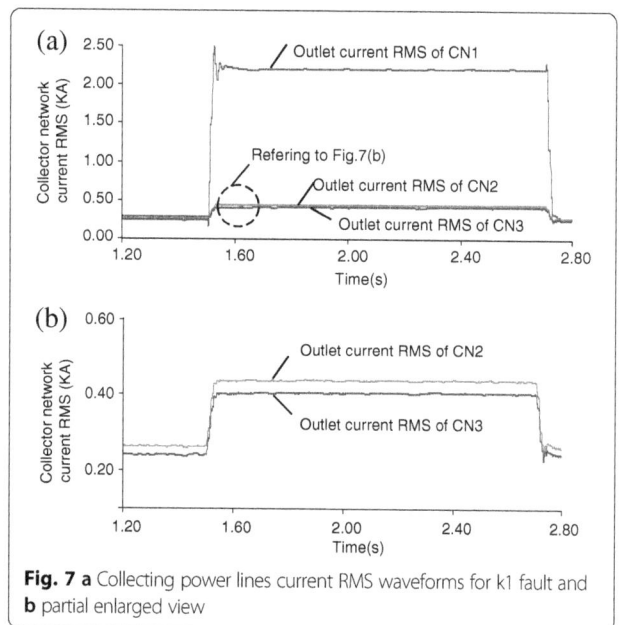

Fig. 7 a Collecting power lines current RMS waveforms for k1 fault and **b** partial enlarged view

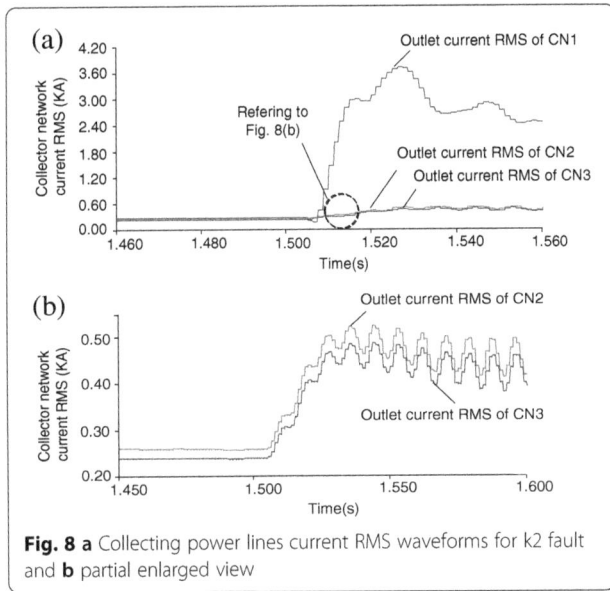

Fig. 8 a Collecting power lines current RMS waveforms for k2 fault and **b** partial enlarged view

capacity accounted for only a smaller proportion of the power system, the positive sequence and negative sequence impedances of the wind farm are far greater than that of the grid. For the unique wind farm grounding mode, its zero sequence impedance is far less than the impedances of its positive and negative sequence. When the wind farm outgoing line has a single phase-to-ground fault, zero sequence current will become the main component and consequently, the three-phase fault currents tend to become similar indicating obvious weak power characteristic (the results of two phase-to-ground faults are similar). This will seriously affect the accuracy of the phase selection element as whose criterion is current overshoot.

5 PMSG based wind farm protection coordination improvement scheme and its validation and analysis

5.1 Protection incoordination of PMSG wind farm with LVRT capability

Based on the researches in Section 4, there are 4 main issues with protection incoordination of PMSG based

Table 3 RMS current of the collector network when single-phase faults occurres at K1 ~ K6

location	Protection 1(KA)	Protection 2(KA)	Protection 3(KA)	Section I setting(A)
k1	0.369	—	—	1700
k2	0.341	—	—	
k3	—	0.352	—	1800
k4	—	0.331	—	
k5	—	—	0.321	1900
k6	—	—	0.313	

Table 4 Setting values of outgoing line protection

Protection type		Section I	Section II	Section III
Zero-sequence protection	Setting(A)	1300	736	150
	Time(s)	0	0.4	0.7
Distance protection	Setting(Ω)	15.193	32.735	56.1
	Time(s)	0	0.4	1.6

wind farms with LVRT capability: ① The current protection zone I for the collector network is in conflict with the fuse at the high voltage side of the turbine transformer; ② The reduced reliability of the non-fault collector's current protection zone III; ③ The reduced reliability due to the incoordination between the collector current protection zone I and zone III; ④ The incoordination caused by wind farm's operation characteristics.

5.2 Improvement for incoordination between collector current protection zone I and fuse at high voltage side of turbine transformer

Considering the structural characteristic of PMSG, zero voltage ride through by combining the hardware and software can be realized. The fuse generally acts within 0.1 s and the generator has to last at least 0.1 s at voltage of 0.05p.u. ~ 0. In order to realize the perfect match between the collector line current protection zone I and the fuse, a short time delay is added to the collector network current protection zone I. Figure 9 shows the improved action curve of the PMSG low voltage protection.

Furthermore, in order to solve the time incoordination problem between the collector protection I and the fuse, the difference of the fault occurred at the high voltage side of the turbine transformer and the collector line should be distinguished. The analysis shows that the fuse current for fault at transformer high voltage side is greater than the corresponding collector line protection setting value, and the current increases with the decrease of the distance from the fault point to the bus. On the contrary, when the fault occurs in the collector network, the fuse current is smaller than the protection setting value of the corresponding collector line and the current increases with the decrease of the distance from the fault point to the bus. Thus, the collector line current protection zone I setting values can be shown as below.

$$\begin{cases} \text{Main creterion}: \ U_{measure} < U_{set} \\ \text{the 1st auxiliary criterion}: \ I_{fuse} > I_{measure}, t = \left(1 + \dfrac{k \times U_C \times I_N}{U_N \times I_C}\right)t(I)s \\ \text{the 2nd auxiliary criterion}: \ I_{fuse} > I_{measure} \ I_{measure} > I_{set}, t = 0 \\ \text{backup protection}: \ I_{measure} > I_{set3} \end{cases}$$

(3)

The analysis shows that with the change of the fault location, the action time of the current protection zone I

Table 5 Maximum and minimum voltages of the wind turbines connected to the collector network during a three-phase fault

Collecting power line	Maximum machine terminal voltage(p.u)	Minimum machine terminal voltage(p.u)
1	0.296	0.220
2	0.262	0.218
3	0.261	0.217

can be adjusted adaptively. To some extent, the LVRT capability can be improved when the fault location is closer to the bus. The smaller the protection time settings are, the quicker the fault can be removed. When the fuse is not reliable, as its back-up, the collector circuit protection with an inverse time-lag characteristic can remove the fault with a certain reliability and sensitivity.

5.3 Coordination improvement of on-fault collector circuits protection zone III

A directional protective relaying is added to the collector network protection. When the current flows towards the bus direction, its output is "0", otherwise, it is "1". In order to avoid maloperation, the non-fault collector protection has the capability of locking itself by judging the current direction regardless the fault location.

5.4 Coordination improvement between protection zone I and zone III of fault collector circuits

The protection incoordination discussed here mainly shows that the single phase-to-ground fault current of the collector network is less than the setting values of its current protection zone I, II and III. The main reason is due to the lack of zero sequence current path in the collector network. In order to overcome this problem, grounding transformer and zero sequence current protection should be installed.

In this case, the grounding transformer capacity is 450kVA and its neutral grounding resistance is 20 Ω. The zero sequence current protection setting values of the respective wind farm collectors are 14.1A, 12.9A and 11.5A, (reliability coefficient is 1.5). Furthermore,

the action time should be 0.7 s, as same as the current protection zone III.

The waveforms in Fig. 10 show the difference between the former protection and the improved protection when a single phase-ground fault occurres in the collector 1 from 1.2 s to 1.5 s. Based on the analysis, the incoordination problem of the network current protection zone I and III can be solved by the grounding transformer and zero sequence current protection configuration.

5.5 Modification of protection incoordination caused by wind farm operating characteristics

The protection incoordination here mainly refers to the inadequate time cooperation among the PMSG low voltage protection, collector protection and the fuse at the turbine transformer high voltage side. Therefore, it can be improved by modifying the generator protection action curve, as shown in Fig. 11.

The simulation results show that when a three-phase fault occurs at the collector outlets (K2, K4, and K6 in Fig. 6), the generator terminal voltage always exceeds 0.05p.u and the turbines can stay connected for at least 0.1 s. These faults will be quickly isolated by the collector current protection whose action time is given in (3).

6 Conclusion

To tackle the existing coordination problem in PMSG based wind farms with LVRT capability, an improved wind farm protection coordination methodology is proposed. The proposed method adds a short time delay to the collector network current protection zone I to improve its coordination with the fuse at the high voltage side of the

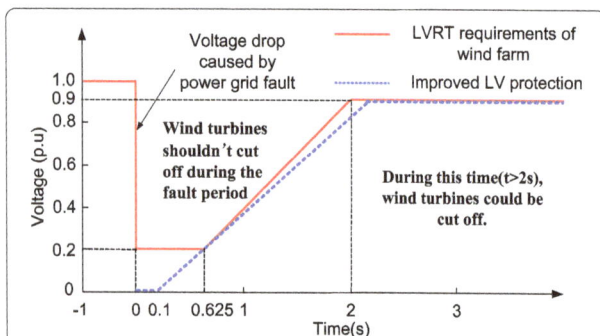

Fig. 9 the improvement low voltage motion curve of the PMSG

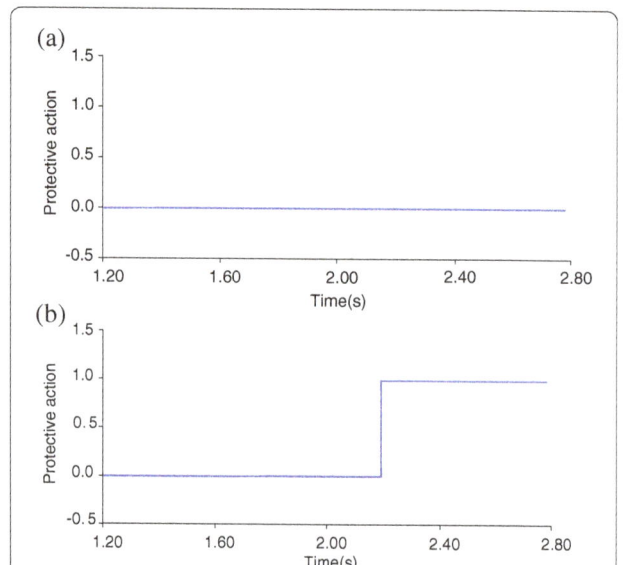

Fig. 10 a Common protective action **b** Improved protective action

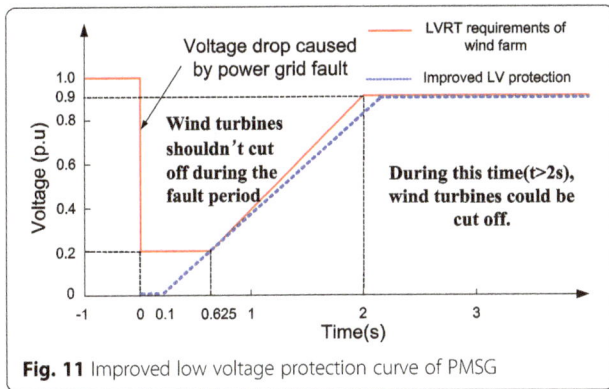

Fig. 11 Improved low voltage protection curve of PMSG

turbine transformer. Reliability improvement of the non-fault collector current protection zone III is achieved by introducing a directional protective relaying to the collector network protection, while reliability improvement of collector current protection zone I and zone III during single-phase faults are accomplished by installing grounding transformer and zero sequence current protection. Furthermore, improvement of low-voltage protection action is also achieved by modifying the generator protection action curve. Taking a PMSG based wind farm in Dabancheng, Xinjiang, China as an example, validity of the proposed method is verified.

Future work will investigate overvoltage problems caused by the turbine transformer winding breakdown, and the high voltage ride through method and protection incoordination modification potentially brought by this method.

Authors' contributions

All authors read and approved the final manuscript.

References

1. Zhou, W., Bi, D., Dai, Y., Cheng, L., & Xiong, S. (2017). Design of VSG testbed for LVRT of renewable energy [J]. *Electric Power Automation Equipment*, 37(1), 107–111.
2. Zhang, Y., Zheng, T. Q., Ma, L., et al. (2013). LVRT of photovoltaic grid-connected inverter adopting dual-loop control [J]. *Transactions of China Electrotechnical Society*, 28(12), 136–141.
3. Li, S., An, R., Xu, Z., et al. (2015). Coordinated LVRT of IG and PMSG in hybrid wind farm [J]. *Electric Power Automation Equipment*, 35(2), 21–27.
4. Zhang, M., Li, N., & Wang, Z. (2014). LVRT ability of PMSG wind power system [J]. *Electric Power Automation Equipment*, 34(1), 128–134.
5. Zhang, X., Hu, M., Wu, Z., et al. (2014). Coordinated LVRT control of wind power generation system based on VSCHVDC [J]. *Electric Power Automation Equipment*, 34(3), 138–143.
6. Khatun MF and Sheikh MRI (2016) Low voltage ride through capability enhancement of DFIG-based wind turbine by a new topology of fault current limiter[C]//2016 2nd International Conference on Electrical, Computer & Telecommunication Engineering (ICECTE), Rajshahi: pp. 1-5.
7. Carnielutti F, de Paris J, Massing JR, Pinheiro H, Tessele B (2016) A human-machine interface applied for a low-voltage-ride-through test system for grid-connected wind turbines[C]//2016 12th IEEE International Conference on Industry Applications (INDUSCON). Curitiba: pp. 1-8.
8. Ananthababu, P., Trinadha, B., & Ramcharan, K. (2009). *Performance of dynamic voltage restorer (DVR) against voltage sags and swells using space vector PWM technique[C]//International Conference on Advances in Computing Control and Telecommunication Technologies* (pp. 206–210). Karnataka: IEEE.
9. Abbey, C., & Joos, G. (2007). Super capacitor energy storage for wind energy applications [J]. *IEEE Transactions on Industry Applications*, 43(3), 769–776.
10. Morren J, Pierik JG (2004) Voltage dips ride-through control of direct-drive wind turbines[C]//39th Inter-national UPEC. Bristol: 934-938.
11. Han, L., Li, F., Wang, C., Wang, H., & Weiwei, C. H. E. N. (2016). A survey on impact of wind farm integration on relay protection [J]. *Power System Protection and Control*, 44(16), 163–169.
12. Zhang, B., Wang, J., Li, G., et al. (2012). Cooperation of relay protection for grid-connected wind power with low-voltage ride-through capability [J]. *Electric Power Automation Equipment*, 32(3), 1–6.
13. Heming, L., Dong, S., Wang, Y., et al. (2013). Coordinated control of active and reactive power of PMSG-basedwind turbines for low voltage ride through [J]. *Transactions of China Electrotechnical Society*, 28(5), 73–81.
14. Xiong, X., Zhang, H., & Ouyang, J. (2011). Simulation analysis on transient output characteristics of DFIG with SVC [J]. *Power System Protection and Control*, 39(19), 89–93. 99.
15. Fan, X., Wang, W., Xie, Y., et al. (2016). Inverter Control Switching Between Grid-connected and Islanding Operating Modes of Permanent Magnet Wind Power Generation System in Low-voltage Microgrid [J]. *Proceedings of the CSEE*, 36(10), 2270–2783.
16. Zhao, Z., Wu, W., & Wang, W. (2009). A Low Voltage Ride Through Technology for Direct-drive Wind Turbines Under Unbalanced Voltage Dips [J]. *Automation of Electric Power Systems*, 33(21), 87–91.
17. Hu, S., Li, J., & Xu, H. (2007). Analysis on the Low- voltage-ride-through Capability of Direct-drive Permanent Magnetic Generator Wind Turbines [J]. *Automation of Electric Power Systems*, 31(17), 73–77.
18. Yao, J., Liao, Y., & Zhuang, K. (2009). A Low Voltage Ride-through Control Strategy of Permanent Magnet Direct-driven Wind Turbine Under Grid Faults [J]. *Automation of Electric Power Systems*, 33(12), 91–96.
19. Lin, H., & Chao, Q. (2010). Simulation study of modeling and control of direct drive wind turbine under grid fault [J]. *Power System Protection and Control*, 38(21), 189–195.
20. Yihui, S. H. I., Zongxiang, L. U., & Yong, M. I. N. (2011). Practical Calculation Model of Three-phase Short-circuit Current for Doubly-fed Induction Generator [J]. *Automation of Electric Power Systems*, 35(8), 38–43.
21. Vicatos, M. S., & Tegopoulos, J. A. (1991). Transient state analysis of a doubly-fed induction generator under three phase short circuit [J]. *IEEE Transactions on Energy Conversion*, 6(1), 62–68.
22. Guan, H., Zhao, H., & Liu, Y. (2008). Symmetrical short circuit analysis of wind turbine generator [J]. *Electric Power Automation Equipment*, 28(1), 61–64.
23. Kawady, T., Feltes, C., Erlich, I., et al. (2010). *Protection system behavior of DFIG based wind farms for grid-faults with practical considerations [C]//IEEE Power and Energy Society General Meeting* (pp. 1–6).
24. Morren, J., Pierik, J. T. G., & de Haan, S. W. H. (2004). *Voltage dip ride-through of direct-drive wind turbines[C]//Proceedings of 39th International Universities Power Engineering Conference* (pp. 934–938).
25. Li, R., Gao, Q., & Liu, W. (2011). Characteristics of direct-driven permanent magnet synchronous wind power generator under symmetrical three-phase short-circuit fault [J]. *Power System Technology*, 35(10), 153–158.
26. Yang, G., Li, X., & Zhou, Z. (2009). Impacts of Wind Farm on Relay Protection for Distribution Network and Its Countermeasures [J]. *Power System Technology*, 33(11), 87–91.
27. Comech, M. P., Montanes, M. A., & Garcia, M. G. (2008). *Overcurrent protection behavior before wind farm contribution[C]//The 14th IEEE Mediterranean Electro technical Conference, MELECON* (pp. 762–767). Acacia: IEEE.
28. Yuling Wen (2009) Study about Wind Power Short Circuit and Impact on Power System Protection[D]. Xinjiang University
29. Su, C., Li, F., & Wu, Y. (2011). An analysis on short-circuit characteristic of wind turbine driven doubly fed induction generator and its impact on relay setting [J]. *Automation of Electric Power Systems*, 35(6), 86–91.
30. Jang, S. I., Choi, J. H., Kim, J. W., et al. (2003). *An adaptive relaying for the protection of a wind farm interconnected with distribution networks[C]//Proceedings of the IEEE Power Engineering Society Transmission and Distribution Conference* (pp. 296–302). Dallas: IEEE.
31. Wen, Y., Chao, Q., & Tuerxun, Y. (2009). Study on adaptive protection of wind farm [J]. *Power System Protection and Control*, 37(5), 47–51.

18

Time series modeling and filtering method of electric power load stochastic noise

Li Huang[1*], Yongbiao Yang[1], Honglei Zhao[2], Xudong Wang[2] and Hongjuan Zheng[1]

Abstract

Stochastic noises have a great adverse effect on the prediction accuracy of electric power load. Modeling online and filtering real-time can effectively improve measurement accuracy. Firstly, pretreating and inspecting statistically the electric power load data is essential to characterize the stochastic noise of electric power load. Then, set order for the time series model by Akaike information criterion (AIC) rule and acquire model coefficients to establish ARMA (2,1) model. Next, test the applicability of the established model. Finally, Kalman filter is adopted to process the electric power load data. Simulation results of total variance demonstrate that stochastic noise is obviously decreased after Kalman filtering based on ARMA (2,1) model. Besides, variance is reduced by two orders, and every coefficient of stochastic noise is reduced by one order. The filter method based on time series model does reduce stochastic noise of electric power load, and increase measurement accuracy.

Keywords: Electric power load, Stochastic noise, ARMA model, Kalman filter

1 Introduction

Power load operation is complex. Accurate power load forecasting has great significance for designing power supply program and making a good power balance between supply and demand. The power load sequence contains relatively obvious white noise. With longer sampling time interval, the white noise becomes more intense [1, 2]. The prediction accuracy of power load is related to the length of historical observation data. With noise and chaos in the observed data, different time series have different upper limit of prediction accuracy [3, 4]. It is important to estimate the noise intensity directly from the observed data and to separate the noise from the observed data, which is very important to improve the accuracy of the power load forecasting result.

To improve the quality of power load data, stochastic noise present in the load data must be identified and filtered out [5, 6]. At present, there are mainly following methods in the power load forecasting field, such as regression analysis, combined forecasting, exponentially smoothing, neural network and wavelet methods, and so on. Moreover, in view of the uncertainties and randomness

of short-term load, innovative data processing strategies are proposed, such as frequency domain decomposition method and property matrix hierarchical analysis method [7–9]. However, the existing time series modeling methods may not meet the requirements of time series stationary. These methods neglect the pretreatment of load data and statistical checking [10]. Independent, steady, normal, zero-mean and trend-item processing of the required data is required, and non-stationary, non-random and non-normal characteristics of power load data are needed to be tested.

Time series method and Kalman filter algorithm are proposed to filter the power load stochastic noise by pretreating and statistically testing of power load data, then, the total variance method is used to evaluate the stochastic errors of the load data before and after filtering effectively.

2 Methods
2.1 Stochastic noise time series method in power load data
2.1.1 Timing sequence processing
Traditional load forecasting method adopts the regression analysis and the least square method. However, this method is difficult to reflect the new information of the load change during the operation of the power system to the model, and the prediction accuracy is low. According to the characteristics of power load data, the statistical

* Correspondence: 15850575576@163.com
[1]NARI Technology Development CO., Ltd, Nanjing, Jiangsu Province 211106, China
Full list of author information is available at the end of the article

parameter model reflecting the running state of the system is established, and the time series of electric load is constructed. Then, the shortcomings of the existing methods can be effectively overcomed [11–13].

The time series model is used to fit the stationary, normal sequence. An auto regressive moving average (ARMA) model ARMA(p, q) with appropriate order can be used to describe the stationary stochastic process of power load. ARMA(p, q) model of a stable normal time series $\{x_k\}(k = 1, \dots, n)$ can be obtained by

$$x_k = \phi_1 x_{k-1} + \phi_2 x_{k-2} + \dots + \phi_p x_{k-p} + a_k - \theta_1 a_{k-1} - \dots - \theta_q x_{k-q} \quad (1)$$

where $\{x_k\}$ is time series, x_k is the value of the time series $\{x_k\}$ at the k-th moment, and x_k can be estimated by the value of the timing in the past periods $x_{k-1}, x_{k-2}, \dots,$ x_{k-p}, ϕ_p is autoregressive coefficient, θ_q is moving average coefficient, a_k is residual, p and q are orders of ARMA model.

The estimated error of x_k is obtained by

$$e = a_k - \theta_1 a_{k-1} - \dots - \theta_a a_{k-a} \quad (2)$$

The prerequisite for establishing the ARMA model is that the load data satisfy the requirements of stationarity and normality. Power load output data usually do not meet these requirements, then, it is necessary to make pre-processing operations and test of the corresponding characteristics for sampled data.

The first step is stationary test. The reverse order test is used to test the stationary state of the power load data sequence. If the stationarity requirement is not satisfied, the trend item extraction is carried out for the stochastic load sequence. The reverse order test method is carried out as following. $\{x_n\}$ is divided as subsequences $\{x_{j,n}\}$ with quantity of l. The mean value μ_i of each subsequence is obtained, and new subsequence is obtained with $\mu_1 \mu_2 \dots \mu_i$. With i > j, the reverse order A_j equals to the amount of $\mu_i > \mu_j$. The total reverse order number of sequences is obtained by

$$A = \sum_{j=1}^{l-1} A_j \quad (3)$$

where $j \in (0, 1, 2, \dots, l)$.

The theoretical mean and variance of the total number of reversal order are obtained as following [14]:

$$E[A] = \sum_{j=1}^{l-1} E[A_j] = \frac{l(l-1)}{4} \quad (4)$$

$$\sigma_A^2 = \frac{l(2l^2 + 3l - 5)}{72} \quad (5)$$

Statistics value u is obtained as following [14]:

$$u = \frac{\left(A + \frac{1}{2} - E[A]\right)}{\sigma_A} \quad (6)$$

If $|u| \leq 1.96$, there is no significant difference between μ_i, and $\{x_n\}$ can be determined to be a stationary sequence.

The second step is trend item extraction. The data sequence is processed by difference to get the new sequence. Data sequence subtracts the mean of the new sequence, then, obtains the mean value of the difference to complete the trend item extraction.

The third step is normality test. The power load data sequence was tested for normality [14], mainly including standard skewness coefficient ξ and standard kurtosis coefficient v.

Mean value is obtained by

$$\bar{x} = \frac{1}{n} \sum_{i=1}^{n} x_i \quad (7)$$

Variance value is obtained by

$$S^2 = \frac{1}{n} \sum_{i=1}^{n} (x_i - \bar{x})^2 \quad (8)$$

Standard skewness coefficient is obtained by

$$\xi = \sqrt{\frac{1}{6n} \sum_{i=1}^{n} \left(\frac{x_i - \bar{x}}{S}\right)^3} \quad (9)$$

Standard kurtosis coefficient is obtained by

$$v = \sqrt{\frac{n}{24} \left[\sum_{i=1}^{n} \left(\frac{x_i - \bar{x}}{S}\right)^4 - 3\right]} \quad (10)$$

$\xi \approx 0$ and $v \approx 0$ indicates stochastic sequence satisfies the normality requirement.

2.1.2 Online timing modeling

After pretreatment and statistical tests of power load data, model order and parameters also need to be calculated. In addition, the applicability of the new model still need to be tested [14, 15]. Based on the new model, the system state equation and output equation can be established, and Kalman filter method can be used to deal with the power load data.

The common method Akaike Information Criterion (AIC) for judging the order of time series models is given by

$$\text{AIC}(p, q) = n \ln \sigma_a^2 + 2(p + q) \quad (11)$$

where p and q are orders of ARMA model, n is the number of data in the sequence, σ_a^2 is the Variance of noise $a(t)$.

The AIC criterion takes into account the interaction between model order and residuals, and the smallest AIC value is to be selected.

The applicability of the model is also a critical task for online modeling of power load data. The criterion is to

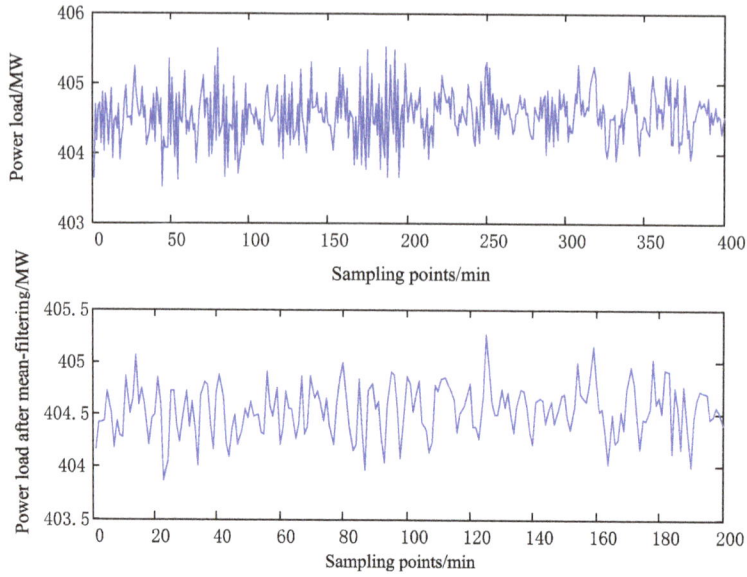

Fig. 1 Raw power load data and the data after mean filtering

check whether the model residuals are white noise. If the model residuals are white noise, the model is available; otherwise, it is not applicable.

2.2 Kalman filtering based on time series model

Kalman filtering method, a kind of effective recursive filtering method, estimates the system state according to a series of measurements including stochastic noise. Kalman filtering selects proper state space, builds state equation and measurement equation, based on the period and characteristic of load prediction. Parameter estimation

and load forecasting are implemented in the filtering, to be an organic whole.

According to the ARMA model, Kalman filtering method is adopted to suppress the stochastic noise of power load. System state equation is built by white noise of the stochastic noise of power load [16, 17].

State equation is as following:

$$X_k = AX_k + Bv_k \tag{12}$$

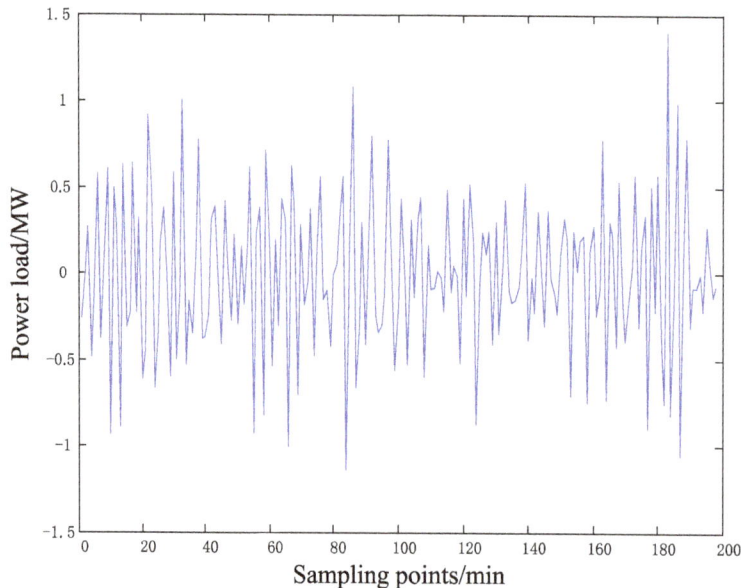

Fig. 2 Stochastic noise of electric power load after one-order differential process

Table 1 AIC values of ARMA model of power load

p	q	AIC value	p	q	AIC value
-	-	-	2	0	0.2145
0	1	−0.0268	2	1	−0.0295
0	2	−0.0265	2	2	−0.0212
0	3	−0.0144	2	3	−0.0273
1	0	0.3546	3	0	−0.0157
1	1	−0.0221	3	1	−0.0219
1	2	−0.0233	3	2	−0.0249
1	3	−0.0275	3	3	−0.0263

Assuming that W_k is estimation error of ARMA model, so there is an equation as following:

$$Y_k = X_k + W_k \tag{13}$$

System output is as following:

$$Z_k = Y_k \tag{14}$$

Output equation is as following:

$$Z_k = CX_k + W_k \tag{15}$$

The mean of both v_k and W_k is zero, white noise with constant autocorrelation function is independent of each other. The statistical properties satisfy the mean equals to zero, $E(W_k) = E(v_k) = 0$. Autocorrelation function $\phi_{vv} = R\delta_{ki}$, $\phi_{vv} - Q\delta_{kj}$, and cross-correlation function $\phi_{vw}(k, j) = 0$.

Kalman filtering equations of power load are built based on state equation and output equation, shown as following:

$$
\begin{cases}
\hat{X}_{k,k} = A\hat{X}_{k-1,k-1} \\
\hat{X}_{k,k} = \hat{X}_{k,k-1} + K_k\left[Y_k - C\hat{X}_{k,k-1}\right] \\
K_k = P_{k,k-1}C^T\left[CP_{k,k-1}C^T + R_k\right]^{-1} \\
P_{k,k-1} = AP_{k,k-1}A^T + BQ_{k-1,k}B^T \\
P_{k,k} = [I - K_k C]P_{k,k-1} \\
\hat{Y}_k = C\hat{X}_{k,k}
\end{cases}
\tag{16}
$$

where, $\hat{X}_{k,k-1}$ is further estimation of filtering state, $\hat{X}_{k,k}$ is the filtering state at the time k, $Y_k - CX_{k,k-1}$ is optimal estimate at the time k being the error between observation estimation and observation value, K_k is gain matrix of filter at the time k, R is error of system measurement noise, and Q is noise variance of system process, and $\hat{Y}_k$ is the output of filter at the time k. Initial values need to be given in advance.

Kalman prediction process is the filtering process of state reconstruction. Known from Eq.(16), estimated information $\hat{X}$ of state phasor X is updated constantly. Considering feedback unit, this part can avoid the effect of dynamic noise v_k. However, for estimation value $\hat{Y}$ of output phasor Y, it can only be approximated owing to the influence of dynamic noise v_k.

3 Result

3.1 Application and analysis of time series model and Kalman filtering

To verify the validity of time series model and Kalman filtering method of power load stochastic noise, 100 power load data of some place in 2015 is analyzed as following. After mean-filtering, power load data is able to

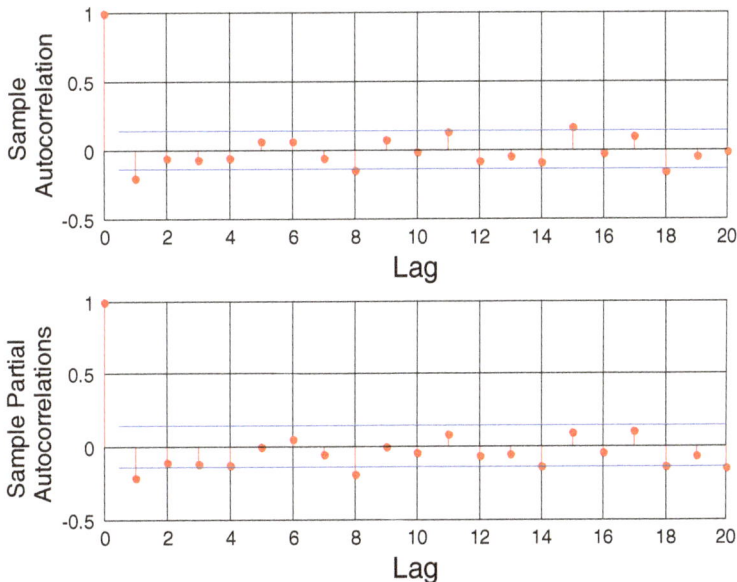

Fig. 3 ACF and PACF of model residual

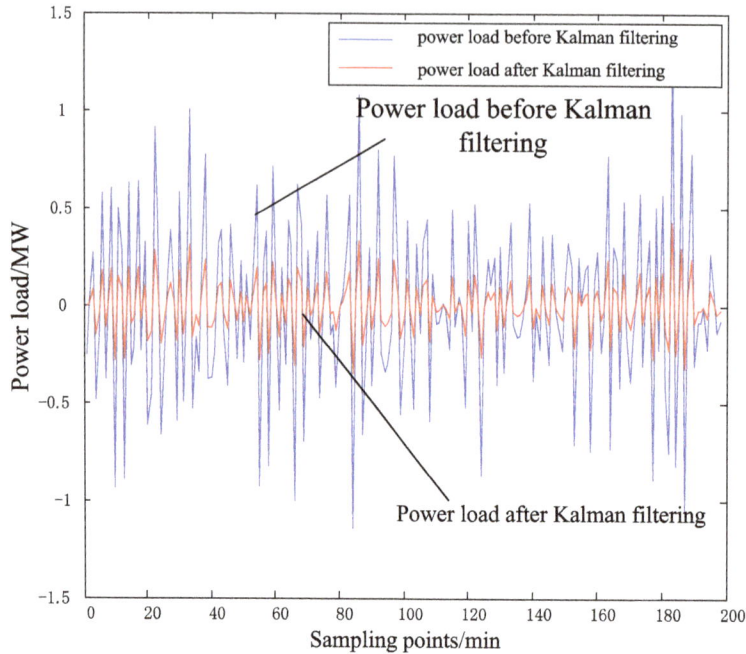

Fig. 4 Electric power load data comparison before and after Kalman filtering

effectively characterize raw power load data, both shown in Fig.1.

The first stationarity test results of power load data is as following: $|u| = 2.54 > 1.96$, which means it doesn't meet the stationarity requirement. After extracting trend, the result is this, namely $|u| = 0.5 < 1.96$, meeting the requirement. The results of normality test are as following: standard skewness coefficient $\xi = 0.0032 \approx 0$, standard kurtosis coefficient $\nu = 4.54 \times 10^{-4} \approx 0$. The new data sequence meets the normality requirements. After trend extraction, stochastic noise of power load is shown in Fig. 2. And new power load data is stationary, zero-mean and normal, satisfying the precondition of online modeling.

As for the power load suitable for modeling, AIC values are calculated. In addition, orders of ARMA model are relatively small, p and q is set to be less than 3. AIC values of chosen model are listed in Table 1.

Table 1 demonsttates that ARMA(2,1) model shall be selected for power load stochastic noise model according to the minimum AIC value, built as following:

$$x_k = \phi_1 x_{k-1} + \phi_2 x_{k-2} + a_k - \theta_1 a_{k-1} \qquad (17)$$

where, x_k is the model output, and a_k is the white noise, of which mean is 0 and variance is σ_a^2. ϕ_1, ϕ_2 and θ_1 is calculated by least square fitting (LSF)

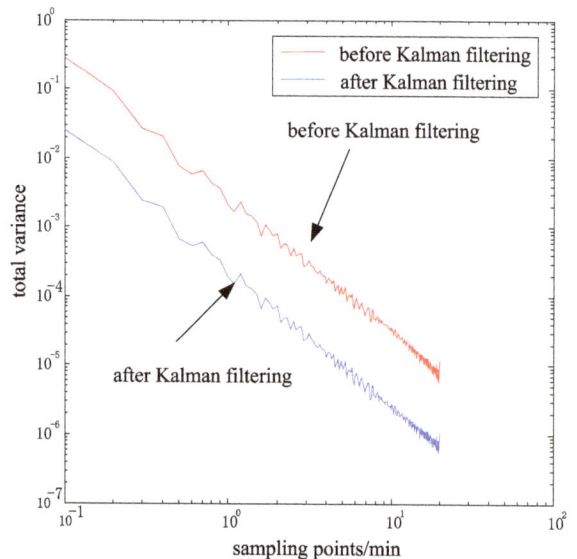

Fig. 5 Total variance analysis before and after Kalman filter

Table 2 Stochastic noise coefficients before and after Kalman filtering

Noise coefficient	Before filtering	After filtering
Q	1.50e-3	1.91e-4
L	1.01e-5	1.26e-6
B	4.74e-4	5.66e-5
K	7.50e-3	8.89e-4
R	4.31e-2	5.10e-3

$$x_k = -0.7072x_{k-1} - 0.1325x_{k-2}$$
$$+ a_k - 0.1242a_{k-1} \quad (18)$$

ACF and PACF of model residual is shown as Fig.3, and both can be regarded as white noise input.

According to established model ARMA(2,1), corresponding system state equation is obtained as following:

$$x_k = -0.7072x_{k-1} - 0.1325x_{k-2}$$
$$+ a_k - 0.1242a_{k-1} \quad (19)$$

System output equation is as following:

$$X_k = AX_{k-1} + BV_k \quad (20)$$

where, $A = \begin{bmatrix} -0.7072 & -0.1325 \\ 1 & 0 \end{bmatrix}$, $B = \begin{bmatrix} 1 & 0.1242 \\ 0 & 0 \end{bmatrix}$, $C = \begin{bmatrix} 1 & 0 \end{bmatrix}$.

Initial value of co-variance matrix P is $\begin{bmatrix} 1 & 0 \\ 0 & 1 \end{bmatrix}$, initial value of matrix X is $\begin{bmatrix} 0 & 0 \end{bmatrix}^T$, value of matrix R is variance of estimation error, and value of progress noise Q equals to $\begin{bmatrix} \sigma_a^2 & 0 \\ 0 & \sigma_a^2 \end{bmatrix}$.

Kalman filtering method is used to denoise the stochastic noise of power load data, the curves before and after filtering shown in Fig. 4. Result demonstrates that noise amplitude in the stochastic noise data is significantly reduced by ARMA model and Kalman filter. Variance before filtering is 1.56×10^{-4}, after filtering it becomes 3.58×10^{-6}, reduced by two orders of magnitude. The filtered stochastic noise is obviously suppressed.

Stochastic noise of power load is presented with different correlation time and power spectral density, and total variance method is effective to evaluate five kinds of stochastic noise of power load data before and after filtering, including load random walk (L), bias instability (B), rate ramp walk (K), rate ramp (R) and quantization noise (Q). Table 2 is each stochastic noise coefficient before and after Kalman filtering, and Fig.5. is the total variance curve of power load before and after Kalman filtering.

Known from Table 2 and Fig.5, selected power load data mainly contains quantization noise, rate random walk and bias instability. However, each stochastic coefficient in the power load data is effectively reduced through time series modeling and Kalman filtering, each coefficient value is reduced by an order of magnitude. The proposed method can eliminate the stochastic noise of power load data and promote power load accuracy.

4 Conclusion

Suppressing power load stochastic noise is one of the important links in power load modeling and forecasting.

Based on the characteristics of power load data, time series analysis is used to model the data of power load on-line, realizing the pretreatment and inspection analysis of power load data. ARMA (2,1) model is established and Kalman filtering method is used to denoise load data. And total variance method is adopted to verify the effect of modeling and filtering, namely the stochastic error coefficients before and after filtering.

The results show that stochastic noise amplitude of power load data after time series modeling and Kalman filtering is significantly reduced, the variance value is decreased by two orders of magnitude, and each stochastic error coefficient of power load is reduced by an order of magnitude. The proposed time series modeling and filtering method can effectively suppress the stochastic noise of power load data and improve the prediction accuracy of power load.

Acknowledgments
This work was financially supported by Science and Technology Project of SGCC (SGTJDK00DWJS1600014).

Authors' contributions
LH, YY and HZ mainly wrote the paper together, LH is responsible for the most of paper, including abstract, Part 1 Stochastic noise time series method in power load data and Part 3 Application and analysis of time series model and Kalman filtering individually. In addition, LH and YY are responsible for the revised manuscript, finishing the reviewers' comments. YY is responsible for the Part 2 Kalman filtering based on time series model and participates in the abstract with LH. HZ writes Part 0. Introduction and participates in the Part 4 Conclusion with YY. And helps to submit the revised manuscript in the web. XW gives advice on the paper structure and helps to check the whole paper's grammar. HZ helps to check the whole paper's grammar and words spelling. All authors read and approved the final manuscript.

Competing interests
The authors declare that they have no competing interests. And the authors certify that none of the material in the paper has been published or is under consideration for publication elsewhere.

Author details
[1]NARI Technology Development CO., Ltd, Nanjing, Jiangsu Province 211106, China. [2]State Grid Tianjin Electric Power Company, Tianjin 300010, China.

References
1. Qian, C. H. E. N., Wenying, H. U. A. N. G., Cheng, L. I., et al. (2008). Component based on-line modeling for electric loads [J]. *Automation of Electric Power Systems, 32*(2), 7–10.
2. Yang, G. A. O. (2009). *The white noise separation in "mechanism model + identification model" strategy for short time micro-grid load forecasting [D]*. Tianjin University.
3. Yundong, G. U., Sujie, Z. H. A. N. G., & Junshu, F. E. N. G. (2015). Multi-model fuzzy synthesis forecasting of electric power loads for larger consumers [J]. *Transactions of China Electrotechnical Society, 30*(23), 110–115.
4. Shuqing, Z. H. A. N. G., Rongyan, S. H. I., & Liguo, Z. H. A. N. G. (2016). Improvement of chaotic application in power forecasting model and its daily load forecasting [J]. *Chinese Journal of Scientific Instrument, 37*(1), 208–214.
5. Chuanping, X. I. O. N. G., Junjie, C. A. O., Qian, C. H. E. N., et al. (2011). A wavelet-based method for de-noising data of electric load modeling [J]. *Journal of Hohai University (Natural Sciences), 39*(4), 470–473.
6. Bai, X. I. A. O., Chao, Z. H. O. U., & Gang, M. U. (2013). Review and Prospect of the spatial load forecasting methods [J]. *Proceedings of the CSEE, 33*(25), 78–92.

7. Bai, X. I. A. O., Xiao, X. U., Kun, S. O. N. G., et al. (2013). Abnormal data identification and treatment in spatial electric load forecasting [J]. *Journal of Northeast China Institute of Electric Power Engineering, 33*(1), 45–50.

8. Shan, G. A. O., & Yuanda, S. H. A. N. (2001). A new method of load data error-correction and smoothing based on wavelet singularity detection [J]. *Proceedings of the CSEE, 21*(11), 105–108 113.

9. Xinyao, S. U. N., Xue, W. A. N. G., Jiangwei, W. U., et al. (2014). Feature weighting based hierarchical probabilistic load forecasting in distributed collaborative network[J]. *Chinese Journal of Scientific Instrument, 35*(2), 241–246.

10. Xiaoling, S. H. E. N. (2009). *Research on power system short-term load forecasting approaches [D]*. Tianjin University.

11. Yuna, Z. H. A. N. G., Quana, Z. H. O. U., Haijunb, R. E. N., et al. (2013). Application and research of improved artificial immune network to power short-term load forecasting [J]. *Journal of Chongqing University (Natural Science Edition), 36*(4), 33–38.

12. Xinran, L. I., Xuejiao, J. I. A. N. G., Jun, Q. I. A. N., et al. (2010). A classifying and synthesizing method of power consumer industry based on the daily load profile [J]. *Automation Of Electric Power Systems, 34*(10), 56–61.

13. Qi, W. A. N. G., Wenchao, Z. H. A. N. G., Yong, T. A. N. G., et al. (2010). A new load survey method and its application in component based load modeling [J]. *Power System Technology, 34*(2), 104–108.

14. LI Xiaojing CHEN Jiabin, YONG Shangguan. A method to analyse and eliminate stochastic noises of fog based on ARMA and kalman filtering method[C]. Intelligent Human Machine Systems and Cybernetics (IHMSC). 2014.

15. Charles, A., Greenhall, A. D., et al. (1999). Total variance, an estimator of long-term frequency stability [J]. *IEEE Transactions on Ultrasonic, Ferroelectrics and Frequency Control, 46*(5), 1183–1191.

16. Chicco, G., Ionel, O. M., & Prumb, R. (2013). Electrical load pattern grouping based on centroid model with ant colony clusting [J]. *IEEE Transactions on Power Systems, 28*(2), 1706–1715.

17 Liqi, W. A. N. G., Huifang, G. H., Guibin, L. I.,et al. (2014). Characteristic wavelength variable optimization of near-infrared spectroscopy based on Kalman filtering [J]. Guang pu xue yu guang pu fen xi = Guang pu, *34*(4), 958.

19

Configuration and operation combined optimization for EV battery swapping station considering PV consumption bundling

Yu Cheng[*] and Chengwei Zhang

Abstract

Integration of electric vehicles (EVs), demand response and renewable energy will bring multiple opportunities for low carbon power system. A promising integration will be EV battery swapping station (BSS) bundled with PV (photovoltaic) power. Optimizing the configuration and operation of BSS is the key problem to maximize benefit of this integration. The main objective of this paper is to solve infrastructure configuration of BSS. The principle challenge of such an objective is to enhance the swapping ability and save corresponding investment and operation cost under uncertainties of PV generation and swapping demand. Consequently this paper mainly concentrates on combining operation optimization with optimal investment strategies for BSS considering multi-scenarios PV power generation and swapping demand. A stochastic programming model is developed by using state flow method to express different states of batteries and its objective is to maximize the station's net profit. The model is formulated as a mixed-integer linear program to guarantee the efficiency and stability of the optimization. Case studies validate the effectiveness of the proposed approach and demonstrate that ignoring the uncertainties of PV generation and swapping demand may lead to an inappropriate batteries, chargers and swapping robots configuration for BSS.

Keywords: Electric vehicle, Battery swapping station, Optimal facilities configuration, Uncertainty, PV consumption bundling

1 Introduction

Electric vehicles (EVs) are being widely concerned as an environmentally friendly transportation compared to internal-combustion-engine vehicles. The government of China introduced related policies to promote the popularization of EVs [1]. Now the popular mode complementary energy for EVs is by charging. But EVs need to wait in queue if the chargers are not enough when they arrive at the charging station. It usually takes less than half an hour for fast charging or $4 \sim 8$ h for normal charging [2, 3]. Battery swapping is much more flexible compared to the battery charging, especially for Taxi and other public transportations.

As for the power system, EV BSS can charge the batteries in advance, even during the valley period, to fulfill swapping peak demand which is highly consistent with peak demand of power system theoretically. In theory, more battery reserves makes BSS much more flexible load to consume clean energy at specific period. BSS can be regard as one kind of demand response resources to bring benefits to both power system and environment. Now in China, relevant practices of battery swapping mode have been carried out in many areas. For example, Hangzhou city launched a pilot project for taxi fleets and placed battery swapping stations into operation on the high way from city center to airport, Sichuan province and Chongqing city which together called Cheng-Yu area are building the first inter provincial travel channel for new energy vehicle in China on the basis of battery swapping mode, trying to set up a green travel

* Correspondence: judychengyu@163.com
School of Electrical and Electronic Engineering, North China Electric Power University, Beijing 102206, China

demonstration zone [4]. Due to energy storage feature, BSS is able to act as supplementary automatic generation control and black-start resource [5, 6]. In spite of that, the most popular and fundamental contribution of BSS is to reducing carbon emissions, facilitate renewable integration.

Optimal strategies and schedules for BSS operation have been wide reported in literatures [7–11]. A charging strategy considering service availability and self-consumption of PV energy is proposed in [7], which proves the performance of PV-based BSS is better for both economic and environment. In [8], a detailed operation model of BSS is presented considering many different energy transfer mode like G2B, B2G, or B2B. A new optimal dispatching strategy for microgrid containing BSS is given in [9], which indicates that BSS is also able to be a kind of controllable resource to support power system. A short-term battery management and market strategy are proposed and a dynamic operation model of BSS in electricity market is built in [11].

Recent studies such as [12–15] have identified the need for combined facility configuration problem and operation strategy of BSS. The investment return of BSS is closely related to the operation strategy and schedule for daily swapping service. Therefore, it is necessary to take into account the operation strategy simulation when study on optimal configuration scheme of BSS. Swapping demand, the cost for battery, other facilities and the cost for charging should be considered synthetically while allocating the number of batteries for BSS [12]. Previous works [16, 17] used queuing theory to solve the EVs' swapping facilities configuration problem, while the expectation result is not applicable for some extreme cases. A multi-objective optimization problem for component capacity of PV based battery swapping station is solved in [13], aiming at maximizing the benefits of economy and environment. A capacity optimization method for PV-based BSS considering second-use of EV batteries is studied in [14]. The EV battery pack quantity planning problem under mode of centralized charging and unified distribution is researched and a comprehensive planning model is proposed in [15]. The power network reinforcement should be taken into consideration during the battery charging or swap station configuration process, and the battery charging/swap station model are developed in [18].

In the particular case of BSS regarded as one kind of demand response resources, various references have demonstrated that BSS has its unique advantages to handle adverse effects on distribution network with high penetration of EVs and distributed renewable resources benefitting both power system and environment [5, 6, 9, 11]. A new supplementary automatic generation control strategy using energy storage in BSSs is proposed in [5], and the

optimization model for power system restoration with support from EVs under battery swapping mode is analyzed. Research results testify it is feasible to include BSS into AGC market and power system restoration process. BSS has its own batteries to buffer the transferring of power from grid to EVs and to balance supply and demand [11], this should be commonly concerned by the system operators and load aggregators in electricity market.

Hence the demand response ability of BSS to further support renewable generation resources integration and thus system operation is closely linked with the facility configuration and operation strategy of BSS. Less attention has been paid to the facility configuration problem for enhancing BSS demand response potential, and theoretical research lags behind application which is not able to meet the demand of the development in this field [3]. Indeed the investment risk for BSS is non-negligible due to large investment scale. As a special electricity provide entity, operating profit of BSS is closely related to both swapping demand and cost for investment and operation. The investment and operation cost of BSS depends on the quantity of the batteries, chargers and the swapping robots, which affect BSS capability to meet swapping demand.

In contrast to the whole operation and capacity cost minimization modelling used to planning the batteries quantity and other facilities of BSS, a profit maximization model is developed in this study with considering how an EV BSS makes planning and operation decision considering the swapping demand, PV consumption demand and other complicated environments. The model aims at deriving the optimal solutions for components configuration in BSS and its operation schedule to recharge EV batteries. In this context, the proposed model is applied to derive evaluate the demand response potential of EV BSS which is much valuable for system operators and load aggregators in electricity market. In addition, the proposed model is able to accurately describe the operation process including issues necessary to be concerned such as customer waiting queue, which is paid less attention in previous work.

To date, there is no modeling for evaluating how facilities planning and operation for BSS can be co-optimized and simulated in order to maximize demand response profitability of BSS. Furthermore, there is currently no formal treatment in modeling the response flexibility of BSS with considering the investment scale, satisfaction of swapping service and batteries charging process. In this context, the paper attempts to contribute those problems modeling. Specifically, this paper's main contributions are:

1) The model combines both long-term facilities configuration and short-term operation strategies optimization with maximizing profit of BSS under different stochastic scenarios of BSS operation environments. Though, coordinated operation and facilities configuration optimization model for EV charging stations are already published on some studies such as [19] and [20], there are few published works specifically on swapping stations. Previous component capacity planning works for swapping station only consider and deal with battery configuration problem. The jointed configuration problem for both batteries and chargers are investigated recently, such as models in [13] and [21]. This paper further conducts on configuration problems for batteries, chargers and swapping robots together.

2) The state flow model for EV battery is proposed to describe a group of batteries status, which is quite different from the modeling method from perspective of each individual battery condition in previous works. The state flow model for EV batteries in BSS used in this paper can identify the charging power and battery refueling process clearly. Therefore, the group of batteries in swapping, charging or idle status can be precisely formulated.

3) The proposed model of BSS considers the time and the capability for swapping operation instead of assumptions that this operation is finished immediately and the swapping capability is large enough in previous works. Consequently, this paper simulates the EVs waiting queue with taking swapping ability of BSS into account, which are more in line with reality as mentioned in [22]. In addition, this paper models the EVs waiting queue constrain as a chance constrain in the optimization model to formulate customer satisfaction with probabilistic approach.

4) The PV consumption bundling requirement is included in proposed model and the profit abilities of BSS are evaluated under different PV capacities and consumption bundling requirement levels in case study. The analysis results are able to assist to instruct a new kind of business pattern for BSS to accomplish clean energy consumption target in electricity market.

The rest of the paper is organized as follows. Section 2 describes the BSS business model, proposes the state flow model to describe battery charging process. Section 3 proposes an optimization model of configuration and operation for BSS considering PV consumption bundling. And case study and some impact discussions are presented in Section 4 and Section 5,

which can provide suggestions for the grid operator to develop and recall BSS as a kind of demand response resources relieve system pressure. Finally, the conclusions are given in Section 6.

2 Problem Descriptions
2.1 Business mode for BSS
Battery swap service is available to EVs in BSS. The BSS could charge batteries by consuming PV power from PV electricity suppliers or power from the main grid during off peak time and offer swapping service by charging for fee to make profit and maintain its business. Considering TOU and the energy of PV electricity supplier can provide, BSS optimizes charging schedule in the case of satisfying customer satisfaction and PV consumption task. BSS is incentive to charge batteries during the lower price period as much as possible and consume enough PV energy to meet PV consumption requirement. This strategy can both save the electricity bills for BSS and utilize more clean energy, bringing economic benefits and increasing the efficiency of environmental reduction of EVs simultaneously.

It is worth noting that the ability BSS can serve at one time interval is influenced by both the number of EV batteries fully charged and the number of swapping robots. During high swapping demand period, EVs should wait in the queue for a short time if BSS can't provide swapping service at once.

The structure of BSS is displayed in Fig. 1 and main facilities are such as follows:

1) EV battery: the key equipment for BSS to serve EVs with swapping serve.
2) EV battery charger: the terminal to charge EV batteries in BSS.
3) Swapping robot: the machine help to swap EV batteries.
4) Control unit for optimization: the key unit for EV BSS, which is responsible for coordinate and control the equipment in BSS to optimize the operation.

2.2 Key problems for business of BSS
The profit level of BSS is mainly influenced by some key factors, they are as follows:

1) Swapping demand: Swapping demand is the key factor to affect BSS's income, including amount level and distribution characteristic. The large amount of swapping demand is, the more income of BSS is under normal condition. The distribution characteristic of swapping demand affects both income and demand response flexibility of BSS, but it is not in our study

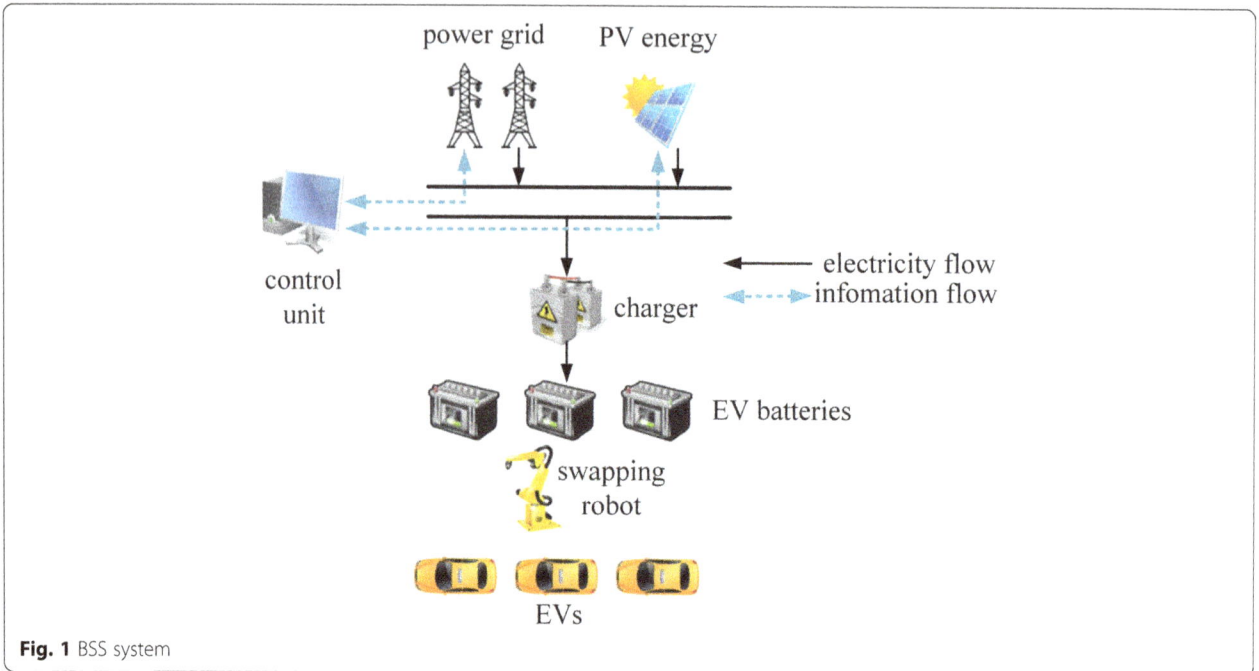

Fig. 1 BSS system

scope. More detailed discussions will be carried out in our future work.

2) Service fee for battery swapping: BSS charges for fee after finishing swapping battery for EVs. Service fee is the key point to influence the income of BSS. Income usually increases as fee increases in the case of unchanged demand. But in reality, the price elasticity of demand is not zero.

3) Electricity price level: it is the main factor influencing BSS's income, as large power consumption for charging. Generally, the lower the electricity price is, the stronger the profitability is for BSS. Considering that the price for renewable energy is usually cheaper than conventional ones, it is obviously that the more renewable energy BSS uses, the electricity bill is low, the stronger profit BSS can get.

4) Costs for facilities: As follows, EV battery, battery charger and the swapping robot are three main facilities of BSS, not a single one can be omitted. In general, the lower the cost for these facilities is, BSS can allocate more facilities to absorb cheaper electricity at the off-peak time. BSS has more decision options under this circumstance, naturally, the more profit it gets.

2.3 Fixed cost calculation model for equipment in BSS

At present, costs for facilities account for a major proportion in investment of BSS, the rationality to invest facilities should be concerned emphatically for EVs swapping demand is undefined nowadays. As for one

specific equipment n in BSS, the annual average fixed cost is as follows:

$$C_{fixed,n}^{year} = b_n \cdot \frac{r_0(1+r_0)^m}{(1+r_0)^m - 1} \tag{1}$$

Where $C_{fixed,n}^{year}$ is the annual fixed cost for facility n; b_n is the cost to buy and install facility n, m is life time of facility n, generally expressed in the number of years; And r_0 is the discount rate.

2.4 BSS customer satisfaction indexes

Customer satisfaction of BSS is closely related with the swapping service ability and fee. The quantity of the batteries, chargers and the swapping robots affect BSS capability to meet the swapping demand. The components configuration strategies model proposed in section 3 aims to enhance the swapping ability and save corresponding investment and operation cost to meet acceptable customer satisfaction.

To characterize customer satisfaction, waiting time for available batteries and EV blocking probability are proposed to evaluate service availability in previous works [21, 22]. Based on the EVs' waiting time for swapping service and waiting queue length in procedure simulation, most works evaluate the BSS operation from the temporal perspective and propose evaluation indexes such as Availability of Battery Swapping Service per Day (ABSSD) [21], queuing time cost per hour and average queuing length [16]. By the model in section 3, this paper can simulate out the length of the waiting queue at each time with maximizing the profit of BSS under acceptable customer satisfaction requirements.

The length of the waiting queue at time t L_t is used as the basic index for characterizing customer satisfaction in this paper, which is generally expressed in the number of cars and contains customers' waiting time cost information either. Based on the basic index L_t, four extended indexes to characterize customer satisfaction multi-dimensionally are defined, including maximum queue length L_{max}, average queue length $\overline{L}$, normalized average queue length for each swapping demand $\overline{L_{unit}}$, and probability of queue length longer than some value P_η, detailed calculation formulations can be got from formula (2), (3), (4), (5). Above mentioned indexes is going to be constrained to acceptable levels in the optimization model in section 3.

$$L_{max} = \max(L_1, L_2, ..., L_t, ..., L_{T-1}, L_T) \qquad (2)$$

Where L_t is the length of the queue at time t, T is the number of time interval in the research period.

$$\overline{L} = \frac{\sum_{t=1}^{T} L_t}{T} \qquad (3)$$

$$\overline{L_{unit}} = \frac{\sum_{t=1}^{T} L_t}{\sum_{t=1}^{T} n_t^d} \qquad (4)$$

Where n_t^d is the swapping demand at time t, which is expressed as the amount of EVs that arrive at BSS for swapping. $\overline{L_{unit}}$ is normalized average queue length for each swapping demand which is obtained by dividing total queue length by total swapping demand. It can characterize the average waiting time for each EV indirectly.

$$P_\eta = p(L_t \geq \eta) \qquad (5)$$

Where η is the estimated acceptable maximum queue length, $p()$ is the probability calculation function.

2.5 State flow model for EV battery

There are three states of batteries in the BSS, which are state of fully charged, being charging and waiting for charging [23]. Assumes that a complete and continuous charging process takes T_{ch} time intervals for an EV battery with capacity of C and the charging power is constant which can be calculated as $p_{ch} = C/T_{ch}$ (including the charging efficiency). A battery can be divided into

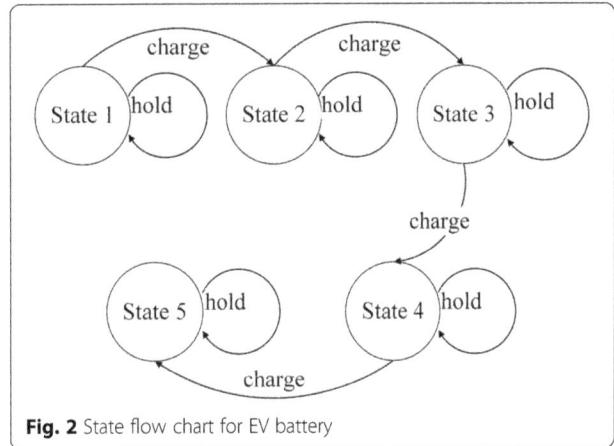

Fig. 2 State flow chart for EV battery

T_{ch} states numbered as $1, 2, 3, ..., T_{ch} - 1, T_{ch}$, according to the SOC of batteries.

With the charging process, the state of the battery changes. The state flow is shown in Fig. 2 when T_{ch} is 5. While being in charge, the state would transfer from state 1 to state 5 constantly, until reaching state 5, means fully charged in this example. The state would continue to be same while in idle state.

3 Methods

3.1 Objective function

The planning model objective for a BSS is to maximize the net profit for providing battery swapping services, which can be calculated by income, fixed cost and variable cost. The income includes the swapping service fees. The fixed cost includes the expenses to allocate facilities like batteries, chargers and swapping robots. The variable cost includes the energy charge paid for the electricity which is used to charge batteries mainly, these electricity can come from power system or the PV supplier.

Due to the uncertainties of future EV swapping demands and output abilities of PV energy, a set of potential future planning scenarios (Ω) should be forecasted and established to ensure the availability and flexibility of planning model strategy.

The aim of BSS is trying to maximize the annual profit. So the objective function is formulated as follows:

$$Max f = Max \left\{ 365 \cdot \frac{\sum_{\omega \in \Omega} (R - C_{variable})}{N_\Omega} - C_{fixed}^{year} \right\} \qquad (6)$$

The first term in objective function is the total annual equivalent operation profit under different stochastic scenarios and the second term is the annual equivalent investment cost.

Where R is the daily income during the specific operation period, C_{variavle} is the daily variable cost such as operation cost to purchase energy and C_{fixed}^{year} is the annual fixed cost of BSS such as investment cost.

The annual fixed cost can be calculated as follows:

$$C_{fixed}^{year} = \text{n}^s \cdot c^{barttry} + n^{\text{charge}} \cdot c^{\text{charge}} + \text{n}^{robot} \cdot c^{robot} \qquad (7)$$

Where n^s is the total number of batteries stored in BSS, $c^{barttery}$ is the annual fixed cost for an EV battery, n^{charge} is the number of chargers in BSS, c^{charge} is the annual fixed cost for a charger, n^{robot} is the number of swapping robots in BSS and c^{robot} is the annual fixed cost for a swapping robot. The annual fixed cost for a battery, a charger and a swapping station can be calculated with formula (1) separately.

The daily income during operation period is as follows:

$$R = \sum_{t=1}^{T} n_t^d \cdot c^{service} \qquad (8)$$

Where $c^{service}$ is the service fee BSS charges for supplying the swapping service.

The daily operation cost can be formulated as follows:

$$C_{\text{variable}} = \sum_{t=1}^{T} \left(P_t^{sys} \cdot c_t + P_t^s \cdot c^{solar} \right) \qquad (9)$$

Where P_t^{sys} is the power purchased from the system at time t, and c_t is the system electricity price at time t, P_t^s is the power purchased from PV supplier at time t, and c^{solar} is the price of solar energy.

3.2 Constrain conditions

Since future EV swapping demand and PV output are both uncertain. Swapping demand can be influenced by various stochastic factors such as work/weekend days, personal travel arrangements, etc. [19]. PV output can be influenced by weather conditions especially solar radiation intensity. Therefore, operation simulation procedure is necessary under enough scenarios and following constrain conditions need to be satisfied under each operation scenario.

1) Swapping demand balance

Typically, EVs battery can be replaced at once if there are full charged batteries. If not, the EVs will wait into the queue until there is a battery ready for swapping.

$$n_t^{service} + n_t^{wait} = n_t^d \qquad (10)$$

Where $n_t^{service}$ is the number of EVs whose battery can be replaces at time t, n_t^{wait} is the number of EVs that should be taken into the queue. It means EVs get into the queue when it is positive, and get out of the queue when it is negative.

2) Number of EV batteries constraint

The number of batteries at different states in BSS is equal to the total number of batteries stored in BSS.

$$\sum_{i}^{T_{ch}} \left(n_t^i + n_t^{i0} \right) + n_t^s = n^s$$

$$\qquad (11)$$

Where n_t^i is the number of charging batteries at state i at time t, n_t^{i0} is the number of idled batteries at state i at time t.

3) Number of EV batteries at each state equation constraint

The following constraint should be satisfied during the swapping process:

$$n_t^{di} + n_t^{di0} + n_t^{wi} = nd_t^i \qquad (12)$$

Where n_t^{di} is the number of batteries in charging at state i replaced from EVs at time t, n_t^{di0} is the number of idle batteries at state i replaced from EVs at time t, n_t^{wi} is the number of batteries at state i that enter into or quit from the queue, nd_t^i is the number of batteries at state i in the swapping demand.

4) Waiting queue length constraint

The length queue at any time can be calculated as follows:

$$n_t^{queue} = \begin{cases} n_{t-1}^{queue} + n_t^{wait} & (2 \leq t \leq T) \\ 0 + n_t^{wait} & (t = 1) \end{cases}$$

$$\qquad (13)$$

$$n_t^{qi} = \begin{cases} n_{t-1}^{qi} + n_t^{wi} & (2 \leq t \leq T) \\ 0 + n_t^{wi} & (t = 1) \end{cases}$$

$$\qquad (14)$$

Where n_t^{queue} is the number of EVs waiting in the queue at time t, n_t^{qi} is the number of batteries at state i in the EVs waiting in the queue at time t.

5) Number of swapped EV batteries into storage constraint

The number of batteries at state i in swapping demand is equals to the number of batteries at state i and will be in charging at next time interval replaced from EVs plus the number of batteries at state i and will be in idle at next time interval replaced from EVs.

$$\sum_{i=1}^{T_{ch}} \left(n_t^{di} + n_t^{di0} \right) = n_t^{service} \qquad (15)$$

6) Number of batteries in charging constraint

Anytime, the number of batteries in charging should be less than the number of chargers.

$$\sum_{i=1}^{T_{ch}} n_t^i + n_t^{di} \leq n^{charge} \qquad (16)$$

7) Number of charging batteries at different state constraint

The number of batteries in charging at state i can be calculated as follows:

$$n_t^i = n_{t-1}^{i-1} + n_{t-1}^{di-1} + nt_{t-1}^{i0} \qquad (2 \leq t \leq T) \qquad (17)$$

Where nt_{t-1}^{i0} is the number of batteries at state i which is in idle at time t-1 and will be in charging at time t.

8) The number of idle batteries constraint

The number of batteries in idle at state i can be calculated as follows:

$$n_t^{i0} = n_{t-1}^{i0} + n_{t-1}^{di0} - nt_{t-1}^{i0} \qquad (2 \leq t \leq T) \qquad (18)$$

9) The number of full charged batteries constraint

The number of full charged batteries can be calculated as follows:

$$n_t^s = n_{t-1}^s + n_{t-1}^{T_{ch}} + n_{t-1}^{dT_{ch}} - n_{t-1}^{service} \qquad (2 \leq t \leq T) \qquad (19)$$

Where n_t^s is the number of full charged EV batteries at time t.

10) EV battery start and end time state balance

The SOC and the states of EV batteries at the end of the day should be the same as it was at the beginning of the day for the cycling operation pattern of BSS.

$$n_1^s = n_T^s \qquad (20)$$

$$n_1^i = n_T^i \qquad (21)$$

$$n_1^{i0} = n_T^{i0} \qquad (22)$$

11) Power balance constraint

The power purchased from the PV electricity supplier plus that one purchased from the grid should equal to the charging power EV batteries need.

$$P_t^s + P_t^{sys} = P_t \qquad (23)$$

Where P_t is the charging power at time t.

12) PV output constraint

The power purchased from the PV electricity supplier should not be greater than the power PV electricity supplier generates.

$$P_t^s \leq P_t^{solar} \qquad (24)$$

Where P_t^{solar} is the output of PV energy at time t.

13) Maximum service ability constraint

The ability BSS can serve at one time interval is influenced by both the number of EV batteries fully charged and the number of swapping robots. And the constraint below should be satisfied.

$$n_t^{service} \leq \min \left(n_t^s, \left[t_{unit} / t_{change} \right] \cdot n^{robot} \right) \qquad (25)$$

Where n^{robot} is the number of swapping robots in BSS, t_{unit} is the time interval in the research, t_{change} is the time to swap for each EV, [ ] is the rounding operator.

14) Variables constraints

Some decision variables of the model proposed should meet constrains such as follows:

$$n_t^i, n_t^{i0}, n_t^{di}, n_t^{di0}, n^s, n^{charge}, n^{robot}, \qquad (26)$$

$$n^{solar}, n_t^{service}, n_t^{queue}, n_t^{wi} \in N$$

$$n_t^{wait}, n_t^{qi} \in Z \qquad (27)$$

15) PV consumption bundling constraint

BSS needs to consume a certain amount of PV output in a certain time period like a month or a year from bundling PV seller.

$$\sum_{\omega \in \Omega} \sum_{t=1}^{T} P_t^s \geq cap^{solar} \cdot \sum_{\omega \in \Omega} \sum_{t=1}^{T} P_t^{solar} \qquad (28)$$

Where cap^{solar} is the PV consumption percentage requirement BSS needs to be satisfied.

16) Customer satisfaction

The probability that customer satisfaction meet the specific level is allowable in reality but should reach an acceptable level. As mentioned in 2.4, four customer satisfaction indexes are set up to the threshold, including maximum queue length L_{max}, average queue length $\overline{L}$, normalized average queue length for each swapping demand $\overline{L_{unit}}$, and probability P_η of queue length longer than some value, see formula (2), (3), (4) and (5).

The above operation condition can be expressed as deterministic and chance constraints as follows:

$$L_{max} \leq \alpha \qquad (29)$$

$$\overline{L} \leq \beta \qquad (30)$$

$$\overline{L_{unit}} \leq \delta \qquad (31)$$

$$P_\eta \leq \varepsilon \qquad (32)$$

Where α, β, δ, ε are the parameters set to meet the expected customer satisfaction requirement.

4 Case study and Results

4.1 Basic stats

This study assesses the individual effect of PV output uncertainty and swapping demand uncertainty. Some critical factors have great influence on battery swapping demand level [24], but it is out of the scope of this paper. We assume that the battery swapping demand is under well forecasting [25]. The typical battery swapping demand curves on workdays, weekends and in festival days are shown in Fig. 3 [26], and the probability for above three typical demand curves are 68.4%, 23.6% and 8.0%. The total daily demand number is around 600. There are two swapping demand peaks on workdays and weekends. In festivals, the swapping demand is concentrated and large during the day time. It is assumed that the energy state-of-charges (SOCs) of EVs that come to BSS for battery swapping service are uniformly randomized between state 2 to state 4 [7], and the total daily energy refueling demand is around 7,500 kWh. The replaced EV batteries will be placed in the storage and fully recharged at a specific time to be available for other EVs. And we assume that only the fully charged EV batteries can be used for swapping.

Typical daily output curves of PV in different seasons are as shown in Fig. 4, which are the typical outputs in Spring, Summer, Autumn, and Winter when sunshine condition is good, and the probability for each seasonal typical outputs are all 25%. It is noted that the short-term fluctuations for both swapping demand and PV output are also taken into account in our study. System parameter values are depicted in Table 1, which are based on the data from open channels or some inquiries. As a main facility in BSS, EV battery accounts for a large part of the cost of one electric vehicle. The key problem while operating BSS is to use and maintenance batteries properly. One simple method is to prevent charging and discharging frequently so as to reduce the number of switching state of the batteries [27], and we assume that the charging process is continuous without interruption. Now China is under the initial

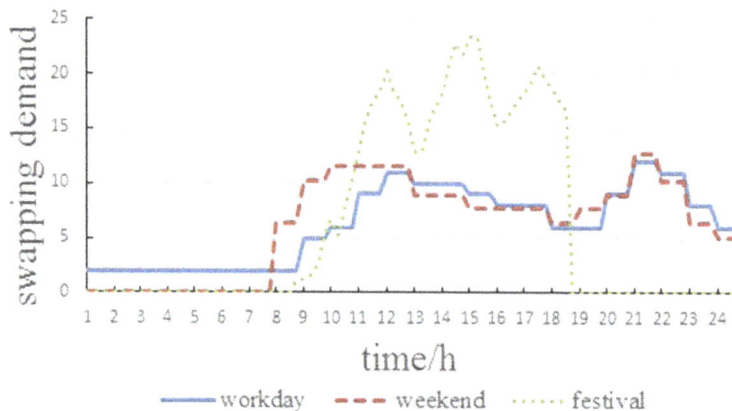

Fig. 3 Typical battery swapping demand curves on workdays, weekends and in festival days

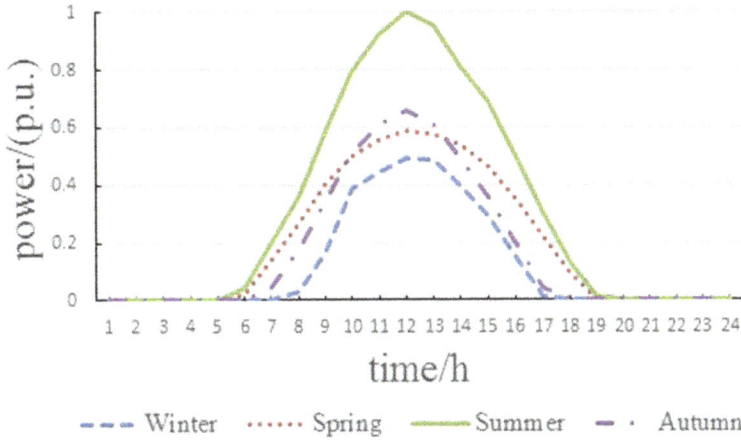

Fig. 4 Typical daily output curves of PV in different seasons

stage of establishing competitive electrical market, real time price mechanism is not yet implemented, and it will need a long time to be perfect. It should be noted that time-of-use price mechanism and related practices were carried out in most cities of China, some experience has been accumulated, so TOU mechanism is adopted in the case study. TOU reflects peak and valley conditions of power system, which encourages EV BSS to make contribution to relieve system pressure. Assuming that the BSS is treated as industry/commerce customer connected to 10 kV distribution network, the electricity price for this kind of electricity customer references to the TOU of Beijing city in China and is provided in Table 2. The daily research period is 24 h, each time interval is 15 min. Customer satisfaction index threshold α =6, β =0.25, δ =0.25, η =3 and the related

probability level ε is 10% respectively. The PV consumption percentage requirement cap^{solar} =80%. Considering that the correlation between PV output uncertainty and swapping demand uncertainty is low, the probability of combined scenarios can be calculated individually. The results of our proposed model are compared with the model without PV consumption bundling. The collaborative optimization configuration and operation decision for BSS in one day can be solved. The model is a mixed-integer linear programming (MILP) model implemented in GAMS 24.4.0 and solved using CPLEX.

4.2 Results

The optimal configuration solutions under scenarios with and without PV bundling are displayed and compared in Table 3. The total energy consumption are

Table 1 Values of system parameters

No.	Parameter	Value
1	PV installed capacity/(kW)	1,000
2	PV energy price/(yuan/kWh)	0.65
3	Battery capacity/(kWh)	20
4	Charging power/(kW)	10
5	Charging efficiency/(%)	88
6	Swapping service fee/(yuan)	25
7	Cost for EV battery/(yuan/each)	83,000
8	Cost for charging charger/(yuan/each)	6,000
9	Cost for swapping robot/(yuan/each)	160,000
10	Robot swapping capability/ (service times/each time interval)	4
11	Life time of each facility/(year)	5
12	Discount rate/(%)	6

Table 2 Time-of-use price

Levels	Time period	Price/(yuan/kWh)
Peak time	10:00–15:00, 18:00–21:00	1.3782
Mean time	7:00–10:00, 15:00–18:00, 21:00–23:00	0.8595
Valley time	23:00–7:00	0.3658

Table 3 The optimal configuration solutions under scenarios with and without PV bundling

PV bundling/(%)	Number of robots	Number of batteries	Number of chargers	Profit/ (yuan)
Without PV bundling (0%)	6	101	125	573,679.7
With PV bundling (90%)	6	101	125	520,679.7

scenario with PV bundling. The charging power at mean time (such as 15:00–18:00) and peak time period (such as 18:00–21:00) is not low enough because it is more economic to pay for high electricity price than configuring more EV batteries to improve its flexibility. BSS needs to absorb PV energy at a certain period due to PV only generates energy at a given period (around 7:00–19:00 in Summer), this sacrifices the profit level.

In the meantime, we could find that the charging power from late night to early morning is still at a high level under scenario without PV bundling. Coincidently, the output of wind power is large in this time period coincidently according to many research reports. This indicates that the BSS also has the potential to absorb wind power.

The customer waiting queue lengths are depicted in Fig. 6. Under both scenarios with and without PV bundling, EVs have to wait in the queue during the second swapping demand peak time at workdays. The queue is due to BSS can charge batteries during mean and valley time price period when second swapping demand peak comes to save electricity payment. But the queue problem can be solved by setting a suitable customer satisfaction levels.

The PV consumption curves are shown in Fig. 7. The BSS consumes 90% energy as PV output (capacity: 1,000 kW) under scenario with PV consumption bundling requirement, that is the majority of power that PV generates. It is noteworthy that BSS indeed has the potential to absorb PV energy in this condition.

The redundancy of the batteries are shown in Fig. 8. It appears that BSS charges the most of batteries in the off-peak time of power system and stores them for the following swapping peak demand especially under scenarios without PV bundling. The number of fully charged batteries declines rapidly as swapping demand rises.

different under scenarios with and without PV bundling, the gap is due to the difference of optimal solutions for battery SOC at initial and end period. The little gap does not affect the validity of our research. The annual profit under scenario with PV bundling is less than ones under scenario without PV bundling, because PV consumption bundling requirement sacrifices the operation flexibility of BSS and makes annual profit lower than optimal strategy without bundling. In addition, more PV electricity consumption can get more environmental benefit, but that has not been included into our model yet. The robot configuration number are same for these 2 scenarios, because the swapping demand level at one time interval is the key impact on deciding how many swapping robots are necessary.

The charging power curves are displayed in Fig. 5. Under scenario without PV bundling, the charging power curve is much smooth than the curve under

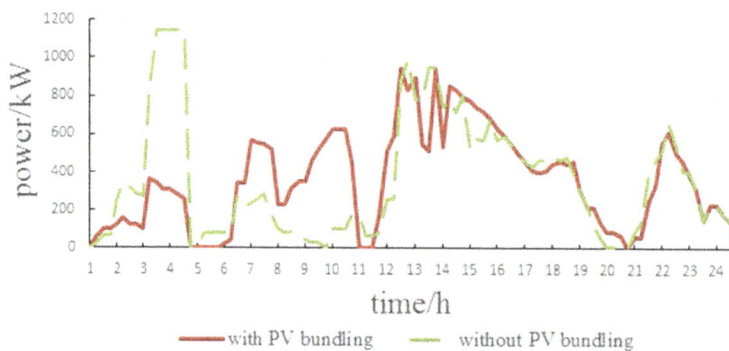

Fig. 5 Charging power curves with and without PV bundling requirements at a workday in Summer

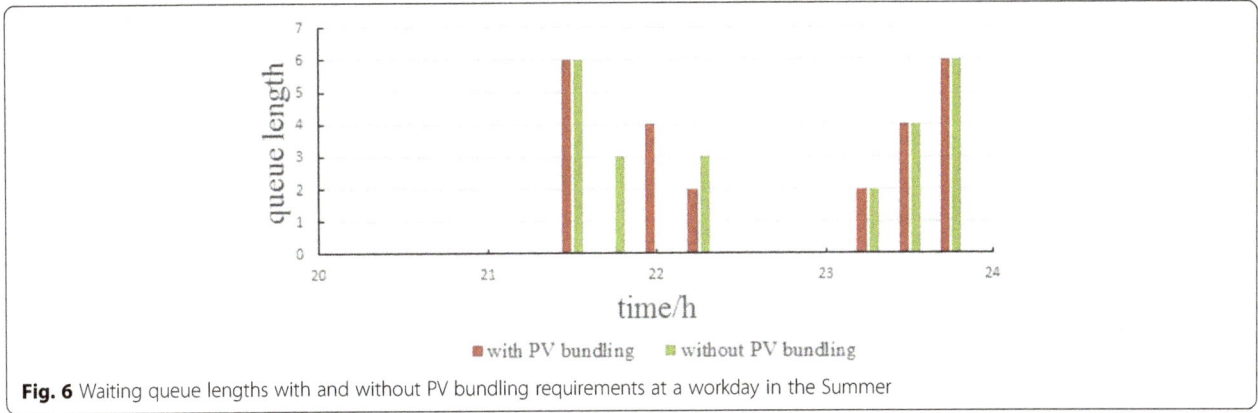

Fig. 6 Waiting queue lengths with and without PV bundling requirements at a workday in the Summer

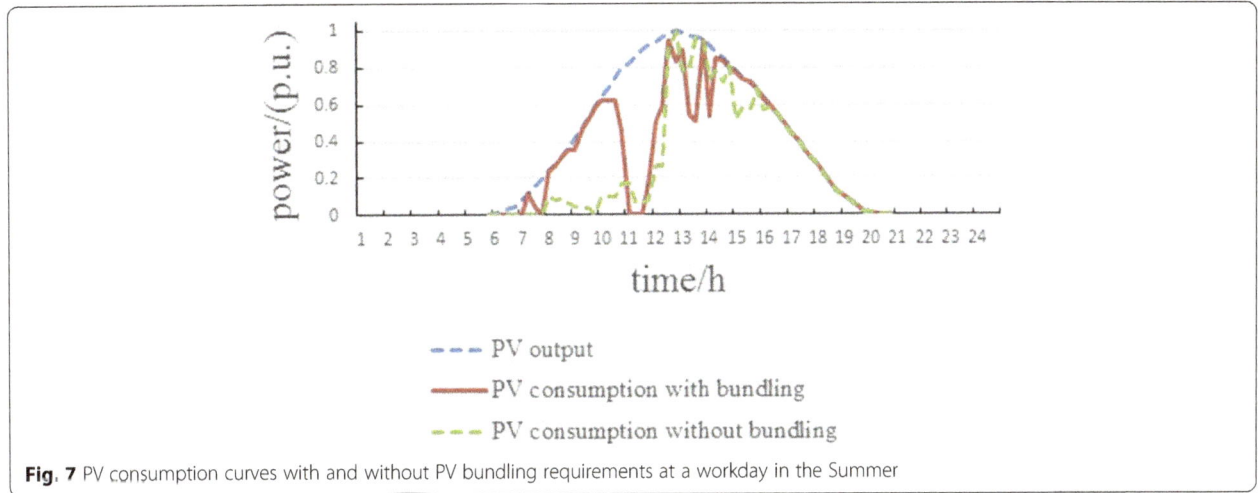

Fig. 7 PV consumption curves with and without PV bundling requirements at a workday in the Summer

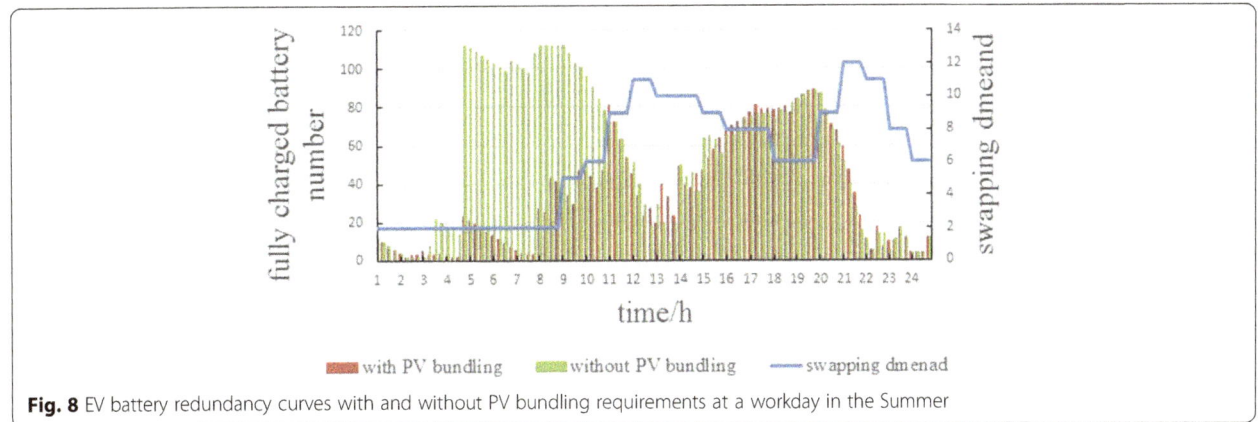

Fig. 8 EV battery redundancy curves with and without PV bundling requirements at a workday in the Summer

5 Discussions

EV BSS is the participator in both electricity system and traffic system, the decision of BSS is influenced by conditions in both two systems. The sensitive analysis is proposed from the aspects of economy and technology.

5.1 Impact analysis to BSS demand response capability with different BSS configuration structures

BSS has two demand response capabilities, one is to respond system TOU price for peak shaving and valley filling, and the other is to consume clean energy that PV or wind power generates. The former is beneficial to power system operation and the latter is environmental. The EVs number coming to refuel is fixed for a specific BSS, which determines daily energy demand. So there is a trade-off between system TOU price response capability and clean energy consumption response capability.

The investment cost accounts for a relatively large expenditure during the life cycle of BSS. Considering the costs for each battery, charger are likely to reduce in the following five years. The impact of this reduction will be discussed in this part. We do not include the swapping robot into discussion because the determinants for BSS to increase the number of robots are the swapping demand level and customer satisfaction level. At the same time, considering that the cost for a battery is much more expensive than the cost for a charger, so only the cost of each battery is selected into discussion here. The charging power curves, system charging curves and PV

consumption power curves with different costs of each battery are depicted in Figs. 9, 10, 11 and the optimal configurations and energy consumptions of BSS are shown in Table 4.

It can be concluded that with the cost of each battery decreases, BSS reschedule the charging arrangement. The charging power is high at valley time period and small at peak time period. The effect of peak shaving and valley filling becomes obvious when battery cost is less than 45,000. While the battery cost is 15,000, the number of batteries and chargers change from 125 and 183 (scenario with 83,000 yuan for each battery) to 101 and 73, which resulting in the great improvement of capability for BSS as demand response resource to response the TOU.

It can be concluded that with the decline rate of battery's price increases, BSS reschedule the charging arrangement. The charging power is high at mean time period and small at peak time period. The effect of peak

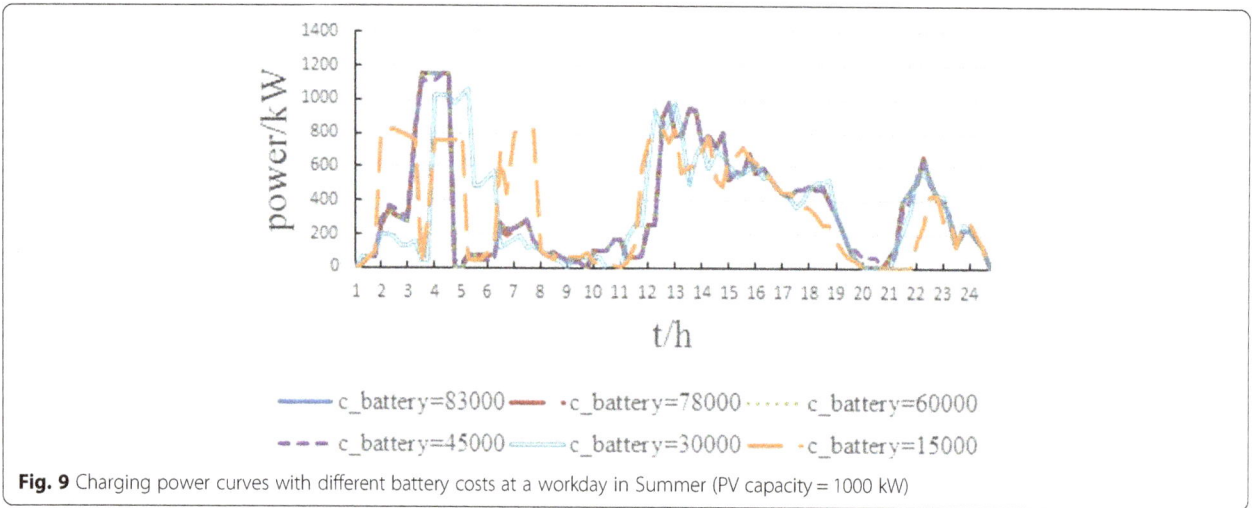

Fig. 9 Charging power curves with different battery costs at a workday in Summer (PV capacity = 1000 kW)

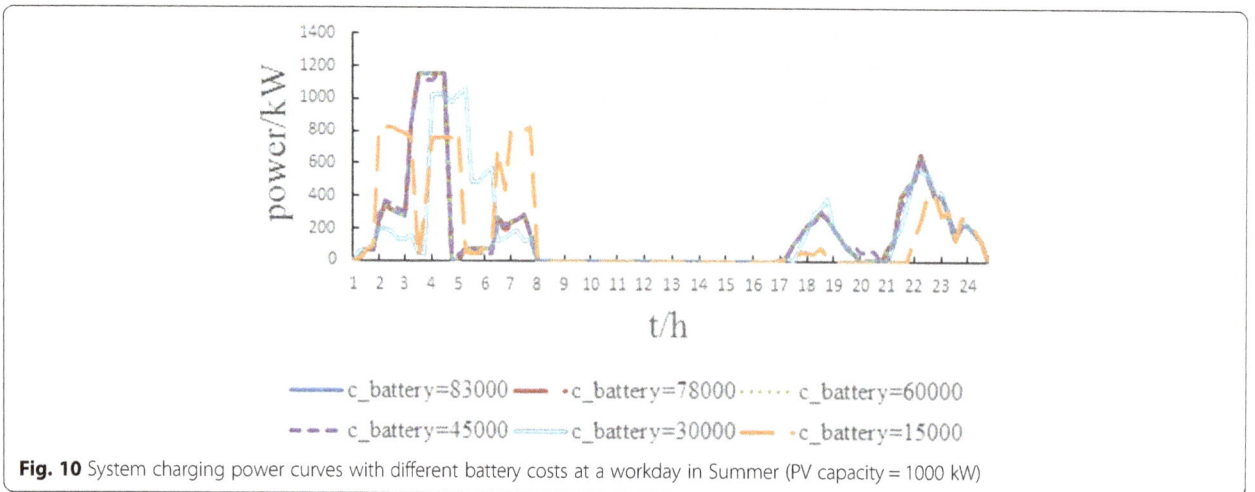

Fig. 10 System charging power curves with different battery costs at a workday in Summer (PV capacity = 1000 kW)

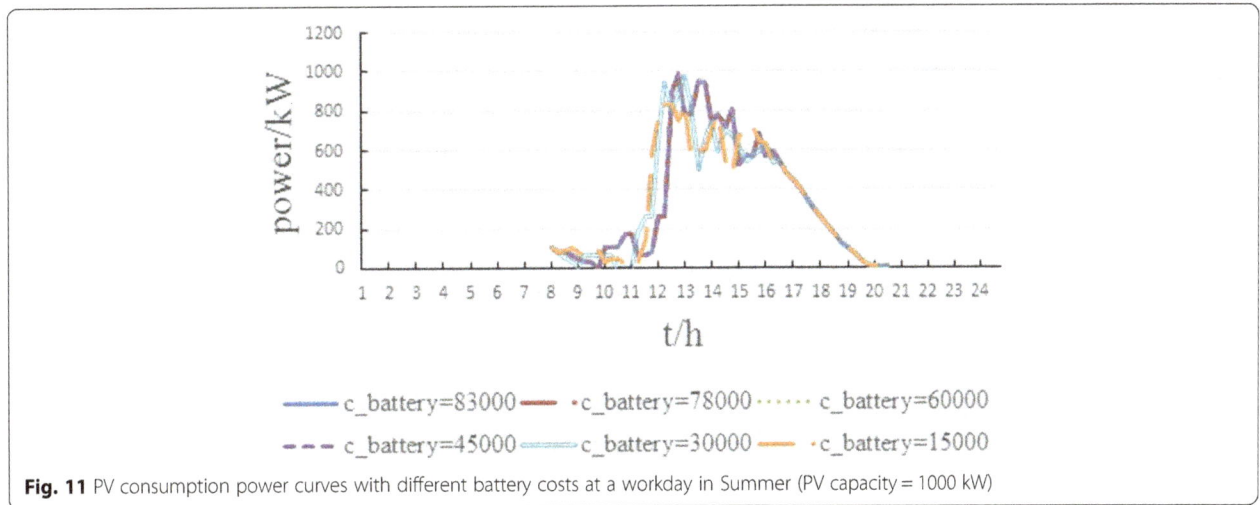

Fig. 11 PV consumption power curves with different battery costs at a workday in Summer (PV capacity = 1000 kW)

Table 4 The optimal configurations and energy consumptions of BSS with different costs of battery

Battery cost/(yuan/each)	83,000	78,000	60,000	45,000	30,000	15,000
Number of robots	6	6	6	6	6	3
Number of batteries	125	125	125	125	132	183
Number of chargers	101	101	101	101	94	73
Valley time energy//(kWh)	2,653.295	2,653.295	2,653.295	2,653.409	2,752.841	3,636.364
Mean time energy//(kWh)	1,140.711	1,140.711	1,140.711	1,140.711	1,140.597	646.3928
Peak time energy//(kWh)	392.5868	392.5868	392.5868	392.5873	293.155	1.6445
Solar energy//(kWh)	4,330.339	4,330.339	4,330.339	4,330.339	4,330.089	4,330.339
Profit/(yuan)	573,679.7	756,179.7	1,212,430	1,668,283	2,131,929	2,674,645

shaving becomes obvious when decline rate is more than 50%. While the decline rate reaches 81%, the number of batteries and chargers are improved from 72 and 50 (base scenario) to 132 and 76, which resulting in the great improvement of ability for BSS as demand response resource to response the TOU and the load peak of charging power is also decreases.

5.2 Impact analysis of different PV capacity and consumption bundling requirement combinations

The emission advantage for EVs is not obvious while the charging power is supplied by coal-fired power plants primarily [7]. The environmental benefit of BSS is focus of study. Analysis of optimal results has indicated BSS indeed has a good quality potential to absorb PV energy while finishing the consumption bundling contract. Fig. 12 shows the annual profit surfaces with different PV capacities and bundling requirements. The annual profit value decreases rapidly when PV bundling requirement and PV capacity increase. The boundary curve between green and red areas at the bottom left of the picture shows the combinations of PV bundling requirements and PV capacities making annual profit of

BSS equals to zero. This curve is an extreme trade-off between PV bundling requirements and PV capacities due to the BSS surrenders all of its profit level to finish PV consumption contract. The surface throws out suggestions for BSS how to sign contract with PV sellers to get specific profit value.

Profit value curves with different PV bundling requirements are depicted in Fig. 13. It is more visualized than Fig. 12. The points whose profits equal to zero is defined as the limitation capability for BSS to accomplish PV consumption task. For instance, the 90% consumption contract with a 1,500 kW capacity PV seller is equivalent to the 95% consumption contract with an around 1,300 kW capacity PV seller. It can be concluded that as PV capacity increases from a low level, the profit value increases and gets a best profit point and then decreases rapidly under scenarios with higher bundling requirements. And the lower the bundling requirement is, the more capacity agreement the BSS can reach with the PV seller.

Considering the profit value curves under different PV bundling requirements which are less than 80%. There is an interesting discovery. The profit curves are inverted U-shaped. There is an optimum point for BSS to get

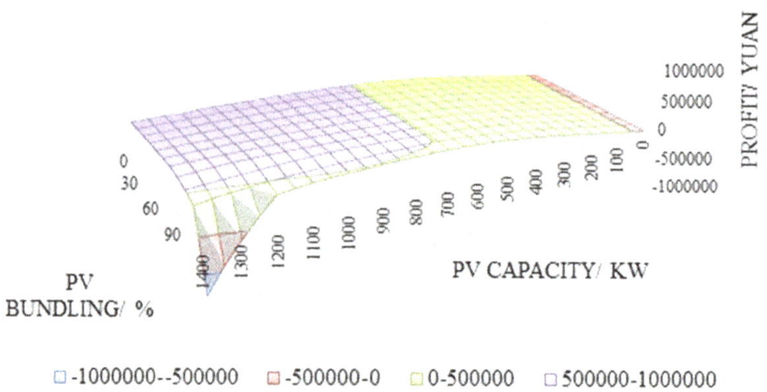

Fig. 12 Profit value surface with different PV capacities and bundling requirements

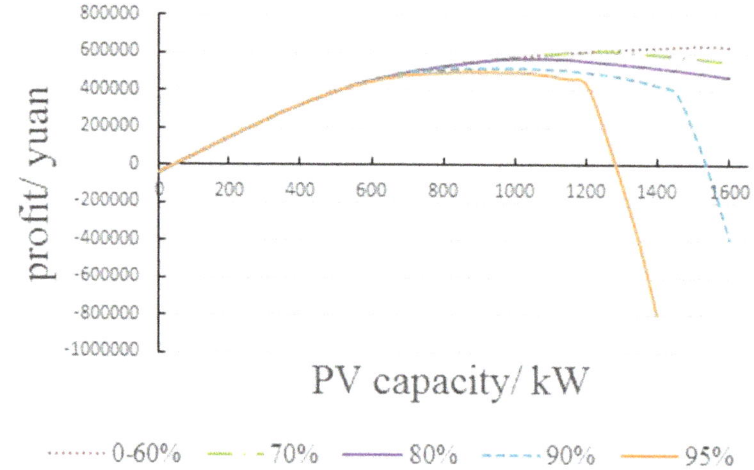

Fig. 13 Profit value curves with different PV bundling requirements

maximal profit value under a certain bundling requirement. As PV capacity increases, the profit value rises at first. And then when the capacity exceeds a certain value, the profit falls as shown in Fig. 13. This discovery indicates there is a natural incentive for the BSS serving a specific amount of swapping demand to consume and absorb a certain amount of PV energy if the price is appropriate. In contrast, the profit levels for BSS will be reduced under excessive PV consumption bundling requirements.

It is attractive that the profit value rises at first and then falls. Table 5 shows the optimal configuration solutions of BSS with different PV capacities when PV consumption bundling requirement is set as 80%. With a fixed bundling requirement, there is no a strict linear pattern of optimal configuration solutions of batteries and chargers as PV capacity increases. It is concluded that BSS overall needs more chargers and then needs more batteries with PV capacity increasing. This phenomenon is much more obvious when capacity is more than 1,200 kW. The annual profit value increases from 146,640.2 yuan to 568,871.9 yuan when capacity increases from 200 kW to 1,000 kW, and then the annual profit decreases to 471,516.5 yuan when capacity

Table 5 The optimal configuration solutions of BSS with different PV capacities (PV bundling requirement = 80%)

PV capacity/(kW)	200	400	600	800	1,000	1,200	1,400	1,600
Number of robots	6	6	6	6	6	6	6	6
Number of batteries	125	125	125	125	125	125	167	249
Number of chargers	102	102	102	102	101	101	77	87
Profit/(yuan)	146,640.2	322,671.8	452,113.7	526,964.5	568,871.9	556,351.1	519,923.2	471,516.5

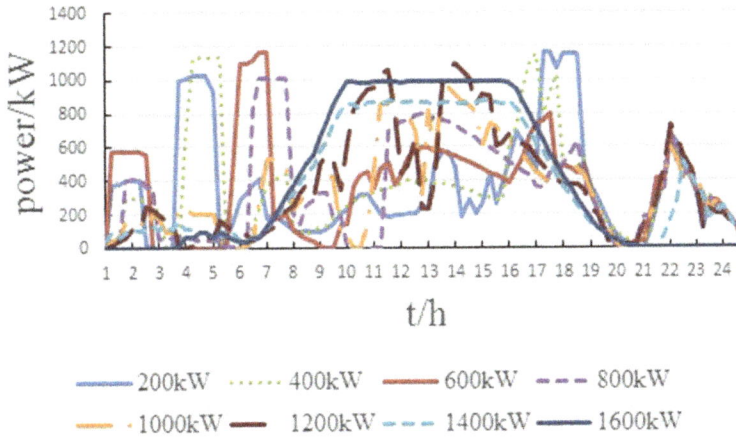

Fig. 14 Charging power curves with different PV capacities (PV bundling requirement = 80%)

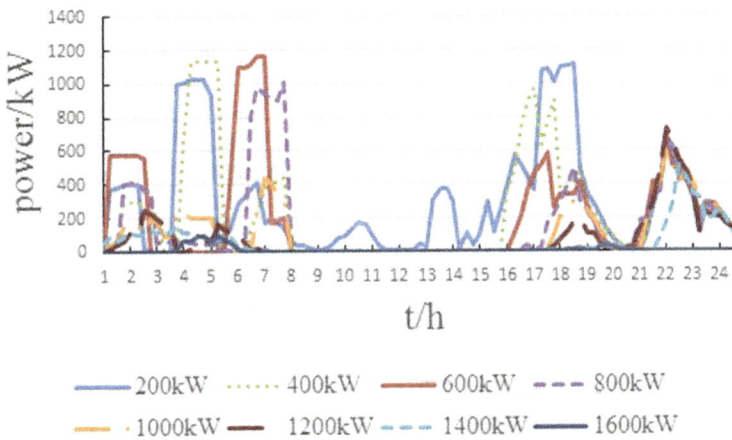

Fig. 15 System charging power curves with different PV capacities (PV bundling requirement = 80%)

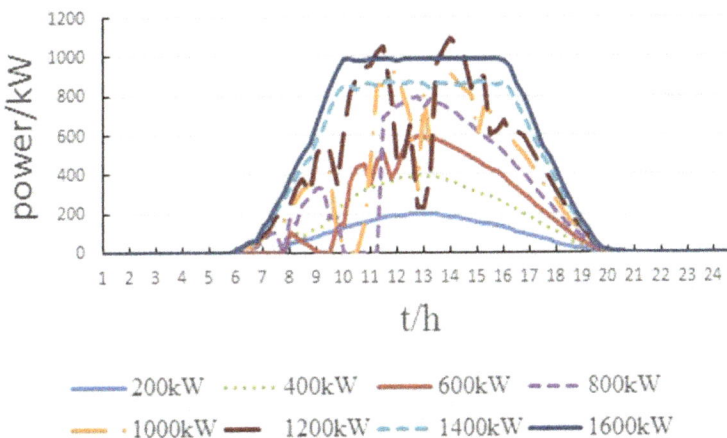

Fig. 16 PV consumption curves with different PV capacities (PV bundling requirement = 80%)

Table 6 The energy consumption distribution of BSS with different PV capacities (PV bundling requirement = 80%)

PV capacity/(kW)	200	400	600	800	1,000	1,200	1,400	1,600
Valley time energy/(kWh)	2,715.909	2,659.092	2,653.409	2,179.297	1,233.234	665.1525	676.2998	165.8008
Mean time energy/(kWh)	3,320.774	2,642.929	1,780.901	1,291.263	1,095.473	966.705	618.705	12.747
Peak time energy/(kWh)	1,085.139	436.8878	418.9328	405.8543	393.9313	360.3088	55.30375	28.80975
Solar energy/(kWh)	1,426.474	2,780.979	3,663.803	4,671.89	5,839.863	7,007.833	8,175.805	9,343.777

continues to increase to 1,600 kW. It is indicated that the optimal PV consumption capability for BSS is around 1,000 kW when requirement is 80%, which can be confirmed in Fig. 13. The numbers of robots are changeless under different scenarios, because the number of swapping robots is influenced by swapping demand at a specific period mainly as mentioned earlier. In our case study, the maximal swapping demand is 23, which can be refueled by 6 robots once at each time interval.

The charging power curves, system charging power curves and PV consumption curves with different PV capacities are shown in Figs. 14 15, 16. Table 6 displays the energy consumption distribution of BSS with different PV capacities. The power curves vary under different scenarios, that means BSS is capable of different operation mode to adapt to external environment and incentive signals. As the PV capacity increases, BSS consumes less and less energy from system at peak time period. The energy consumption from system also decreases because BSS needs to accomplish the PV consumption bundling requirement. It can be concluded that BSS indeed has capability to achieve peak shaving even without sacrificing much profit level when incentives are appropriate.

5.3 Impact analysis of configuration considering multi-swapping demand scenarios and single swapping demand scenario

There are many uncertainties in BSS operation external environment such as PV outputs and EVs swapping demands, et.al. The difference of optimal configurations between considering multi-swapping demand scenarios and only single typical daily swapping demand are compared in this section. The different solutions are displayed and compared in Table 7.

It can be concluded that the configuration number of robots, batteries and chargers are all larger under multi-swapping demand scenarios and the annual profit when

considering multi-scenarios is lower compared to typical daily scenario configuration solution. But it should be concerned that the optimal configuration while considering multi-swapping demand scenarios in our model is much more flexible because the extreme swapping demand like more concentrated in the temporal dimension and larger swapping demand in festival days is taken into account. The difference of configuration means that in order to meet the demand in extreme scenarios, more facilities should be equipped in BSS which results in there are a large number of batteries and chargers idled on other days like workdays. Therefore, how to use a good mechanism such as time of use price for swapping service fee to shape swapping demand more smooth to avoid these low-utilized facilities investment and improve BSS economic is an essential and meaningful problem needed to be investigated.

5.4 Summary

The results of case study prove the effectiveness of proposed model. According to the results of case study and impact analyses, some summaries and suggestions can be concluded as follows:

1) BSS has the potential to be a kind of demand response resources in order to absorb clean energy and shave charging power peak. BSS would be a participator in electricity market to provide many kind of services like G2V, V2G, and other kind of demand response services [28]. However, it needs more incentives for BSS to change the configuration decision and reschedule the charging arrangement, including the decline of battery's price, increasing the price difference between peak and valley price in grid power TOU price and increasing allowance for PV system to reduce price for purchasing PV energy.

2) The optimal operation strategy and configuration decision of BSS are both greatly influenced by the PV consumption bundling requirement. BSS needs

Table 7 The optimal configuration solutions under different scenarios

Scenarios	Number of robots	Number of batteries	Number of chargers	Profit/(yuan)
Multi-swapping demand	6	101	125	573,679.7
Single swapping demand	3	51	72	1,511,402.5

to find an appropriate combination of PV capacity and PV consumption bundling requirement in contract with PV sellers to keep its profit at a satisfaction level.

3) The distribution characteristics of swapping demand have great influence on the equipment configuration planning for BSS. It is necessary to anticipate the fluctuation of BSS profit under multi-swapping demand scenarios at planning stage.

6 Conclusions

In this paper, we have proposed an optimization model of configuration and operation for BSS considering PV consumption bundling. There are several optimal solutions for components of BSS, including EV batteries, swapping robots and battery charging machines, and operation schedule suggestions can be derived. The scheme of proposed model is also a guidance to dig out demand response potential of BSS, in order to satisfy the expectation of multi-stakeholders in the market and is also a toolbox for BSS investor to configure facilities for future's operation with considering the uncertainties of PV generation and swapping demand. The case study results demonstrate that as the penetration of EVs increases, EV batteries are focused on as a kind of natural storages. It is reasonable and effective to excavate the potential of EV batteries to offer demand response by various technical means and incentive mechanism after satisfying swapping demand. What needs to be stressed is that the electricity price in the market and the SOCs of arriving EVs are also uncertain. Moreover, several types of EVs all need to be recharged at one EVs battery swapping station, one BSS should be able to provide different types of batteries [13]. The comprehensive planning model should be established considering different kinds of incentive mechanisms for BSS which may guide and change swapping demand in the market. It will be one of our future's work.

Abbreviations
BSS: Battery swapping station; EV: Electric vehicle

Funding
This work is supported by the National Natural Science Foundation of China (Grant No. 51207050).

Authors' Contributions
YC helped performing the case study and revised the manuscript. CZ contributed to the conception of this study and significantly to the modeling establishment process. Both authors read and approved the final manuscript.

Authors' information
Yu CHENG (1978–), female, Ph. D. degree. She is currently an Associate Professor with the School of Electrical and Electronic Engineering, North China Electric Power University. Her research interests include demand response, the theory and methods of electricity pricing, power system economy analysis.

Chengwei ZHANG (1992-), male, he is currently working toward the M.S. degree at the School of Electrical and Electronic Engineering, North China Electric Power University, Beijing, China. His research interests include demand response, electric vehicles.

Competing interests
The authors declare that they have no competing interests.

References
1. National Development and Reform Commission, National Energy Board, Ministry of industry and Information Technology, the Ministry of housing and urban construction ministry of housing (2015) Guidefor the development of electric vehicle charging infrastructure. Available via DIALOG http://www.sdpc.gov.cn/zcfb/zcfbtz/201511/t20151117_758762.html. Accessed 9 Oct 2015.
2. Yang, S. J., Yao, J. G., Kang, T., & Zhu, X. Q. (2014). Dynamic operation model of the battery swapping station for EV (electric vehicle) in electricity market. *Energy, 65*, 544–549.
3. Gao, C. W., & Wu, X. (2013). A survey on battery-swapping mode of electric vehicles. *Power Syst Technol, 37*(4), 891–898 (in Chinese).
4. Rao, R., Zhang, X. P., Xie, J., & Ju, L. W. (2015). Optimizing electric vehicles users' charging behavior in battery swapping model. *Applied Energy, 155*, 547–559.
5. Xie, P. P., Li, Y. H., Zhu, L., Shi, D. Y., & Duan, X. Z. (2016). Supplementary automatic generation control using controllable energy storage in electric vehicle battery swapping stations. *IET Generation, Transmission & Distribution, 10*(4), 1107–1116.
6. Sun, L., Wang, X. L., Liu, W. J., Lin, Z. Z., Wen, F. S., Ang, S. P., et al. (2016). Optimisation model for power system restoration with support from electric vehicles employing battery swapping. *IET Generation, Transmission & Distribution, 10*(3), 771–779.
7. Liu, N., Chen, Q. F., Lu, X. Y., Liu, J., & Zhang, J. H. (2015). A charging strategy for PV-based battery switch stations considering service availability and self-consumption of PV energy. *IEEE Trans Industrial Electronics, 62*(8), 4878–4889.
8. Sarker, M. R., Panzic, H., & Oretga-Vazquez, M. A. (2015). Optimal operation and services scheduling for an electric vehicle battery swapping station. *IEEE Trans Power Systems, 30*(2), 901–910.
9. Miao, Y. Q., Jiang, Q. Y., & Cao, Y. J. (2012). Operation strategy for battery swap station of electric vehicles based on microgrid. *Autom Electr Power Syst, 36*(15), 33–38 (in Chinese).
10. Tian, W. Q., & He, J. X. (2012). Research on dispatching strategy for coordinated charging of electric vehicle battery swapping station. *Power Syst Protect Control, 40*(21), 114–119 (in Chinese).
11. Yang, S. J., Yao, J. G., Kang, T., & Zhu, X. Q. (2014). Dynamic operation model of the battery swapping station for EV(electric vehicle) in electricity market. *Energy, 65*, 544–549.
12. Yang, Y. X., Hu, Z. C., & Song, Y. H. (2012). Research on optimal operation of battery swapping and charging station for electric buses. *Proc CSEE, 32*(31), 35–42 (in Chinese).
13. Liu, N., Chen, Z., Liu, J., Tang, X., Xiao, X. N., & Zhang, J. H. (2014). Multi-objective optimization for component capacity of the photovoltaic-based battery switch stations: Towards benefits of economy and environment. *Energy, 64*, 779–792.
14. Nian, L., Tang, X., Duan, S., & Zhang, J. H. (2013). Capacity Optimization method for PV-based battery swapping stations considering second-use of electric vehicle batteries. *Proc CSEE, 33*(4), 34–44 (in Chinese).
15. Gao, C. W., Wu, X., Xue, F., & Liu, H. C. (2013). Demand Planning of Electric Vehicle Battery Pack Under Battery Swapping Mode. *Power Syst Technol, 37*(7), 1783–1791 (in Chinese).
16. Jing Z, Fang L, Lin S, Shao W. (2014). Modeling for Electric Taxi Load and Optimization Model for Charging/Swapping Facilities of Electric Taxi. *Transportation Electrification Asia-Pacific IEEE*, 1–5
17. Wang L, Wang X, Sun Q, Lai L. (2015). The Service Modeling of EV Charging/swapping Station and its Cost Benefit Analysis. *10th International Conference on Advances in Power System Control, Operation & Management (APSCOM)*, 31–37
18. Zheng, Y., Dong, Z. Y., Xu, Y., Meng, K., Zhao, J. H., & Qiu, J. (2013). Electric Vehicle Battery Charging/Swap Stations in Distribution Systems: Comparison Study and Optimal Planning. *IEEE Trans Power Systems., 29*(1), 221–229.
19. Zhang, H., Hu, Z., Xu, Z., & Song, Y. (2016). Optimal Planning of PEV Charging Station With Single Output Multiple Cables Charging Spots. *IEEE Trans Smart Grid., 99*, 1–10.

20. Gharbaoui M, Martini B, Bruno R. (2013). for efficient usage of an EV charging infrastructure deployed in city parking facilities. *International Conference on ITS Telecommunications*, 384–389

21. Lu X, Liu N, Huang Y, Zhang J, Zhou N. (2014). Optimal Configuration of EV Battery Swapping Station Considering Service Availability. *International Conference on Intelligent Green Building and Smart Grid IEEE*, 1–5

22. Sun B, Tan X, Tsang D. Optimal Charging Operation of Battery Swapping and Charging Stations with QoS Guarantee. *IEEE Trans Smart Grid*, 99:1–13

23. Zhang L, Lou S, Wu Y, Lin Y, Bin H. (2014). Optimal Scheduling of Electric Vehicle Battery Swap Station Based on Time-of-Sse Pricing. *Power and Energy Engineering Conference IEEE*, 1–6

24. Zhang CH., Meng JS, Cao YX, Cao X, Huang Q, Zhong, QC. (2012). The Adequacy Model and Analysis of Swapping Battery Requirement for Electric Vehicles. *IEEE Power and Energy Society General Meeting*, 1–5

25. Dai, Q., Cai, T., Duan, S., & Zhao, F. (2014). Stochastic Modeling and Forecasting of Load Demand for Electric Bus Battery-Swap Station. *IEEE Trans Power Delivery*, 29(4), 1909–1917.

26. Zhang, Q. D., Huang, X. L., Chen, Z., Chen, L. X., & Xu, Y. P. (2015). Research on control strategy for the uniform charging of electric vehicle battery swapping station. *Transactions of China Electrotechnical Society, 30*(12), 447–453 (in Chinese).

27. Chen, Y. Z., Zhang, B. H., Wang, J. H., Mao, B., Fang, R. C., Mao, C. X., et al. (2011). Active control strategy for microgrid energy storage system based on short-term load forecasting. *Power Syst Technol, 35*(8), 35–40 (in Chinese).

28. Yao, W. F., Zhao, J. H., Wen, F. S., Xue, Y. S., & Xin, J. B. (2012). A charging and discharging dispatching strategy for electric vehicles based on bi-level optimization. *Autom Electr Power Syst, 36*(11), 30–37 (in Chinese).

20

A novel protection scheme for synchronous generator stator windings based on SVM

Magdi El-Saadawi* and Ahmed Hatata

Abstract

This paper proposes a novel scheme for detecting and classifying faults in stator windings of a synchronous generator (SG). The proposed scheme employs a new method for fault detection and classification based on Support Vector Machine (SVM). Two SVM classifiers are proposed. SVM1 is used to identify the fault occurrence in the system and SVM2 is used to determine whether the fault, if any, is internal or external. In this method, the detection and classification of faults are not affected by the fault type and location, pre-fault power, fault resistance or fault inception time. The proposed method increases the ability of detecting the ground faults near the neutral terminal of the stator windings for generators with high impedance grounding neutral point. The proposed scheme is compared with ANN-based method and gives faster response and better reliability for fault classification.

Keywords: Support vector machine, Artificial neural networks, Synchronous generator, Differential protection, Fault detection, Fault classification

1 Introduction

Synchronous Generators are the majority source of commercial electrical energy. The failure of SGs cause severe damage to the machine, interruption of electrical supply, and ensuing economic loss. Therefore, it is essential to have a protection system that is able to detect and diagnose all credible faults and provide effective protection for the SG to increase their useful life and reliability. Internal faults present a real challenge for the protection of electrical machines; especially ground faults in case of high impedance grounding as they are not detectable by differential relays, the most commonly devices used for generator protection. Hence, a reliable and accurate diagnosis of internal faults is still a challenging problem in the area of fault diagnosis of electrical machines [1]. This fact has motivated many works over long period to develop various protection techniques [2–13] include digital, Artificial Intelligence (AI) and other machine learning techniques.

A digital computer technique for the protection of a generator against internal asymmetrical faults has been introduced in [2]. The technique relies on the detection of a second harmonic component in the field current at the onset of a fault in the armature windings. The discrimination against external faults is achieved by monitoring the direction of the negative sequence power flow at the machine terminals. In [3], a digital technique for detecting internal faults in the stator windings of synchronous machine, using positive- and negative-sequence models of the SG, has been introduced. The performance of that technique was evaluated using fault data generated by applying electromagnetic transient program (ETP). Authors in [4, 5] introduced a power-based protection algorithm to provide protection for non-utility generation units against islanding. Another power-based algorithm was introduced in [6] for detecting pole-slipping conditions using three phase power measurements taken at the generator's terminals.

AI techniques have been increasingly used for fault diagnosis of SGs. Among various AI techniques, Artificial Neural Network (ANN) method has become the most widely used tool for solving complex electric power system problems. Fault detection in SGs is one of the areas of intensive application of ANN because of their superior learning, generalization capabilities, and fault-tolerance capabilities [8–13]. Unfortunately, the diagnostic accuracy of ANN cannot be high enough due to the limitations of 'over-fitting' and slow convergence velocity.

Rapid development of the Support Vector Machine (SVM) [14] and its capability to successfully solve the

* Correspondence: saadawi1@gmail.com
Mansoura University Faculty of Engineering, Mansoura, Egypt

problems with ANN [15] resulted in a wide use of this tool in many applications in recent years [16–23]. These applications include: fault detection in transformers during impulse test [16], prediction and control of fuel cell operating conditions [17], load forecasting process in a deregulated power system [18], fault detection [19, 20], modeling of nonlinear dynamic systems [21], power quality events recognition [22] and voltage disturbance classification [23].

In this paper a new scheme for protecting the stator windings of a SG using SVMs is introduced. A multilayer SVM classifier is proposed to detect and classify the internal faults on stator windings of a SG using the instantaneous values of three phase currents on both sides of the SG stator windings. The algorithm is verified using simulation results to evaluate the performance of the proposed method in terms of accuracy and speed. A typical 100 MVA, 13.8 kV, 50 Hz synchronous generator is simulated using Matlab/Simulink software to generate the required test data. The test data obtained from simulation covers all possible states of the SG; namely normal state, internal fault state and external fault state at various conditions such as: fault type, fault location, fault resistance and fault inception angle.

The rest of the paper is organized as follows. Following this introduction, a brief description of the SVM is presented in Section 2, whereas in section 3 the proposed protection scheme is presented. The test system and results are presented in section 4.

2 Classifications with SVM

SVM is a machine learning based method that is applied to classify input data patterns into predefined classes. SVM is based on the idea of a hyper plane classifier, where data space is divided into classes separated by a hyper plane. The goal of SVM is to find a linear optimal separating hyper plane so that the margin of separation between the two classes is maximized [24–27]. Figure 1 illustrates the SVM classification principles where it shows the support vectors and the training data.

Assuming that the data space X consists of several input vectors, x_i :(i = 1, 2,.... N), where x_i are components belong to one of two different classes, and N is the number of samples. The class label of x_i is y_i = {1, -1} which indicates the class to which xi belongs.

The principle of operation of SVMs classifier will be modified according to the type of the data samples. For linearly separable data the hyper-plane satisfies the following equality equation:

$$w^T x + b = \sum_{i=1}^{N} w_i x_i + b = 0 \qquad (1)$$

where w is a normal vector on the hyper-plane, and b is a bias representing the distance from the origin.

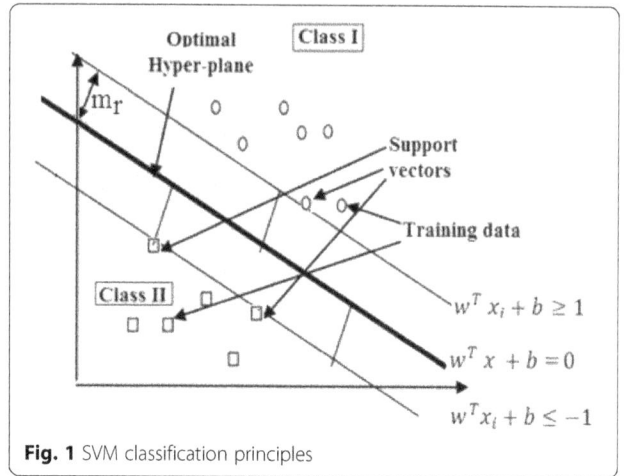

Fig. 1 SVM classification principles

The separating hyper-plane that creates the maximum distance between the plane and the nearest data is called the optimal separating hyper-plane as shown in Fig. 1. The maximum-margin classifier is the discriminate function that maximizes the geometric margin $1/||w||$, which is equivalent to minimizing $||w||^2$. This leads to the following constrained optimization problem [28, 29]:

Minimize: $m = \frac{1}{2}||w||^2$

Subject to:

$$y_i(w^T x_i + b) \geq 1 \text{ for } i = 1, 2, ... N \qquad (2)$$

The constraints in this formulation ensure that the maximum margin classifier classifies each example correctly, which is possible since we have assumed that the data are linearly separable.

The objective of maximizing the margin i.e., minimizing $||w||^2$, will be augmented with a term $C\sum_{i=1}^{N} \zeta_i$ to penalize misclassification and margin errors, where ζ_i are slack variables that allows an example to be in the margin and is called a margin error ($\zeta_i \geq 1$). The optimization problem now becomes [28]:

Minimize:

$$\frac{1}{2}||w||^2 + C\sum_{i=1}^{N} \zeta_i \qquad (3)$$

Subject to:

$$y_i(w^T x_i + b) \geq 1 - \zeta_i \text{ for } i = 1, 2, ... N$$
$$\zeta_i \geq 0 \quad \text{for all } i \qquad (4)$$

With the help of Lagrange multipliers, the dual form of the above minimization problem is:

Minimize:

$$W(\alpha) = \sum_{i=1}^{N} \alpha_i - \frac{1}{2}\sum_{i,k=0}^{N} \alpha_i \alpha_k y_i y_k x_i^T x_k \qquad (5)$$

Subject to:

$$\sum_{i=1}^{N} y_i \alpha_i = 0, \quad 0 \le \alpha_i \le C, \quad i = 1, 2, \ldots .N \quad (6)$$

The number of variables of the dual problem is the number of training data. Let us denote the optimal solution of the dual problem with α^* and w^*. The training examples x_i is a support vector (SV). The number of SVs is considerably lower than the number of training samples making SVM computationally very efficient. The value of the optimal bias b^* is found from the geometry,

$$b^* = -\frac{1}{2} \sum_{SVs} y_i \alpha_i^* (S_1^T x_i + S_2^T x_i) \quad (7)$$

where S_1 and S_2 are the arbitrary SVs for class-I and class-II, respectively.

Only the samples associated with the SVs are summed because the other elements of the optimal Lagrange multiplier α^* are equal to zero. The final decision function is given by:

$$f(x) = \sum_{SVs} \alpha_i y_i x_i^T x + b^* \quad (8)$$

The unknown data sample x are then classified as:

$$x = \in \begin{cases} Class-I & \text{if} \quad f(x) \ge 0 \\ Class-II & \text{otherwise} \end{cases} \quad (9)$$

The nonlinear classification problems can be solved by applying Kernel functions. The classified data is mapped onto a high-dimensional feature space where the linear classification is possible. Kernel functions are defined as follows:

$$K(x_i, x_k) = \varphi(x_i).\varphi(x_k) \quad (10)$$

substituting by (10) into Kuhn-Tucker conditions (5) results in:

Maximize:

$$W(\alpha) = \sum_{i=1}^{N} \alpha_i - \frac{1}{2} \sum_{i,k=0}^{N} \alpha_i \alpha_k y_i y_k K(x_i, x_k) \quad (11)$$

Subject to:

$$\sum_{i=1}^{N} y_i \alpha_i = 0, \quad 0 \le \alpha_i \le C, \quad i = 1, 2, \ldots .N \quad (12)$$

Different types of kernels can be used to train SVMs. The most commonly used kernel functions in power system applications are Linear, Polynomial and Gaussian radial basis functions (RBF) [22, 23, 26–33], they can be defined as:

Linear:

$$K(x_i, x_k) = x_i.x_k \quad (13)$$

Polynomial:

$$K(x_i, x_k) = (x_i^T.x_k + 1)^n, \quad n > 0 \quad (14)$$

Gaussian (RBF):

$$K(x_i, x_k) = \exp\left(\sigma \|x_i^T - x_k\|^2\right), \quad \sigma > 0 \quad (15)$$

where n is the degree of kernel inner product, $\sigma = -\frac{1}{2\gamma^2}$ and γ is the kernel width parameter.

3 SVM Methodology for Generator Protection

This section describes the generalization of SVMs classification technique to the protection of a SG stator windings. Firstly, the proposed scheme is introduced followed by the details of each of its parts. The heart of the proposed scheme composed of two SVM classifiers that receive instantaneous 3 phase current measurements and based on these current measurements the classifiers decide whether the SG is in normal or fault condition. The classifiers also determine whether the fault is internal or external. The main purpose of using SVM classifiers is to free the proposed protection scheme of the over fitting and slow convergence limitations experienced by other techniques, and hence to improve the diagnostic accuracy. The diagnostic accuracy of the proposed scheme, as all other machine learning based techniques, depends on its training in terms of how sufficient and representative the training data sets are.

3.1 Proposed protection algorithm

Figure 2 presents a combined functional block diagram and a flow chart of the proposed protection scheme. The proposed procedure for fault classification can be described in the following steps:

1- The three phase currents at both ends of the SG stator windings are sampled at a sampling frequency of 1 kHz.
2- An anti-aliasing filter eliminates the current samples corrupted by high frequency transients to ensure effective and accurate discrimination of internal faults.
3- A moving window current waveform of 4 samples width, less than a quarter of a cycle, is used as the input pattern to the SVM1 classifier. The output will be -1 in case of a fault occurrence (either internal or external), and will be +1 for normal state.
4- In case of a fault occurrence, SVM2 discriminates between internal and external faults.
5- The step length of moving window is taken as one sample, where the window is moved one sample ahead in each calculation step. The steps 1 to 4 are repeated unless a fault is detected and the SG is tripped.

3.2 SVM classifiers

As stated earlier and illustrated in Fig. 2, the proposed protection scheme employs two SVM classifiers to discriminate between the normal, internal fault and external fault states of the SG. SVM1 is trained to distinguish between normal and fault states. If the input pattern to

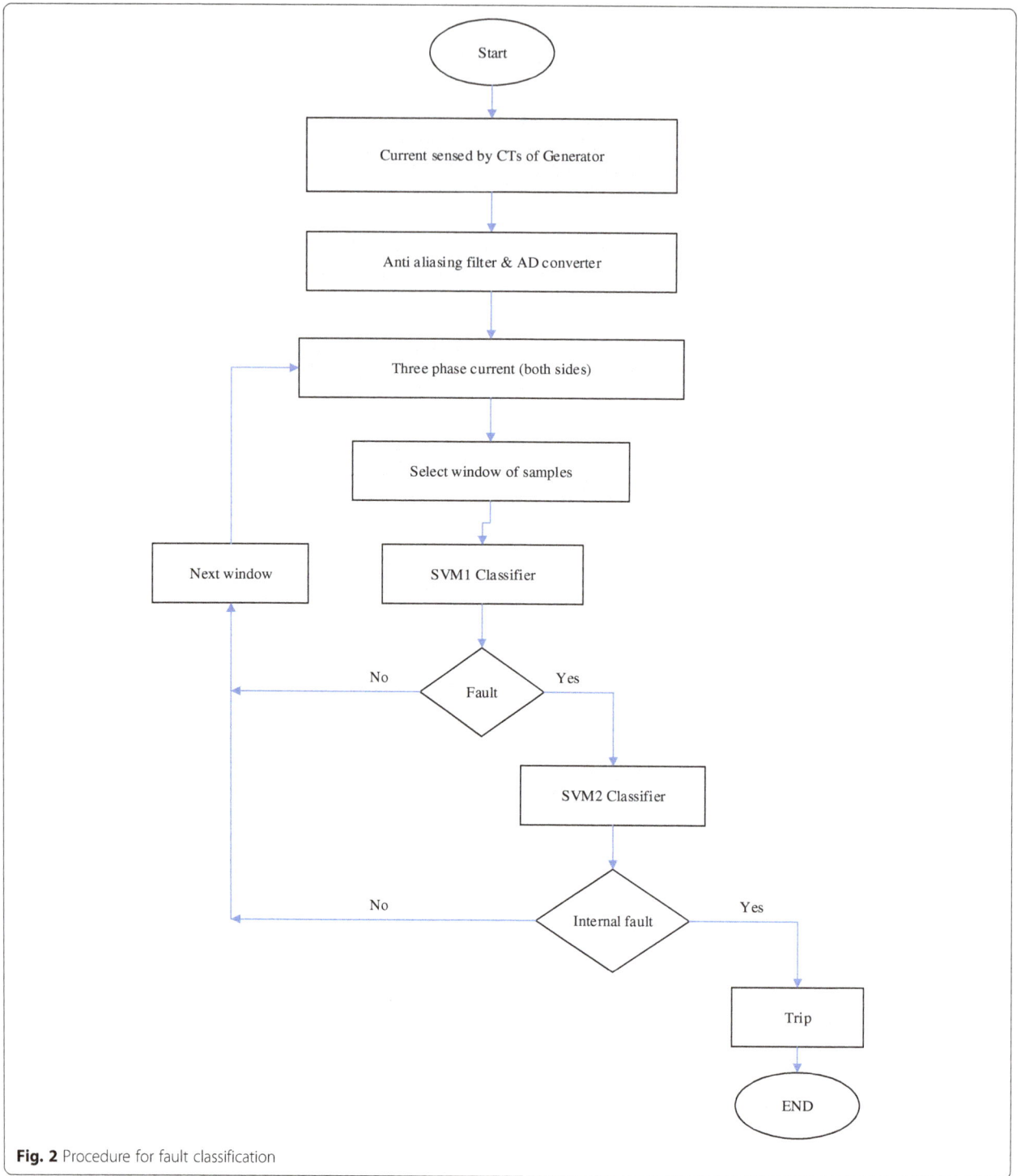

Fig. 2 Procedure for fault classification

SVM1 represents a normal state, its output is set to +1; otherwise it is set to - 1. On other hand, SVM2 is trained to discriminate between internal fault state and external fault state. Its output will be +1 for input patterns representing internal faults and -1 otherwise.

In the training process of SVMs, the three prescribed kernel functions are tried and tested. The two functions that give the desired performance are used for the test cases. These two functions are the polynomial kernel function and the kernel radial basis function (RBF). Selection of the optimum parameters for SVMs is done during the training process using training data. The SVM classifier with the best performance is obtained by testing different values of the kernel parameters. These parameters are varied in the following manner; kernel width parameter γ is varied in the range [0.1, 5] with a step of 0.05. The order of the polynomial kernel n is varied in the range [1, 10] with a step of 0.5. The penalty due to the error C is varied in range [1, 1000] with a step of 1. The values of the used kernel parameters are shown in Table 1. The performances of the two SVMs are assessed for each combination of these values by calculating the classification accuracy (CA) defined as:

$$CA\% = \frac{Correctly\ classified\ patterns}{Total\ patterns} \times 100 \quad (16)$$

From these results, the SVM classifier with the highest percentage classification accuracy is selected. The testing and generalization processes are then performed.

3.3 Training and testing patterns for SVMs
Since the classification technique is based on a supervised learning mechanism, it needs to be trained for possible normal and faulty states of SG. In this paper, training and testing data sets are obtained through excessive simulations for different operating conditions.

In general, a large number of cases that covering all the credible conditions and representing the whole data space are required for training. In this work, a total training and testing set of 4600 patterns representing different cases of the generator states are generated via simulation of the SG. Each training or testing pattern consists of 24 elements; these elements represent 4 samples of each of the three phase currents at the two sides of the SG stator windings. Details of these training and testing patterns are presented in section 4.

3.4 Data preprocessing
Data preprocessing is implemented for the purpose of correcting errors that may present in the raw data. To improve generalization capability of SVM, the preprocessing of training data may include data smoothing, excluding odd values, and normalizing experimental data. Also, unequal interval data series are transformed into equal interval series by linear interpolation to train SVMs.

4 Results and Discussion
This section presents the test system and the details of the training and testing data sets as well as the results of applications of the proposed scheme.

Table 1 Values of the used kernel parameters

Parameter	Values
Kernel width parameter, γ	0.1:0.05: 5
Order of polynomial kernel, n	1:0.5: 10
Penalty due to the error, C	1: 1000

4.1 System under study
The data required for training and testing SVMs are obtained through Matlab/Simulink simulations. These training cases were selected so that the obtained training data should contain necessary information to generalize the problem. A suitable SG model is required to characterize the different operating and fault conditions. The fault conditions include fault type, location, and resistance and inception angle. In a previous work [34], the authors have developed a dynamic model to simulate generator states (internal fault, external fault, normal states). Three phase power system was simulated and the input/output pair patterns were generated. That developed model will be used in this paper to represent the SG.

The test power system consists of a three phase SG connected to an infinite bus through a transmission line. The measured devices are located at the two ends of the generator. A single line diagram of the modeled power system is shown in Fig. 3.

The training cases include: five generator loading conditions (50, 60, 70, 80 and 90% of rated MVA), three values of fault resistance (0 Ω, 10 Ω and 20 Ω), ten fault locations on stator windings (10, 20, 30, 40, 50, 60, 70, 80, 90, and 100% from the neutral point), four fault types (single line to ground (SLG), Double line to ground (DLG), line to line (LL), and three phase fault (3 ph)), seven fault incipient time instants (0.07, 0.072, 0.075, 0.078, 0.08, 0.085 and 0.088 s) and current transformer saturation condition.

The testing cases include: three generator' loading conditions (55, 65, 75% of the rated MVA), two values of fault resistance (5 Ω, 15 Ω), ten fault locations on stator windings (5, 15, 25, 35, 45, 55, 65, 75, 85, and 95% from the neutral point), four fault types (SLG, DLG, LL, and 3 ph) and current transformer saturation condition.

The total training and testing patterns are classified as:

- 1200 patterns represent the normal operation state. They are generated by applying three phase balanced operation at different loading conditions.

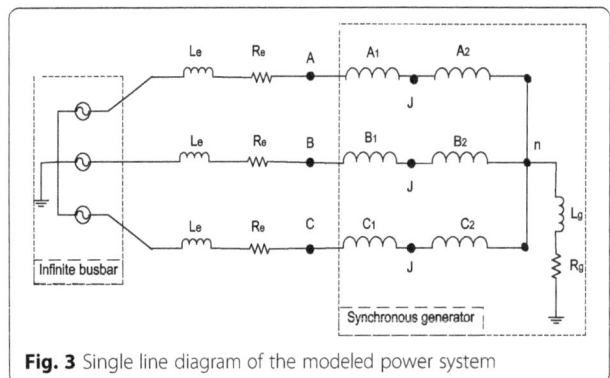
Fig. 3 Single line diagram of the modeled power system

- 800 patterns represent the external fault state. They are generated by applying different types of external faults at various locations along a transmission line.
- 2600 patterns represent the internal fault state. They are generated by applying different types of internal faults at different locations on the stator windings.

4.2 SVMs training and testing results

Many SVMs with different values of kernel parameters were trained and tested using Matlab toolbox. The classification accuracy of the SVMs is determined by applying (16) to all of the 3800 training patterns. Tables 2 and 3 show the SVMs with the highest classification accuracies for both the Polynomial kernel function; and the Gaussian kernel function respectively.

From the above results, it can be noticed that the highest training efficiency obtained with the polynomial kernel function is 99.8% for SVM1 and 99.67% for SVM2. The highest accuracy for SVM1 has been achieved with the value of $C = 883$, and $\gamma = 0.1$, whereas for SVM2: parameter values for the highest accuracy are $C = 725$, and $\gamma = 0.15$. These parameters are used for "learning" the two proposed SVMs. Once the training is completed, the trained SVMs are used for testing the new patterns. Table 4 illustrates the classification accuracy of the designed SVMs for testing data.

To help judging the validity and accuracy of the proposed system, the SVMs results have compared with those obtained using an ANN-based system [13]. In that study the ANN-based method was applied to detect and classify different types of SG faults. The same training and testing data described above have been used to train the ANN-based system. A comparison between the two methods is depicted in Tables 5 and 6. It can be noticed that the proposed SVMs-based technique is faster and more accurate (both for training and testing patterns) than ANN-based method.

Table 6 lists the time taken by both the ANN-based and the proposed methods to detect a fault. It is clear

Table 2 SVMs for the Polynomial kernel function during the training process

SVM		kernel width parameter γ	Penalty due to the error C	Accuracy of SVM%
1	SVM1	0.55	231	98.601
	SVM2	0.40	437	96.455
2	SVM1	0.50	404	95.745
	SVM2	0.35	578	88.461
3	SVM1	0.3	77	99.117
	SVM2	0.8	85	98.096
4	SVM1	0.1	883	99.80
	SVM2	0.15	725	99.67

Table 3 SVMs for the Gaussian kernel function during the training process

SVM		Order of polynomial kernel n	Penalty due to the error C	Accuracy of SVM%
1	SVM1	1.5	265	95.143
	SVM2	1	104	96.257
2	SVM1	2	513	95.328
	SVM2	6	682	81.250
3	SVM1	3	610	92.341
	SVM2	2	595	96.861
4	SVM1	8.5	490	94.654
	SVM2	9	463	86.095

that proposed SVMs-based method takes shorter time to detect a fault for all of the tested fault types.

4.3 Case studies

This section presents the details of four test cases to illustrate the ability and accuracy of the proposed method to detect and diagnose a fault.

4.3.1 Case I: External fault

An external SLG fault is simulated at the middle of the transmission line. The pre-fault power flow from generator to infinite bus is $P = 1$p.u. at p.f of 0.85 lag. Figure 4-a shows the three-phase currents at the two ends of the stator windings for a fault incident at $t = 0.077$ s. It can be noticed that currents at both ends of the stator windings are identical as the fault is external. Figure 4-b shows the classification label generated by SVM1 and SVM2. It can be seen that the output of SVM1 is +1 indicating no fault condition until $t = 0.077$ s, time of fault occurrence, then it changes to -1 indicating a fault condition. The output of the SVM2 become +1 at $t = 0.077$ s detecting the external fault.

4.3.2 Case II: SLG fault at 55% of stator windings

This is a case of an internal SLG fault occurred at 55% of stator windings away from the neutral point at $t = 0.077$ s. The fault resistance is 5 Ω, the pre-fault power flow from the generator to infinite bus is $P = 1$ p.u. at p.f 0.85 lag. Figure 5-a shows the waveform of the three-phase currents. It can be observed that the currents of phase (a) at both ends of the stator windings are now different ia1 and ia2. The current ia2 has a larger value due to short circuit occurrence (this is the sum of the ground fault current and the load current).

Table 4 Classification accuracy of the designed SVMs

SVM	No. of testing patterns	Correct patterns	Incorrect patterns	Classification accuracy %
SVM1	800	758	42	94.75
SVM2	600	553	47	92.17

Table 5 Comparison between accuracy and training time for ANN and SVMs methods

Method	Accuracy for training patterns %	Accuracy for testing patterns %	Training time (s)
SVM1	99.80	94.75	<1
SVM2	99.67	92.17	<1
ANN [13]	93.7	89.1	10

Figure 5-b illustrates the response of the SVM1 and SVM2 as a function of the time (sec). The output of SVM1 is almost the same as in case I, that is +1 indicating no fault until $t = 0.077$ s and then -1 indicating fault incidence at $t = 0.077$ and afterwards. The output of the SVM2 become -1 at $t = 0.077$ s indicating that fault detected by SVM1 is an internal fault. The response of SVMs in this case and in the previous case proved the ability of the proposed technique to discriminate between healthy and faulty conditions and also between internal and external faults.

4.3.3 Case III: Three phase to ground fault at 75% of stator windings

A three phase to ground fault is simulated at 75% of stator windings away from the neutral point. The fault resistance is 15 Ω. The inception fault time is $t = 0.075$ s. The pre-fault power flow from generator to the infinite bus is $P = 0.9$ p.u. at power factor of 0.75 lag. Figure 6-a shows the input current waveforms of the SVM fault classifier and Fig. 6-b illustrates the response of the SVM1 and SVM2 as a function of the time (sec). It can be noticed that despite the big difference in fault currents, the SVMs have given the right classification of the fault condition as in case II.

4.3.4 Case IV: SLG fault at 5% of stator winding

In this case a single line to ground fault is placed at 5% of stator windings away from the neutral point. The fault resistance is 15 Ω. The inception fault time is $t = 0.073$ s. The pre-fault power flow from generator to infinite bus is $P = 1$ p.u. at power factor of 0.8 lag.

Figure 7-a shows the input current waveforms of the SVM fault classifier and Fig. 7-b illustrates the response of the SVM1 and SVM2 as a function of the time (sec). It can

Table 6 Comparison between fault detection time for ANN and SVMs methods

Type of fault	Time to detect the fault (ms)	
	ANN [13]	SVM
Single line to ground	10.9	6.4
Phase to phase	12.8	8.5
Double line to ground	10.6	6.8
Three phase	8.3	5.4

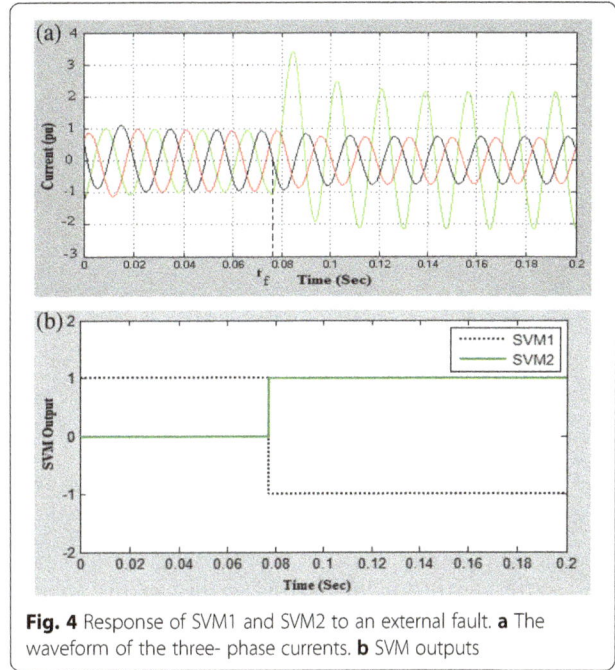

Fig. 4 Response of SVM1 and SVM2 to an external fault. **a** The waveform of the three-phase currents. **b** SVM outputs

be noticed that the SVMs have correctly classified the fault incident as an internal fault.

From the above case studies, it can be observed that the proposed method succeeded to detect and classify the internal faults in the stator windings whatever the changes

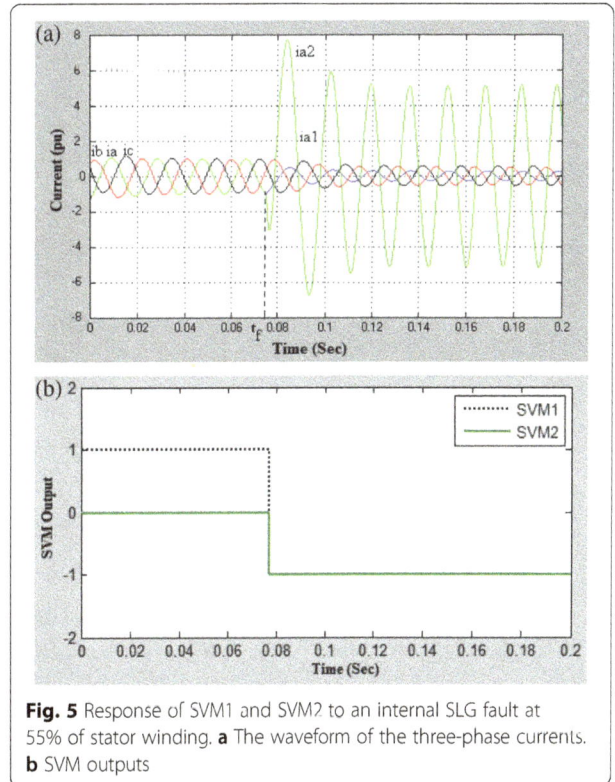

Fig. 5 Response of SVM1 and SVM2 to an internal SLG fault at 55% of stator winding. **a** The waveform of the three-phase currents. **b** SVM outputs

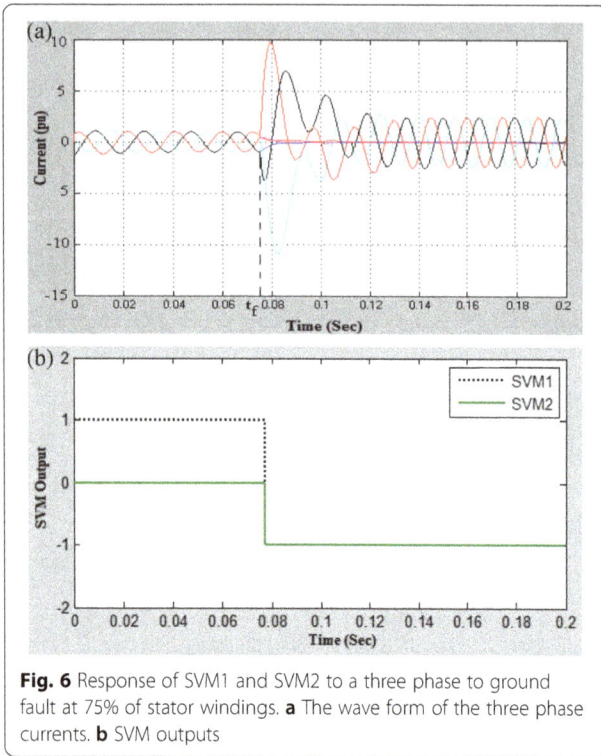

Fig. 6 Response of SVM1 and SVM2 to a three phase to ground fault at 75% of stator windings. **a** The wave form of the three phase currents. **b** SVM outputs

in the prescribed conditions i.e. the fault location, resistance, inception angle or the pre-fault loading conditions.

Extensive simulation results show that the SVM based fault classifier gave excellent predictive ability in all simulation tests. The results show also the stability of the SVM

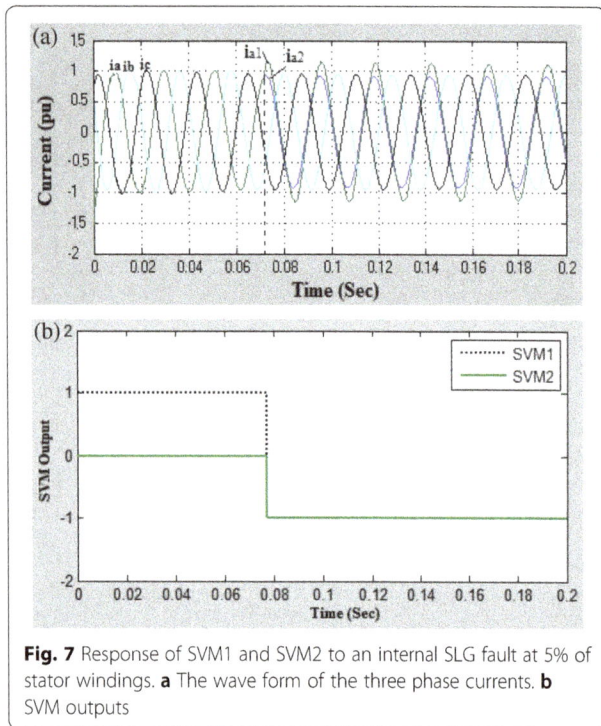

Fig. 7 Response of SVM1 and SVM2 to an internal SLG fault at 5% of stator windings. **a** The wave form of the three phase currents. **b** SVM outputs

outputs under normal steady state conditions and rapid convergence of the output variables to the expected values under fault conditions. This clearly confirms the effectiveness of the proposed SVM based fault detector.

4.4 Verification of synchronous generator internal fault model

The internal fault simulation was verified experimentally (A detailed description of this experimental work is explained in the Additional file 1). A three phase synchronous machine was physically modeled in electric machine lab. at Mansoura university, faculty of engineering. Figure 11 (in Appendix) shows some photos of the tested system. The overall schematic diagram of the experimental system setup is shown in Fig. 8.

The current signals from the power system were obtained through a current transformer. The data acquisition board received the analog signals through an Analog Input Card. The analog filtered signal was then transferred to personal computer, and was converted to a digital one by internal Analog to Digital (A/D) converters. Details of these components are presented in the following items.

i. Universal laboratory machine
   The B.K.B. universal laboratory machine set consists of a two-pole uniform air gap universal machine coupled to a dc dynamometer. The nominal rating of the universal machine is 2 kVA as a three phase 50 Hz, 3000 r.p.m. The dynamometer is rated at 3 kW, 220 V, 2000/3000 r.p.m. The parameters of the machine are shown in Appendix (Table 7). All of the winding connections are brought out to the large terminal panel. The stator connections are also terminated at a 24-way socket.
ii. Current Transformer
   The current signals are obtained by 100/5 A current transformer. The CT secondaries are connected to shunt resistances to obtain an equivalent voltage signals. The low-level output voltage signals out from transformers are filtered and then are passed as an input to the data acquisition card.
iii. Analog filter
   A 2nd-order Butter worth band-pass filter is used to attenuate the dc component and high frequency components. The filter is centered at the nominal system frequency and its pass-band is chosen to be 80 Hz. An array of six filter circuits simultaneously filters the current signals before being fed to the data acquisition system. The filter unit circuit is preceded by an amplifier circuit.
iv. Data acquisition board
   The objective of the DAQ is to convert the analog signals into a digital one so that it can be used by the computer. A 1 kHz sampling rate implies 1 ms time

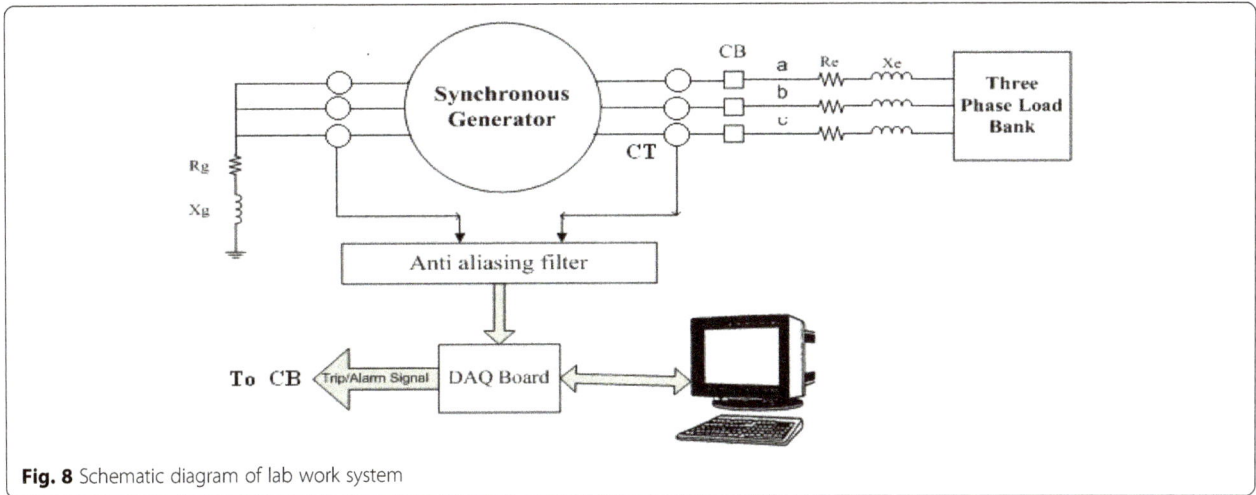

Fig. 8 Schematic diagram of lab work system

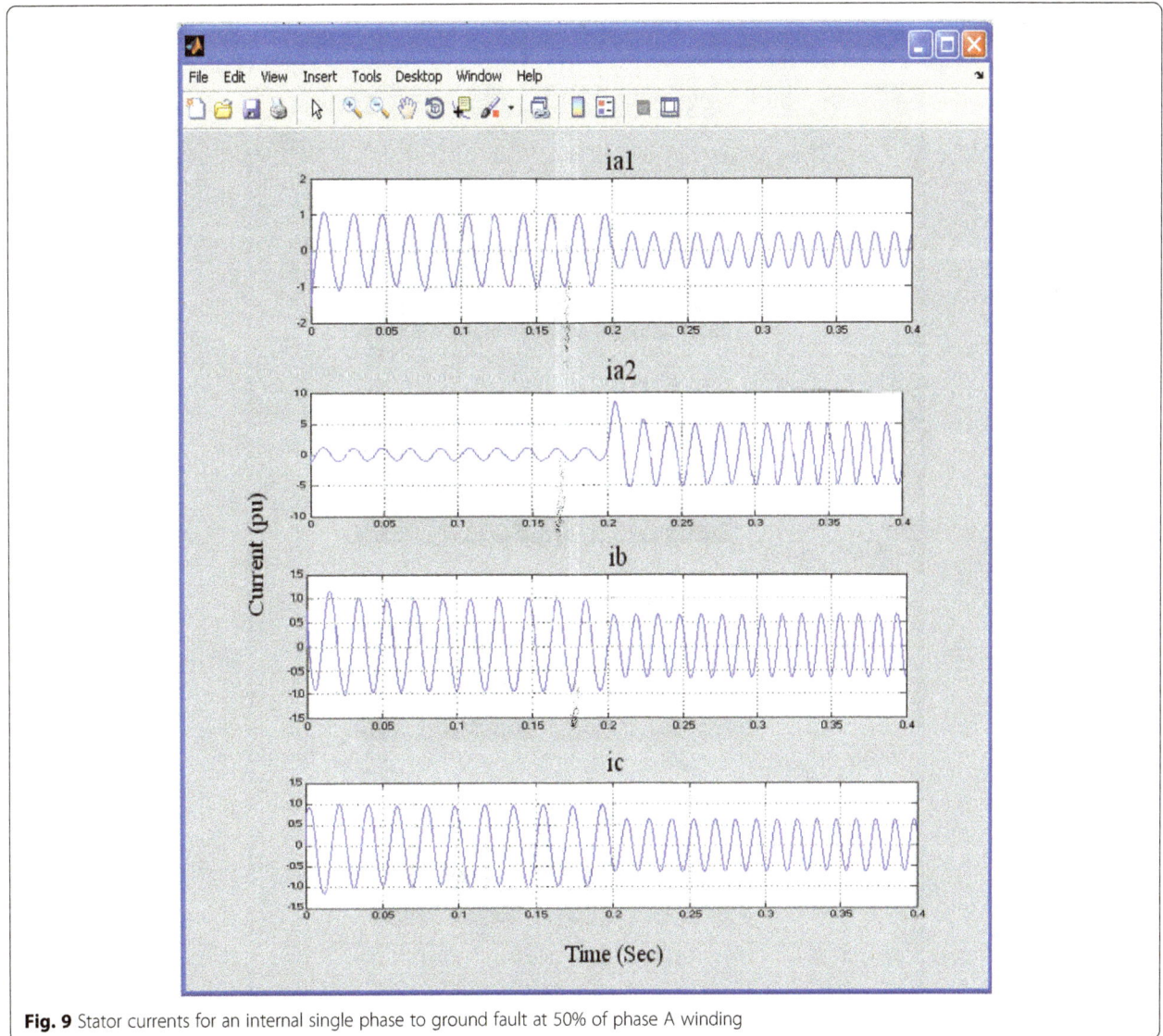

Fig. 9 Stator currents for an internal single phase to ground fault at 50% of phase A winding

interval between samples which is needed for an appropriate software and hardware setup to accomplish protecting relay task within this time interval. To acquire voltage and current signals, a national instruments NI USB-6008 multifunction I/O device is connected to the computer.

Numerous internal fault types such as single phase to ground fault, phase to phase fault, double phase to ground fault, and three phase fault at different locations, inception time and pre-fault loading were applied using the prescribed laboratory system.

For example, simulation and laboratory results for internal single phase to ground faults at 50% of phase A winding are shown in Figs. 9 and 10 respectively. The currents on terminal sides of phase B and C are equal to their counter parts on neutral side, so it is sufficient to show only the currents of one side of each healthy phase. Figure 10 shows the laboratory fault currents for the same case. From

the two figures it can be concluded that the current waveforms of the simulated and laboratory are very close in shape and magnitude. There is a small difference in the curves of simulated cases with respect to the experimental ones. These differences are due to the operating environment of physical system, which is not ideal due to the existence of noise, CT errors and CT saturation.

5 Conclusion

This paper proposed a novel scheme based on SVMs for detecting and classifying faults in stator windings of a synchronous generator. Two SVM classifiers have been proposed. SVM1 was used to identify the fault occurrence in the system and SVM2 was used to determine whether the fault, if any, was internal or external. In the proposed scheme, fault detection and identification was performed in less than a quarter cycle of the 3-phases current at the two ends of stator windings. The detection and classification of

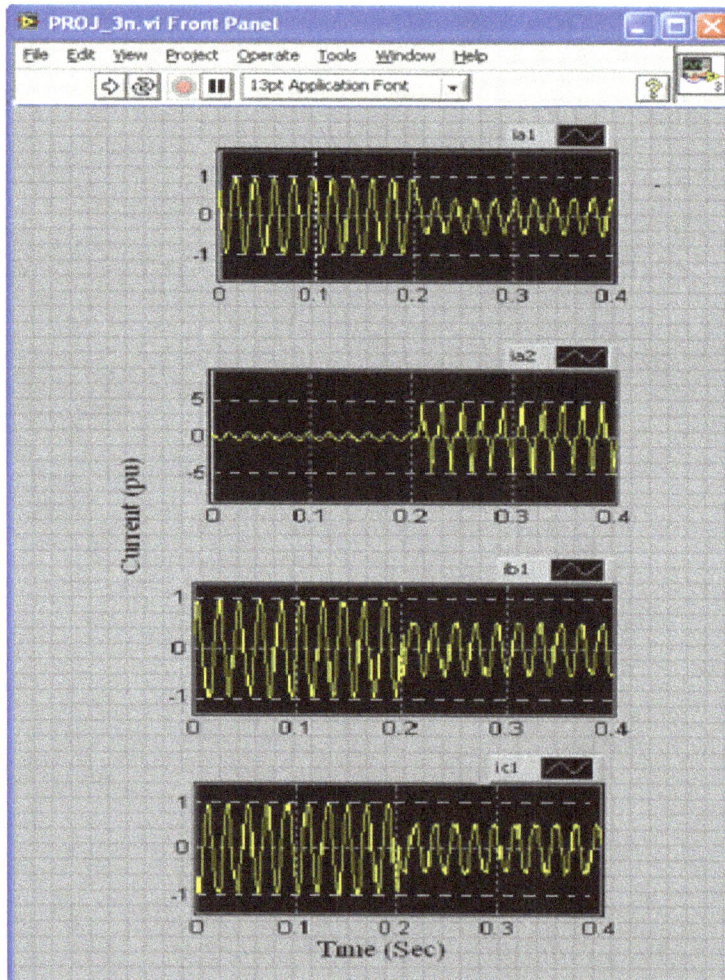

Fig. 10 Recorded stator currents for an internal single phase to ground fault at 50% of phase A winding

faults were not affected by the fault type and location, pre-fault power, fault resistance or fault inception time.

Comparing the proposed SVMs with the ANN-based method prove that the proposed SVM-based method is faster and more accurate. Applying the proposed method to several case studies has shown that the SVM based fault classifier has consistently accurate detection and discrimination in almost all operating conditions and for the expected ranges of different parameters.

The proposed SVMs classification technique has been proven to be highly reliable and very fast in detecting and classifying faults with an accuracy of 99.7% average for all the test cases. This makes the SVM a good candidate to compete with, or even replace, conventional methods. The SVM is proven to be able to generalize the situation from the provided patterns and accurately indicate the presence or absence of a fault. Its fast response compared to other techniques makes it more advantageous for on-line fault detection. In a future work the authors will develop a prototype using Digital signal processing (DSP) to test the ability of applying the proposed system in real life.

6 Appendix

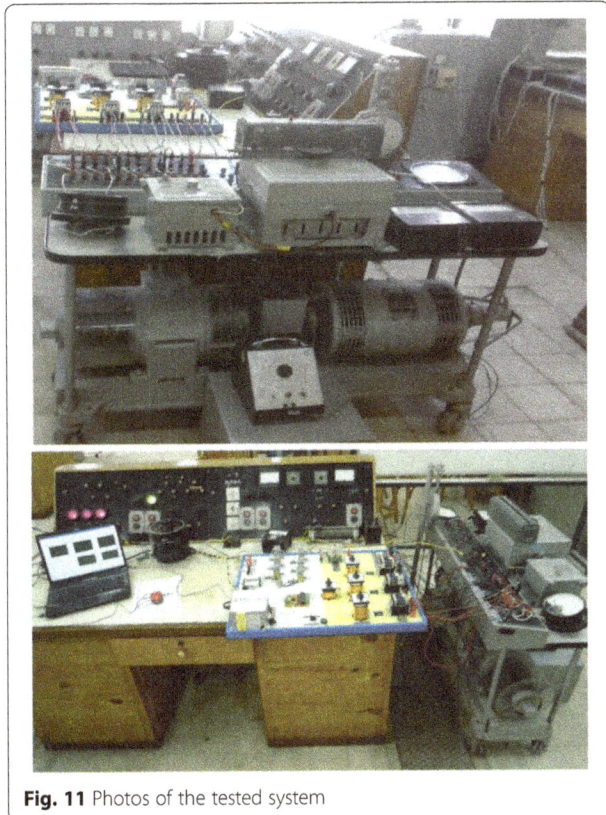

Fig. 11 Photos of the tested system

Table 7 Principal data of the machine

Parameter	Value
Stator resistance/coil (R_s)	0.03
Rotor resistance/coil (R_f)	0.0057
Total moment of inertia of rotors plus coupling	6.2×10^{-2} kg.m^2
Self-inductance of stator winding (L_s)	2
Self-inductance of field winding (L_f)	1.37
Self-inductance of d-axis damper winding (L_D)	1.344
Self-inductance of q-axis damper winding (LQ)	1.357
Mutual inductance between stator and field windings (L_{sf})	1.34
Mutual inductance between stator and d-axis damper windings (L_{sD})	1.264
Mutual inductance between stator and q-axis damper windings (L_{sQ})	1.264
Mutual inductance between field and d-axis damper windings (L_{fD})	1.325

Authors' contributions
The first author suggested the paper topic, planed the structure of the paper and reviewed it. The second author developed the computer program, implemented it and analyzed the results. He prepared the required figures and table and wrote the paper in its preliminary form. Both authors read and approved the final manuscript.

Competing interests
The authors declare that they have no competing interests.

References
1. Yaghobi, H., Ansari, K., & Mashhadi, H. R. (2013). Stator turn-to-turn fault detection of synchronous generator using total harmonic distortion analyzing of magnetic flux linkage. *IJST, Transactions of Electrical Engineering, 37*(E2), 161–182.
2. Dash, P. K., Malik, O. P., & Hope, G. S. (1977). Fast generator protection against internal faults. *IEEE Transactions on Power App. Syst., 96*, 1498–1505.
3. Sidhu, T. S., Sunga, B., & Sachdev, M. S. (1996). A digital technique for stator winding protection of synchronous generators. *Electric Power System Research, 36*, 45–55.
4. Usta, Ö. (1993). *A power based digital algorithm for the protection of embedded generators. Ph.D. dissertation.* Bath: University of Bath.
5. Redfern, M. A., Usta, Ö., Fielding, G., & Walker, E. P. (1994). Power based algorithm to provide loss of grid protection for embedded generation. *IEE Proceedings on Generation, Transmission and Distribution, 141*(6), 640–646.
6. Redfern, M. A., & Checksfield, M. J. (1995). A new pole-slipping protection algorithm for dispersed storage and generation using the equal area criterion. *IEEE Transactions on Power Delivery, 10*, 194–202.
7. Asfani, D., Muhammad, K., Syafaruddin, Purnomo, M., & Hiyama, T. (2012). Temporary short circuit detection in induction motor winding using combination of wavelet transform and neural network. *Expert System with Applications, 39*(5), 5367–5375.
8. Megahed, A. I., & Malik, O. P. (1999). An artificial neural network based digital differential protection for synchronous generator stator winding protection. *IEEE Transactions on Power Delivery, 14*(1), 86–93.
9. Taalab, A. I., Darwish, H. A., & Kawady, T. A. (1999). ANN-based novel fault detector for generator stator winding protection. *IEEE Transactions on Power Delivery, 14*(3), 824–830.
10. Gketsis, Z. E., Zervakis, M. E., & Stavrakakis, G. (2009). Detection and classification of winding faults in windmill generators using wavelet transform and ANN. *Electric Power Systems Research, 79*(11), 1483–1494.
11. Fang, H., & Xia, C. (2009). *A Fuzzy Neural Network based fault detection scheme for synchronous generator with internal fault. 6th International Conference on Fuzzy Systems and Knowledge Discovery* (Vol. 4). NJ, USA: IEEE Press Piscataway.

12. Kamaraj, N. (2008). Diagnosis of inter turn fault in the stator of synchronous generator using wavelet based ANFIS. *International Journal of Mathematical, Physical & Eng. Sciences, 2*(2), 68–74.

13. Hatata, A., Helal, A., El Dessouki, H., El-Saadawi, M., & Tantawy, M. (2011). Neural network based fault detector and classifier for synchronous generator stator windings. *Mansoura Engineering Journal (MEJ), 36*(4), E19–E28.

14. Cortes, C., & Vapnik, V. (1995). Support-vector networks. *Machine Learning, 20,* 273–297.

15. Heisele, B., Serre, T., Prentice, S., & Poggio, T. (2003). Hierarchical classification and feature reduction for fast face detection with support vector machines. *Pattern Recognition, 36*(9), 2007–2017.

16. Koley, C., Purkait, P., & Chakravorti, S. (2007). SVM classifier for impulse fault identification in transformers using fractal features. *IEEE Transactions on Dielectric Electrical Insulation, 14*(6), 1538–47.

17. Li, X., Cao, G., & Zhu, X. (2006). Modeling and control of PEMFC based on least squares support vector machines. *Energy Conversion and Management, 47*(7–8), 1032–1050.

18. Pai, P. F., & Hong, W. C. (2005). Support vector machines with simulated annealing algorithms in electricity load forecasting. *Energy Conversion and Management, 46*(17), 2669–2688.

19. Yan, W. W., Shao, H.H. (2002). Application of support vector machine nonlinear classifier to fault diagnoses. Fourth World Congress Intelligent Control and Automation. IEEE, Shanghai, 10–14 June 2002. pp. 2697–2670.

20. Jack, L. B., & Nandi, A. K. (2002). Fault detection using support vector machines and artificial neural networks: augmented by Genetic algorithms. *Mechanical Systems and Signal Processing, 16*(2–3), 373–390.

21. Chan, W. C., Cheung, K. C., & Harris, C. J. (2001). On the modeling of nonlinear dynamic systems using support vector neural networks. *Engineering Applications of Artificial Intelligence, 14*(2), 105–113.

22. Moravej, Z., Banihashemi, S., & Velayati, M. (2009). Power Quality events classification and recognition using a novel support vector algorithm. *Energy Conversion and Management, 50*(12), 3071–3077.

23. Axelberg, P. G. V., Gu, I. Y. H., & Bollen, M. H. J. (2007). Support vector machine for classification of voltage disturbances. *IEEE Transactions on Power Delivery, 22*(3), 1297–303.

24. Vanpik, V. (1998). *Statistical learning theory.* New York: Wiley.

25. Duda, R. O., Hart, P. E., & Stork, D. G. (2001). *Pattern Classification* (2nd ed., p. 2001). New York: Wiley.

26. Elsamahy, M., Faried, S., Sidhu, T., & Ramakrishna, G. (2011). Enhancement of the coordination between generator phase backup protection and generator capability curves in the presence of a midpoint STATCOM using support vector machines. *IEEE Transactions on Power Delivery, 26*(3), 1841–1853.

27. Ziani, R., Felkaoui, A., Zegadi, R., (2007). Performances de la classification par les séparateurs à Vaste Marge (SVM): application au diagnostic vibratoire automatisé. 4th International Conference on Computer Integrated Manufacturing CIP'2007, 03–04 Nov. Sétif. http://www.univ-setif.dz/cip2007/.

28. Boley, D., & Cao, D. (2004). *Training support vector machine using adaptive clustering.* Lake Buena Vista: SIAM International Conference on Data Mining.

29. Carugo, O., Eisenhaber, F., (2010). Data Mining techniques for the life sciences: methods in molecular biology. Humana Press, 223–239, LLC. https://link.springer.com/content/pdf/10.1007%2F978-1-60327-241-4.pdf.

30. Moulin, L., Silva, A., El-Sharkawi, M., & Marks, R. (2004). Support Vector Machines for transient stability analysis of large-scale power systems. *IEEE Transactions on Power Systems, 19*(2), 818–825.

31. Kaytez, F., Taplamacioglu, M., Cam, E., Hardalac F. (2015). Forecasting electricity consumption: A comparison of regression analysis, neural networks and least squares support vector machines. *International Journal of Electrical Power & Energy Systems, 67,* 431–438.

32. Parikh, U. B., Biswarup, D., & Maheshwari, R. P. (2008). Combined wavelet-SVM technique for fault zone detection in a series compensated transmission line. *IEEE Transactions on Power Delivery, 23*(4), 1789–1794.

33. Ravikumar, B., Thukaram, D., & Khincha, H. P. (2009). An Approach using support vector machines for distance relay coordination in transmission system. *Transactions on Power Delivery, 24*(1), 1789–1794.

34. Helal, A., El-Saadawi, M., & Hatata, A. (2010). Modeling of synchronous generators for internal faults simulation using Matlab/Simulink. *Mansoura Engineering Journal (MEJ), 35*(3), E10–E22.

k-NN based fault detection and classification methods for power transmission systems

Aida Asadi Majd, Haidar Samet*⍟ and Teymoor Ghanbari

Abstract

This paper deals with two new methods, based on k-NN algorithm, for fault detection and classification in distance protection. In these methods, by finding the distance between each sample and its fifth nearest neighbor in a pre-default window, the fault occurrence time and the faulty phases are determined. The maximum value of the distances in case of detection and classification procedures is compared with pre-defined threshold values. The main advantages of these methods are: simplicity, low calculation burden, acceptable accuracy, and speed. The performance of the proposed scheme is tested on a typical system in MATLAB Simulink. Various possible fault types in different fault resistances, fault inception angles, fault locations, short circuit levels, X/R ratios, source load angles are simulated. In addition, the performance of similar six well-known classification techniques is compared with the proposed classification method using plenty of simulation data.

Keywords: Short circuit faults, Fault detection, Fault classification, K nearest neighbor algorithm

1 Introduction

Distance protection is one of the major protections of power systems, utilized for detection, classification, and location of short circuit faults. In the detection stage, any change caused by different normal and abnormal conditions is recognized. Then in the classification stage, the type of faults (Ag, Bg, Cg, ABg, BCg, CAg, AB, BC and CA) is determined.

In the fault location stage, the distance between the fault and the relay is determined. Due to importance of speed and accuracy of fault detection and classification units, too many investigations have been dedicated to these fields.

When a fault occurs in the power system, variables such as current, power, power factor, voltage, impedance, and frequency change. Many detection techniques detect fault occurrence by comparing the post-fault values of these variables with their values during system normal operation. Some of fault detection methods are based on Kalman filter [1], first derivative method, Fourier transform (FT), and least squares [2]. Some other methods are based on differential equations [2], travelling waves [3, 4], phasor measurement [5], discrete wavelet transform [6], fuzzy logic, genetic algorithm [7] and neural network [8].

Also, many efforts have been made in the field of fault classification, which can be broadly categorized in two main groups. First, methods that are based on signatures of the signals and definition of some criteria such as: discrete wavelet transform (DWT) [9–13], Fourier transform (FT), S-transform [14], adaptive Kalman filtering [15], sequential components [16, 17], and synchronized voltage and current samples [18]. The second group includes the methods based on artificial intelligence techniques such as: Artificial Neural Networks (ANN) [19–21], fuzzy logic [22, 23], Support Vector Machine (SVM) [24–26], and decision-tree [27].

In this paper, two new methods are presented for detection and classification of faults. A moving window with the length of half cycle of power frequency is considered and the RMS value of the current samples is computed in the window. The RMS value obtained in the last window before fault, in which the fault instant is the last sample, is saved. The current waveforms are divided by the saved RMS value. Then, k-NN algorithm is applied to these normalized waveforms and their squares in classification and detection methods, respectively.

* Correspondence: samet@shirazu.ac.ir
School of Electrical and Computer Engineering, Shiraz University, Shiraz, Iran

In the detection method, a moving window with the length of half cycle is considered. In the window, besides finding the fifth nearest neighbor for each point of the squared normalized currents, the distance between each point and its corresponding neighbor is found. By comparing the maximum distance in each window with an adaptive threshold, the fault is detected.

The classification method has a similar trend, but the k-NN algorithm is applied to the instantaneous values of normalized three-phase currents and length of the window is three quarters of a cycle.

Various scenarios including different fault types, fault inception angles, fault resistances, fault locations, sources phase angles, X/R ratios, and short circuit levels are used to evaluate the performance of the methods in a simulated typical five-bus power system. Also, in order to evaluate the performance of the proposed classification method, it is compared with six other similar methods. The methods are compared in terms of delay time and accuracy using a data set including 450 different cases. Beside the simplicity, the proposed techniques have small calculation burden and high accuracy. Moreover, the methods performance is preserved in different conditions.

The remainder of this paper is organized as follows: Section 2 presents the under-study power system. In Section 3, basis of k-NN and its application for fault detection as well as an improved fault detection algorithm are presented. In Section 4, the proposed classification algorithm is introduced. The simulation results are presented in Section 5. A comparison between the performance of the proposed method and some other similar methods is presented in Section 6. Finally, the main conclusions are presented in Section 7.

2 Simulated power system

A five-bus power system is modeled in MATLAB Simulink. A schematic single line diagram of the under study system is presented in Fig. 1. The modeled system comprises of two generators, four transformers and active and reactive loads connected to buses 4 and 5. Detailed specification of the system components are as follows:

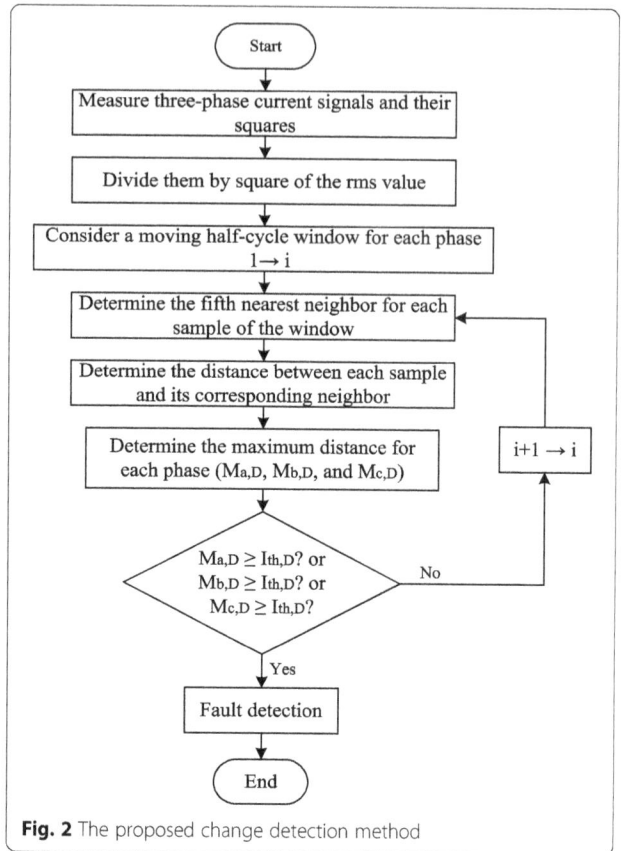

Fig. 2 The proposed change detection method

- Generators: Rated line to line voltage is 20 kV, three-phase short-circuit power is 1000 MVA, frequency is 50 Hz, X/R ratio is 10. Also it is assumed that the angles of sources 1 and 2 are 0 and −10 degree, respectively.
- Transformers: Rated power is 600 MVA, voltage ratio is 20/230 kV with delta-star-grounded connection, its primary and secondary impedances are $0.06 + j0.3\ \Omega$ and $0.397 + j2.12\ \Omega$.
- Lines: All of line impedances are $0.02 + j0.15\ \Omega/km$. Lines 1–2, 2–3, 3–4, 4–1, and 5–2 are 200, 70, 120, 40, and 50 km, respectively.
- Loads: The active and reactive powers of load 1 are 400 MW and 100 MVAr, respectively. The active

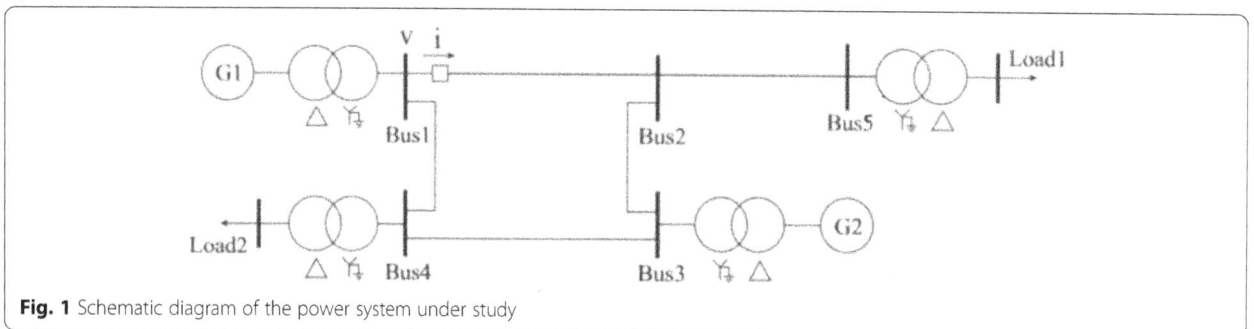

Fig. 1 Schematic diagram of the power system under study

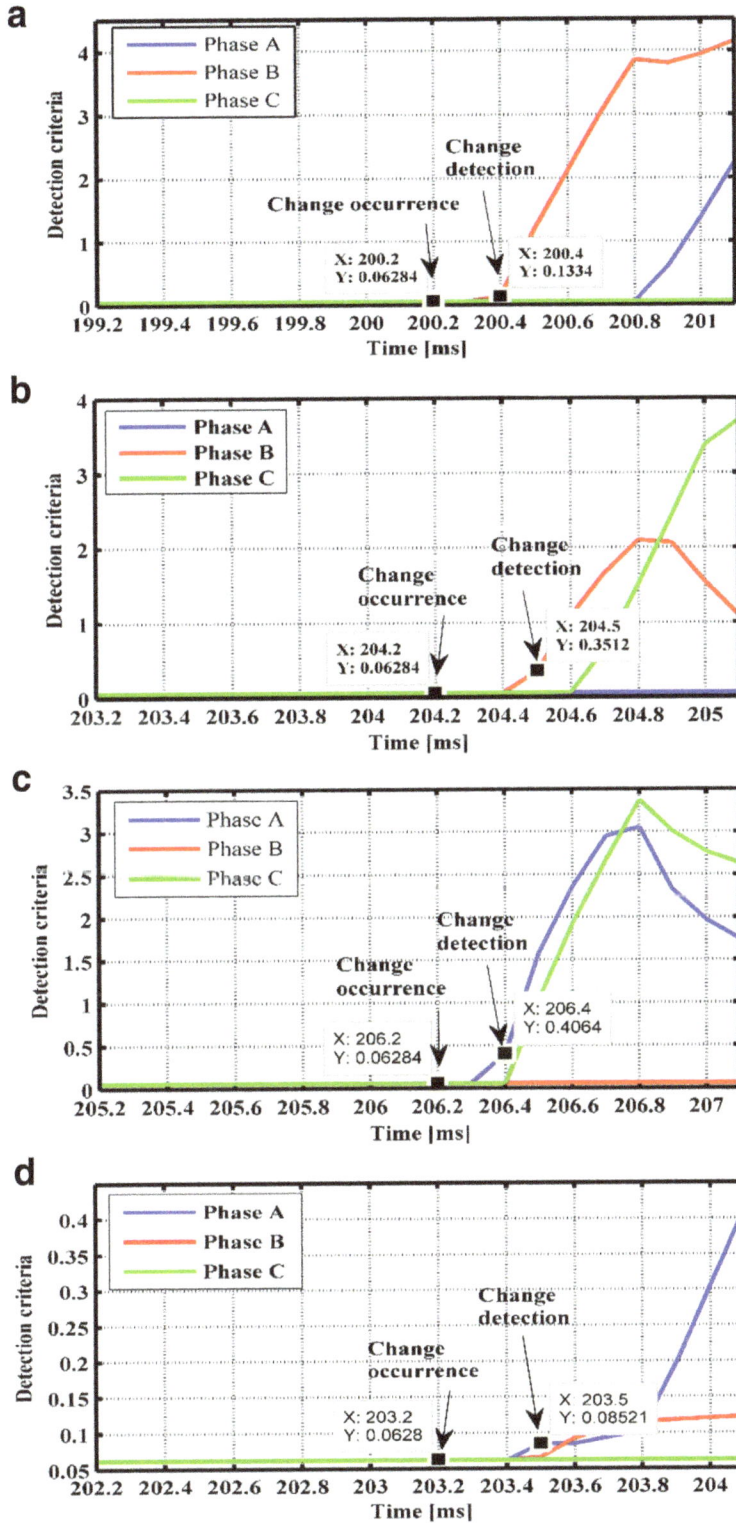

Fig. 3 The proposed criterion. **a** Fault AB, negligible resistance, t0 = 0.2002 s. **b** Fault BCg, R_f = 10 Ω, t0 = 0.2042 s **c** Fault AC, R_f = 40 Ω, t0 = 0.2062 s **d** Switching of load 200 MW, t0 = 0.2032 s

and reactive powers of load 2 are 100 MW and 50 MVAr, respectively.

Sampling frequency: It is equal to 10 kHz.

3 The proposed change detection scheme
3.1 k-Nearest Neighbor algorithm (k-NN)
The k-NN algorithm is a nonparametric classification method that can achieve high classification accuracy in problems with non-normal and unknown distributions. For a particular sample, k closest points between the data and the sample are found. Usually, the Euclidean distance is used, where one point's components are utilized to compare with the components of another point.

The basis of k-NN algorithm is a data matrix that consists of N rows and M columns. Parameters N and M are the number of data points and dimension of each data point, respectively. Using the data matrix, a query point is

provided and the closest k points are searched within this data matrix that are the closest to this query point.

In general, the Euclidean distance between the query and the rest of the points in the data matrix is calculated. After this operation, N Euclidean distances which symbolize the distances between the query with each corresponding point in the data set are achieved. Then, the k nearest points to the query can be simply searched by sorting the distances in ascending order and retrieving those k points that have the smallest distance between the data set and query.

3.2 The proposed fault detection algorithm
Considering fixed sampling frequency, Euclidean distance between each sample and other samples of a considered sliding window varies when a change occurs. In fact, Euclidean distance represents differences between the samples values. k-NN algorithm can derive variation of the Euclidean distance for change

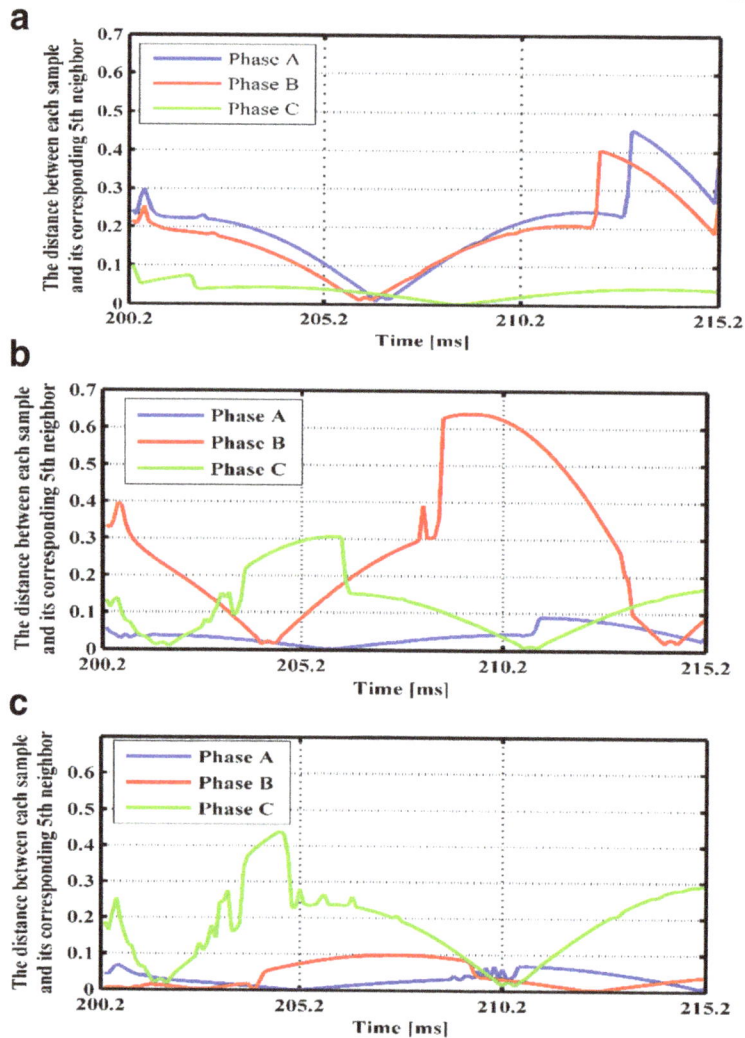

Fig. 4 The distance between each sample and its corresponding neighbor in the analysis window. **a** Fault AB, negligible resistance, t0 = 0.2002 s. **b** Fault BCg, negligible resistance, t0 = 0.2002 s. **c** Fault Cg, negligible resistance, t0 = 0.2002 s

detection. In this work, a sliding window with length of half cycle of power frequency is moved on squared normalized current waveform of each phase. Then, k-NN algorithm is applied to the samples of each

window and the fifth nearest neighbor for each sample and the distance between them is obtained. Finally, the maximum distance is selected for each phase named $M_{a,D}$, $M_{b,D}$, and $M_{c,D}$. Based on

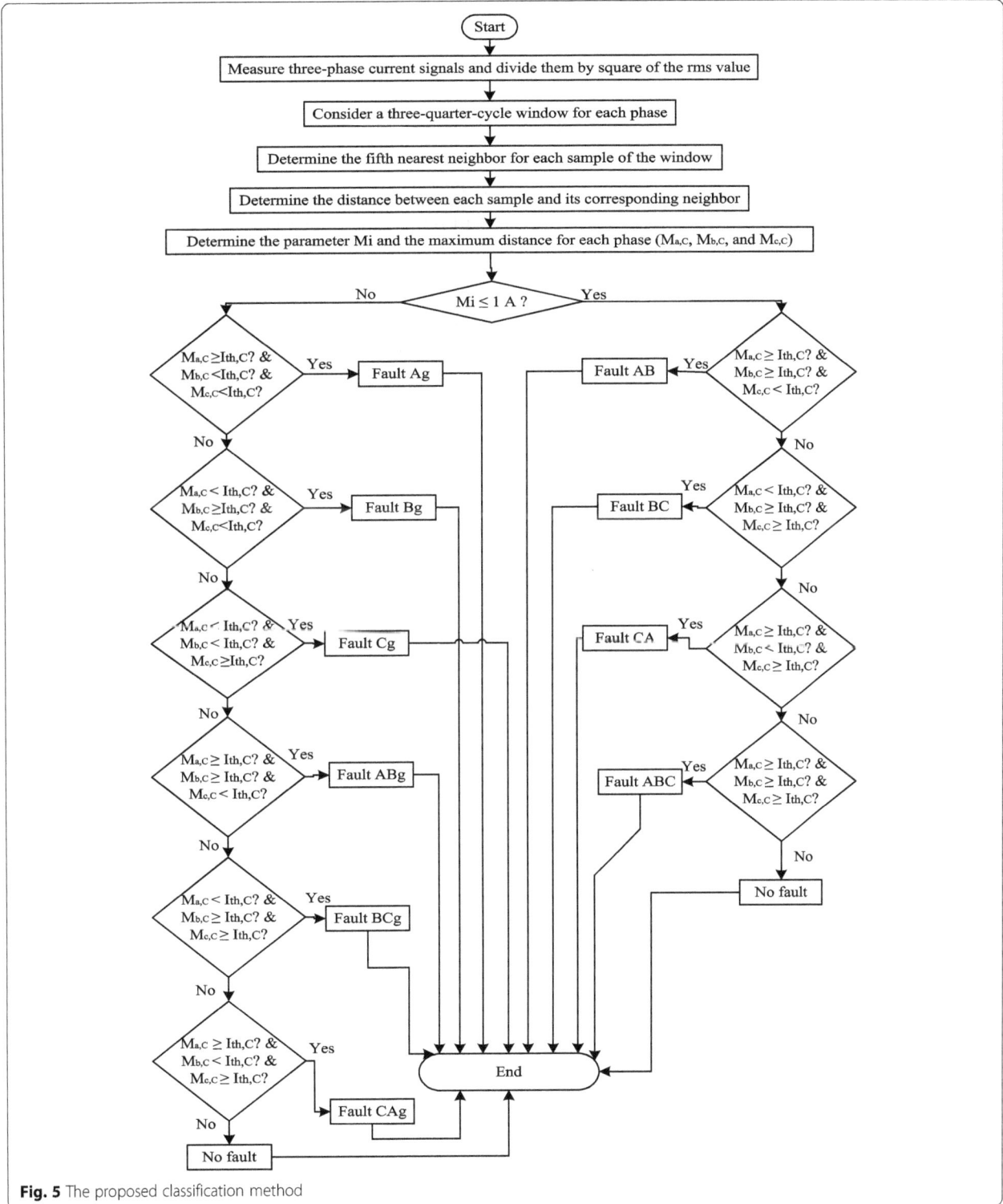

Fig. 5 The proposed classification method

different simulations, it is confirmed that the fifth nearest neighbor gives the best accuracy. In addition to the derived fifth neighbor, the distance between each sample and its corresponding fifth neighbor is derived. Considering sampling frequency 10 kHz, there are 100 samples in each half cycle, result in 100 different distances. Among them, the maximum distance is compared with a certain threshold value to detect fault condition.

In case of change occurrence, the sample corresponding to the change enters the end of the window. It is observed that after three or four samples, the maximum distance of some or all of the phases exceed the threshold value. By considering an appropriate value for the threshold, it is possible to detect the fault after 0.2 ms to 0.4 ms. In this study, $I_{th,D} = 0.0667$ is selected for fault detection threshold. Flowchart of the proposed algorithm for change detection is shown in Fig. 2.

In Fig. 3, the proposed criterion for some different fault cases is presented. The instants of change occurrence and the relevant detection times, are shown.

4 The proposed fault classification scheme

The general approach for fault classification is the same as detection method. However, in the classification method the k-NN algorithm is implemented in a window applied to normalized current waveforms with length of three quarters of a cycle, called analysis window. The considered k value and length of analysis window are selected based on different simulations to achieve the best accuracy and speed for the classification.

In Fig. 4, three-phase distances values for some different fault types with negligible resistance and inception instant equal to 0.2002 s are presented. In these figures, the fifth nearest neighbor for each sample of the analysis window is shown.

It is obvious, the distance between each sample of current and its fifth neighbor is a suitable criterion for fault classification. By choosing the maximum distance for each phase ($M_{a,C}$, $M_{b,C}$, and $M_{c,C}$) and comparing it with a threshold value, the type of fault can be determined. It is obvious that the values of $M_{a,C}$, $M_{b,C}$, and $M_{c,C}$ are obtained exactly the same as detection method, but in a window with the length of three quarters of a cycle. The best threshold value is selected using different simulations.

Some other considerations are taken into account for the classification method, which are as follows:

1. For discrimination between two phase faults (LL) and grounded two phase faults (LL-g), the means of three phases' corresponding current samples in the analysis window is obtained and the maximum mean is utilized as follows:

$$Mi = \max\left(\frac{ia + ib + ic}{3}\right) \; in \; the \; analysis \; window$$

In case of grounded faults (LL-g), Mi > 100 A and Mi < 1 A for two phase faults (LL). This criterion can discriminate between LL and LL-g with a very high accuracy.

2. In order to omit the initial transient behavior of the signal, twenty first samples of the window are not considered.

The flowchart of the classification method is presented in Fig. 5. Threshold $I_{th,C}$ is set to 0.1108.

5 Test cases and simulation results

5.1 Case 1: Various fault types

Different fault types are applied at the middle of line 1–2 of the power system shown in Fig. 1. The results are shown in Table 1. The faults are solid and applied at an identical inception instant 0.2002 s. Results including the discrimination criteria (Mi) and the maximum distance of each phase are presented in Table 1. From the results, one can conclude that the proposed method is able to classify different faults using the mentioned rules.

The results for each group of phase-to-ground, phase-to-phase-to-ground, and phase-to-phase faults are similar. Therefore, hereafter only four types of faults including: Ag, ABg, AB, and ABC are considered.

5.2 Case 2: Various inception instants

In Table 2, the results for different inception instants are presented for the mentioned faults. The inception instant is varied by step 3 ms. Faults are also considered solid type. The results confirm that the proposed method is able to classify faults at different inception instants.

5.3 Case 3: Various fault resistances

In Table 3, the results of this case study for fault resistances 10, 30, 50,70, and 90 Ω, are shown. The

Table 1 Results of various fault types

Type	Mi	$M_{a,C}$	$M_{b,C}$	$M_{c,C}$
Ag	1.2652e + 03	0.5853	0.0711	0.0824
Bg	1.0727e + 03	0.1017	0.5539	0.0342
Cg	310.2327	0.0698	0.0986	0.4405
ABg	1.0528e + 03	0.7518	0.3433	0.0580
BCg	994.4663	0.0903	0.6393	0.3064
CAg	878.8438	0.3729	0.0881	0.5971
AB	0.0286	0.4575	0.4040	0.0444
BC	0.1351	0.0888	0.4539	0.3813
CA	0.0488	0.2934	0.0889	0.4018
ABC	0.0065	0.6779	0.5013	0.4630

Table 2 Results of various fault inception instants

Inception instant (sec)	Type	Mi	$M_{a,C}$	$M_{b,C}$	$M_{c,C}$
0.2032	Ag	993.4711	0.5603	0.0610	0.1014
0.2062	Ag	436.8110	0.4895	0.0658	0.0992
0.2092	Ag	535.8503	0.7499	0.0713	0.0952
0.2032	ABg	1.0660e + 03	0.6429	0.3094	0.0894
0.2062	ABg	944.6056	0.6065	0.2122	0.0889
0.2092	ABg	512.7830	0.9243	0.3425	0.0805
0.2032	AB	0.0454	0.4573	0.3848	0.0878
0.2062	AB	0.0449	0.4300	0.3439	0.0889
0.2092	AB	0.0450	0.5665	0.4345	0.0860
0.2032	ABC	0.0059	0.5048	0.4663	0.2638
0.2062	ABC	0.0047	0.4698	0.2639	0.5052
0.2092	ABC	0.0040	0.2640	0.5038	0.4723

faults are applied at an identical inception instant 0.2002 s. From the results, it is confirmed that the proposed method has acceptable performance for fault resistance up to 90 Ω. Although the technique can also classify the faults with resistances more than 90 Ω, the performance may be less than the acceptable value.

Table 3 Results of various fault resistances

Type	Resistance (Ω)	Mi	$M_{a,C}$	$M_{b,C}$	$M_{c,C}$
Ag	10	898.5760	0.4789	0.0753	0.0771
Ag	30	561.8625	0.3472	0.0804	0.0493
Ag	50	405.9855	0.2790	0.0830	0.0482
Ag	70	316.9368	0.2381	0.0844	0.0474
Ag	90	259.7276	0.2114	0.0853	0.0469
ABg	10	779.8930	0.5540	0.3489	0.0568
ABg	30	508.6675	0.3957	0.2961	0.0538
ABg	50	375.8039	0.3096	0.2524	0.0456
ABg	70	297.5452	0.2593	0.2221	0.0454
ABg	90	246.1880	0.2269	0.2006	0.0453
AB	10	0.0536	0.4328	0.3798	0.0444
AB	30	0.0889	0.3280	0.2957	0.0444
AB	50	0.1022	0.2637	0.2435	0.0444
AB	70	0.1076	0.2242	0.2110	0.0444
AB	90	0.1101	0.1981	0.1893	0.0444
ABC	10	0.0231	0.4989	0.4472	0.3738
ABC	30	0.0651	0.3610	0.3447	0.2130
ABC	50	0.0862	0.2903	0.2817	0.1453
ABC	70	0.0970	0.2471	0.2420	0.1236
ABC	90	0.1031	0.2185	0.2151	0.1193

5.4 Case 4: Various fault locations

One of the other challenges that should be considered for a fault identification technique is location of the fault in the transmission lines. In this test case, the system is analyzed with a fault applied at 0%, 20%, 40%, 60%, 80%, and 100% of the transmission line 1–2. Results of the four fault types are shown in Table 4. The faults are solid type and applied at an identical inception instant 0.2002 s.

In addition, several faults for locations more than 100% are simulated. The faults are applied at 105%, 110%, and 120% of the transmission line 2–5 at an identical inception instant 0.2002 s. The results are tabulated in Table 5.

From the results, it can be concluded that the performance of the proposed method is preserved even for locations more than 100%. It should be mentioned that the performance of the proposed method degrades for locations more than 120%.

5.5 Case 5: Various sources load angles

The results for various angles, according different inception instant, fault resistances, and fault types verify that

Table 4 Results of various fault locations

Type	location (%)	Mi	$M_{a,C}$	$M_{b,C}$	$M_{c,C}$
Ag	0	1.8356e + 03	0.7890	0.0715	0.0821
Ag	20	1.5959e + 03	0.7035	0.0711	0.0825
Ag	40	1.3724e + 03	0.6222	0.0710	0.0821
Ag	60	1.1608e + 03	0.5493	0.0712	0.0811
Ag	80	957.3303	0.4802	0.0719	0.0560
Ag	100	758.6445	0.4136	0.0732	0.0590
ABg	0	1.5206e + 03	1.0038	0.4373	0.0521
ABg	20	1.3260e + 03	0.8987	0.3990	0.0543
ABg	40	1.1418e + 03	0.7994	0.3615	0.0566
ABg	60	965.7605	0.7050	0.3253	0.0593
ABg	80	795.2633	0.6144	0.2899	0.0598
ABg	100	628.0395	0.5265	0.2549	0.0621
AB	0	0.0251	0.5906	0.5384	0.0444
AB	20	0.0268	0.5338	0.4784	0.0444
AB	40	0.0281	0.4819	0.4270	0.0444
AB	60	0.0289	0.4337	0.3787	0.0444
AB	80	0.0293	0.3884	0.3327	0.0444
AB	100	0.0292	0.3450	0.2898	0.0444
ABC	0	0.0077	0.8967	0.6611	0.6104
ABC	20	0.0077	0.8034	0.5927	0.5474
ABC	40	0.0071	0.7180	0.5305	0.4900
ABC	60	0.0057	0.6387	0.4729	0.4368
ABC	80	0.0036	0.5640	0.4187	0.3868
ABC	100	0.0025	0.4926	0.3668	0.3391

Table 5 Results of fault locations more than 100%

Fault Type	Resistance	Location (%)	Mi	$M_{a,C}$	$M_{b,C}$	$M_{c,C}$
ABC	negligible	105	6.1112e-08	0.4899	0.3648	0.3372
ABg	90	105	144.2950	0.1833	0.1639	0.0473
AC	negligible	110	0.0081	0.1963	0.0888	0.3032
Ag	90	110	150.6250	0.1693	0.0879	0.0480
AB	negligible	120	0.0080	0.3385	0.2827	0.0444
Bg	90	120	144.2091	0.0961	0.1671	0.0666

proposed method classify the faults in different values of sources load angles. For abbreviation, the results relevant to this case are not presented.

5.6 Case 6: Various X/R ratios

Different X/R ratios impact on the performance of the proposed method is also investigated, considering different inception instant, fault resistances, and fault types. From the results, it can be concluded that accuracy of the proposed method is preserved for different values of X/R ratios.

5.7 Case 7: Various short circuit levels

The performance of the proposed method is also evaluated for various sources short circuit levels. The algorithm also has desirable performance for these cases.

5.8 Case 8: Various load levels

In Table 6, the results of some simulated cases for no-load and loads with fraction of the nominal value are shown. It should be noted that for each load, different load values are considered in the condition of no-load of the other one. All the faults are applied in the location

of 80% of the transmission line 1–2. From the results, one can observe that the performance of the proposed method is preserved in different load levels.

5.9 Case 9: Current transformer saturation

The performance of the method is also evaluated during current transformer saturation. Two typical cases are considered. The faults are solid type and applied at an identical inception instant 0.2345 s. The classification criteria for both cases are shown in Fig. 6 and Table 7. It is observed that the proposed method is able to classify the faults during current transformer saturation.

6 A comparison with other techniques

The performance of the proposed method is compared with six other similar approaches in this Section. All of the methods are evaluated using an identical data set in similar conditions. The six methods are briefly reviewed as follows:

a. Sequence Component [16]: This technique classifies the faults using the phase differences between positive and negative sequences. Also, relative magnitudes of negative and zero sequences from pre-fault to the fault stage are used to distinguish between phase-to-phase (LL) and phase-to-phase-to-ground (LLg) faults.

b. Alienation Coefficients [28]: In this algorithm, alienation technique is applied to two half successive cycles with the same polarity. The alienation coefficients of the successive cycles as two dependent variables are calculated. This technique is capable of classification using only three-phase current waveforms and its delay time is half cycle of power frequency. Also, another version of this approach is presented in [29].

Table 6 Results of various load levels

Fault Type	resistance	Inception instant		No-load	Load1 100 MW 25 MVAr	Load1 200 MW 50 MVAr	Load1 300 MW 75 MVAr	Load1 400 MW 100 MVAr	Load2 50 MW 25 MVAr	Load2 100 MW 50 MVAr
AB	Negligible	0.2002	Mi	0.0128	0.0124	0.0120	0.0116	0.0112	0.0125	0.0123
			Ma	1.8524	0.8232	0.5531	0.4304	0.3605	2.1232	2.3803
			Mb	1.7653	0.7522	0.4856	0.3686	0.3023	2.0399	2.3048
			Mc	0.0773	0.0921	0.0444	0.0444	0.0444	0.0822	0.0869
ABg	30	0.2032	Mi	467.5109	450.9092	434.2540	418.0157	402.4154	457.0304	446.9282
			Ma	1.6118	0.7110	0.4746	0.3673	0.3067	1.8174	2.0083
			Mb	0.6028	0.2936	0.2268	0.1966	0.1796	0.6865	0.7655
			Mc	0.1003	0.1102	0.1066	0.0993	0.0946	0.1024	0.1018
ABC	90	0.2062	Mi	5.1811e-09	1.3579e-08	1.7955e-08	2.7285e-08	2.4108e-08	2.5092e-08	2.5051e-08
			Ma	1.1341	0.2924	0.1651	0.1534	0.1483	1.2205	1.1439
			Mb	0.8183	0.3847	0.2699	0.2172	0.1868	0.9138	1.0000
			Mc	0.8122	0.3815	0.2678	0.2161	0.1868	0.9061	0.9907

Table 7 Results of two fault cases during current transformer saturation

	Fault Type	Mi	$M_{a,C}$	$M_{b,C}$	$M_{c,C}$
a	AB	9.2980e-04	1.0595	1.1337	0.0915
b	ABC	0.0074	1.2737	1.2458	1.4452

c. Discrete Wavelet Transform [23]: Daubechies family of wavelet transform is used in this technique. Third level output among different decomposed levels is used and the summation of detailed current signals for each phase (S_a, S_b, and S_c) is obtained. If the summation of S_a, S_b, and S_c is equal to zero, then the fault type is either three-phase or LL, otherwise, it is phase-to-ground (Lg) or LLg fault.

d. Fuzzy Logic [22]: The prerequisite of this technique is fault occurrence time. In this algorithm, using measured current samples, some specific characteristics for the samples are defined for the fault classification. The technique takes three quarters of a cycle to classify the fault.

e. Using RMS Values of current: A simple approach to classify the faults is based on comparing the RMS values of three-phase current waveforms with a certain threshold. The RMS values of the phases are obtained using Fourier transform in a half cycle window after fault occurrence. Discrimination between LL and LLg is determined using zero sequence component of current, which is large for LLg and zero for LL.

f. Using RMS Values of Voltage: This technique is exactly the same as previous method for three-phase voltage signals. Type of fault is determined when the RMS values of the voltages become less than a certain threshold.

The performance of the proposed method is compared with the above-mentioned methods based on following factors; the results are tabulated in Table 8:

- Fault resistances
- Fault inception instants
- Fault locations
- Generators X/R ratios
- Phase difference between two generators
- Generators short circuit levels
- Delay operation time
- Error percentage

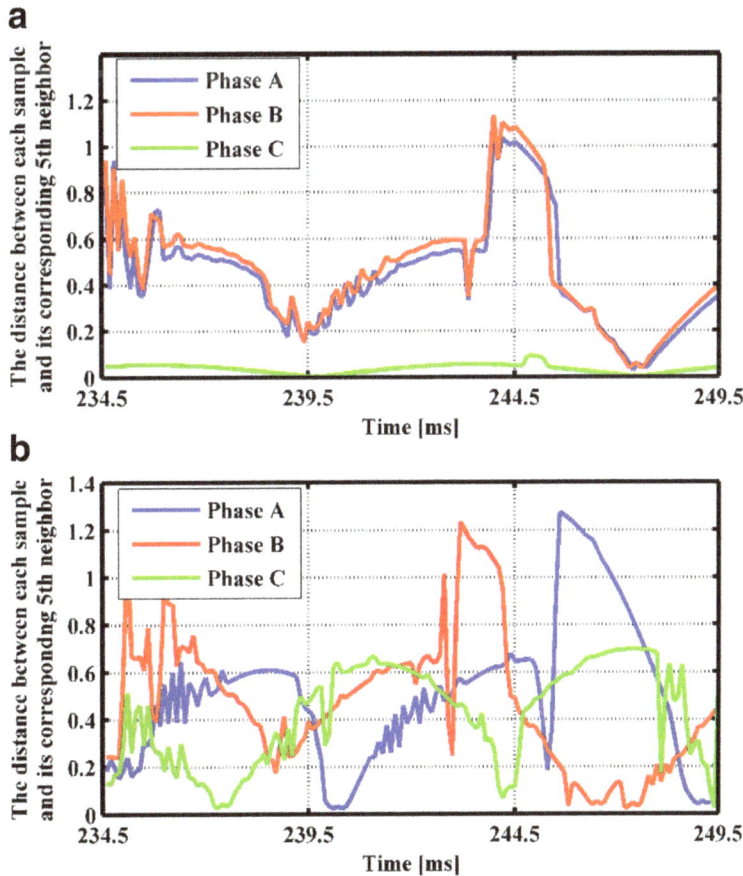

Fig. 6 The distance between each sample and its corresponding neighbor in the analysis window. **a** Fault AB, negligible resistance, t0 = 0.2345 s. **b** Fault ABC, negligible resistance, t0 = 0.2345 s

Table 8 Comparison between the different methods

Technique	Error Percentage of different fault resistance and occurrence time (%)	Error Percentage of different fault location (%)	Error Percentage of different X/R ratio of sources (%)	Error Percentage of different phase angle of sources (%)	Error Percentage of different short circuit level of sources (%)	Total error percentage (%)	Mean delay time (ms)
a	10.5	12	10	14	10	10.98	15
b	22.5	30	22.86	26	25	24.15	10
c	26	20	25.71	32	37.5	27.07	10
d	0	0	0	2	2.5	0.49	15
e	13	10	7.14	46	32.5	17.56	10
f	28	20	24.29	34	17.5	26.10	10
Proposed technique	0.5	0	0	8	7.5	1.95	15

The number of the whole cases considered in this Section is 410; 200 cases for different fault resistances and inception instants, 50 cases for different fault locations, 70 cases for different sources X/R ratios, 50 cases for different sources angles, and 40 cases for different short circuit levels.

In Table 8, error percentages for the above mentioned factors are calculated as the ratio of number of malfunction operations to number of the relevant cases. Then, total error percentage for each method is calculated as ratio of number of whole mal-function operations to number of whole the cases.

Techniques a and d have a delay time 15 ms and techniques b, c, e, and f have a delay time 10 ms. Among the methods with delay time 15 ms, fuzzy logic has a very good performance with only 0.49% error.

The proposed technique has a good performance with error percentage of 1.95% and average delay time of 15 ms. Based on the calculated total error percentage and delay time, it is confirmed that the proposed method has acceptable performance in comparison with other methods.

7 Conclusion

Two simple methods for fault detection and classification are presented in this paper. The methods are based on k-NN algorithm. Plenty of simulations were used in order to evaluate the performance of the methods. The performance of the proposed classification method is compared with six other similar methods. From the results, the good accuracy and speed of the methods are confirmed. The classification technique has accuracy about 98% for the considered data set with 15 ms average delay time.

Authors' contributions
All authors read and approved the final manuscript.

Competing interests
The authors declare that they have no competing interests.

References
1. Chowdhury, F. N., Christensen, J. P., & Aravena, J. L. (1991). Power system fault detection and state estimation using Kalman filter with hypothesis testing. *IEEE Transactions on Power Delivery, 6*(3), 1025–1030.
2. Öhrström, M., & Söder, L. (2002). Fast fault detection for power distribution systems. Power and energy systems (PES), Marina del Rey, USA, may 13–15.
3. Magnago, F. H., & Abur, A. (1999). A new fault location technique for radial distribution systems based on high frequency signals. *IEEE in Power Engineering Society Summer Meeting, 1,* 426–431.
4. Xiangjun, Z., Yuanyuan, W., Yao, X. (2010). Faults detection for power systems. INTECH Open Access Publisher. In W. Zhang (E.d.), Fault Detection (pp. 512). InTech. ISBN 978-953-307-037-7. doi:10.5772/56395. https://www.intechopen.com/books/fault-detection
5. Gopakumar, P., Reddy, M. J. B., & Mohanta, D. K. (2015). Transmission line fault detection and localisation methodology using PMU measurements. *Journal of IET, Generation, Transmission & Distribution, 9*(11), 1033–1042.
6. Bezerra Costa, F. (2014). Fault-induced transient detection based on real-time analysis of the wavelet coefficient energy. *IEEE Transactions on Power Delivery, 29*(1), 140–153.
7. Haghifam, M. R., Sedighi, A. R., & Malik, O. P. (2006). Development of a fuzzy inference system based on genetic algorithm for high-impedance fault detection. *Journal of IEE Proceedings-Generation, Transmission and Distribution, 153*(3), 359–367.
8. Baqui, I., Zamora, I., Mazón, J., & Buigues, G. (2011). High impedance fault detection methodology using wavelet transform and artificial neural networks. *Journal of Electric Power Systems Research, 81*(7), 1325–1333.
9. Shaik, A. G., & Pulipaka, R. R. V. (2015). A new wavelet based fault detection, classification and location in transmission lines. *International Journal of Electrical Power & Energy Systems, 64,* 35–40.
10. Torabi, N., Karrari, M., Menhaj, M. B., Karrari, S. (2012). 'Wavelet Based Fault Classification for Partially Observable Power Systems. IEEE, In Asia-Pacific Power and Energy Engineering Conference (APPEEC) (pp. 1–6).
11. Usama, Y., Lu, X., Imam, H., Sen, C., & Kar, N. (2013). Design and implementation of a wavelet analysis-based shunt fault detection and identification module for transmission lines application. *IET Journal of Generation, Transmission & Distribution, 8*(3), 431–444.
12. Guillen, D., Arrieta Paternina, M. R., Zamora, A., Ramirez, J. M., & Idarraga, G. (2015). Detection and classification of faults in transmission lines using the maximum wavelet singular value and Euclidean norm. *IET Journal of Generation, Transmission & Distribution, 9*(15), 2294–2302.
13. Liu, Z., Han, Z., Zhang, Y., & Zhang, Q. (2014). Multiwavelet packet entropy and its application in transmission line fault recognition and classification. *IEEE Transactions on Neural Networks and Learning Systems, 25*(11), 2043–2052.
14. Dash, P. K., Das, S., & Moirangthem, J. (2015). Distance protection of shunt compensated transmission line using a sparse S-transform. *IET Journal of Generation, Transmission & Distribution, 9*(12), 1264–1274.
15. Girgis, A., & Makram, E. B. (1988). Application of adaptive Kalman filtering in fault classification, distance protection, and fault location using microprocessors. *IEEE Transactions on Power Systems, 3*(1), 301–309.
16. Adu, T. (2002). An accurate fault classification technique for power system monitoring devices. *IEEE Transactions on Power Delivery, 17*(3), 684–690.
17. Rahmati, A., & Adhami, R. (2014). A fault detection and classification technique based on sequential components. *IEEE Transactions on Industry Applications, 50*(6), 4202–4209.

18. Esmaeilian, A., & Kezunovic, M. (2014). Transmission-line fault analysis using synchronized sampling. *IEEE Transactions on Power Delivery, 29*(2), 942–950.

19. Butler, K. L., Momoh, J. (1993). Detection and classification of line faults on power distribution systems using neural networks. IEEE Proceedings of the 36th Midwest Symposium, In Circuits and Systems. (pp. 368–371).

20. Upendar, J., Gupta, C. P., Singh, G. K. (2008). ANN based power system fault classification. IEEE, In Region 10 Conference (TENCON), November, (pp. 1–6).

21. Tayeb, E. B. M., Rhim, O. A. A. A. (2011). Transmission line faults detection, classification and location using artificial neural network. IEEE, international conference, utility exhibition on power and energy systems: Issues & prospects for Asia (ICUE), September.

22. Mahanty, R. N., & Gupta, P. D. (2007). A fuzzy logic based fault classification approach using current samples only. *Journal of Electric power systems research, 77*(5), 501–507.

23. Reddy, M. J., & Mohanta, D. K. (2007). A wavelet-fuzzy combined approach for classification and location of transmission line faults. *International Journal of Electrical Power & Energy Systems, 29*(9), 669–678.

24. Shahid, N., Aleem, S. A., Naqvi, I. H., Zaffar, N. (2012). Support vector machine based fault detection & classification in smart grids. IEEE, In Globecom Workshops (GC Wkshps), December, (pp. 1526–1531).

25. Livani, H., Evrenosoğlu, C. Y. (2012). A fault classification method in power systems using DWT and SVM classifier. IEEE PES, In Transmission and Distribution Conference and Exposition (T&D), May, 1–5.

26. Moravej, Z., Pazoki, M., & Khederzadeh, M. (2015). New pattern-recognition method for fault analysis in transmission line with UPFC. *IEEE Transactions on Power Delivery, 30*(3), 1231–1242.

27. Swetapadma, A., & Yadav, A. (2015). Data-mining-based fault during power swing identification in power transmission system. *Journal of IET Science, Measurement & Technology, 10*(2), 130–139.

28. Masoud, M. E., & Mahfouz, M. M. A. (2010). Protection scheme for transmission lines based on alienation coefficients for current signals. *IET Journal of Generation, transmission & distribution, 4*(11), 1236–1244.

29. Samet, H., Shabanpour-Haghighi, A., & Ghanbari, T. (2017). A fault classification technique for transmission lines using an improved alienation coefficients technique. doi:10.1002/etep.2235. http://onlinelibrary.wiley.com/doi/10.1002/etep.2235/abstract.

Permissions

All chapters in this book were first published in PCMPS, by Springer International Publishing AG.; hereby published with permission under the Creative Commons Attribution License or equivalent. Every chapter published in this book has been scrutinized by our experts. Their significance has been extensively debated. The topics covered herein carry significant findings which will fuel the growth of the discipline. They may even be implemented as practical applications or may be referred to as a beginning point for another development.

The contributors of this book come from diverse backgrounds, making this book a truly international effort. This book will bring forth new frontiers with its revolutionizing research information and detailed analysis of the nascent developments around the world.

We would like to thank all the contributing authors for lending their expertise to make the book truly unique. They have played a crucial role in the development of this book. Without their invaluable contributions this book wouldn't have been possible. They have made vital efforts to compile up to date information on the varied aspects of this subject to make this book a valuable addition to the collection of many professionals and students.

This book was conceptualized with the vision of imparting up-to-date information and advanced data in this field. To ensure the same, a matchless editorial board was set up. Every individual on the board went through rigorous rounds of assessment to prove their worth. After which they invested a large part of their time researching and compiling the most relevant data for our readers.

The editorial board has been involved in producing this book since its inception. They have spent rigorous hours researching and exploring the diverse topics which have resulted in the successful publishing of this book. They have passed on their knowledge of decades through this book. To expedite this challenging task, the publisher supported the team at every step. A small team of assistant editors was also appointed to further simplify the editing procedure and attain best results for the readers.

Apart from the editorial board, the designing team has also invested a significant amount of their time in understanding the subject and creating the most relevant covers. They scrutinized every image to scout for the most suitable representation of the subject and create an appropriate cover for the book.

The publishing team has been an ardent support to the editorial, designing and production team. Their endless efforts to recruit the best for this project, has resulted in the accomplishment of this book. They are a veteran in the field of academics and their pool of knowledge is as vast as their experience in printing. Their expertise and guidance has proved useful at every step. Their uncompromising quality standards have made this book an exceptional effort. Their encouragement from time to time has been an inspiration for everyone.

The publisher and the editorial board hope that this book will prove to be a valuable piece of knowledge for researchers, students, practitioners and scholars across the globe.

List of Contributors

Qian Ai, Songli Fan and Longjian Piao
Department of Electrical Engineering, Shanghai Jiao Tong University, Shanghai 200240, China

Anamika Yadav, Yajnaseni Dash and V. Ashok
National Institute of Technology Raipur, Raipur, Chhattisgarh, India

Zhiqing Yao
Xuchang KETOP Electrical Research Institute, Xuchang, China

Qun Zhang, Peng Chen and Qian Zhao
XJ Electric Co., Ltd, Xuchang, China

Baohui Zhang and Zhiguo Hao
School of Electrical Engineering, Xi'an Jiaotong University, Xi'an 710049, China

Zhiqian Bo
XUJI Group Corporation, Xuchang 461000, China

Qingping Wang, Zhiqian Bo, Xiaowei Ma, Ming Zhang, Yingke Zhao, Yi Zhu and Lin Wang
Corporate R&D Center, XJ group, State Grid Corporation of China, Beijing, China

Wenming Guo and Longhua Mu
Department of Electrical Engineering, Tongji University, Shanghai, China

Md Habibur Rahman and Lie Xu
University of Strathclyde, Glasgow, UK

Liangzhong Yao
China Electric Power Research Institute, Beijing, China

Jalal Khodaparast and Mojtaba Khederzadeh
Electrical Engineering Department, Shahid Beheshti University, Tehran 165895371, Iran

Yuzheng Xie, Hengxu Zhang and Changgang Li
Key Laboratory of Power System Intelligent Dispatch and Control of the Ministry of Education (Shandong University), 17923 Jingshi Road, Jinan, Shandong 250061, China

Huadong Sun
China Electric Power Research Institute, 15 Xiaoying East Road, Qinghe, Beijing 100192, China

Xianggen Yin, Zhe Zhang, Zhenxing Li, Xuanwei Qi, Wenbin Cao and Qian Guo
State Key Laboratory of Advanced Electromagnetic Engineering and Technology, Huazhong University of Science and Technology, Wuhan 430074, China

Zhengming Li, Wenwen Li and Tianhong Pan
School of Electrical and Information Engineering, Jiangsu University, Zhenjiang 212000, Jiangsu Province, China

Saeed Roostaee, Mini S. Thomas and Shabana Mehfuz
Department of Electrical Engineering, Faculty of Engineering and Technology, Jamia Millia Islamia, New Delhi, India

Jinghan He, Lin Liu, Wenli Li and Ming Zhang
School of Electrical Engineering, Beijing Jiaotong University, Beijing 100044, China

Junhui Huang, Jun Han and Hu Li
State Grid Jiangsu Economic Research Institute, Nanjing, Jiangsu province, 210008, China

Shaoyun Ge, Xiaomin Zhou, Hong Liu and Bo Wang
Key Laboratory of Smart Grid of Ministry of Education, Tianjin University, Tianjin 300072, China

Zhengfang Chen
State grid Suzhou Power Supply Company, Suzhou, Jiangsu province, 215000, China

Chao Li, Jianzhao Geng, Xiuchang Zhang and Tim Coombs
EPEC Superconductivity Group, University of Cambridge, Cambridge CB3 0FA, UK

Bin Li
Key Laboratory of Smart Grid of Ministry of Education, Tianjin University, Tianjin 300072, China

Fengrui Guo
State Grid Tianjin Electric Power Company, Tianjin 300010, China

Lin Feng, Jingning Zhang and Guojie Li
School of Electrical Information and Electronic Engineering, Shanghai Jiao Tong University, Shanghai, China

Bangling Zhang
Shanghai Power & Energy Storage Battery System Engineering Tech. Co. Ltd., Shanghai, China

Renfeng Tao, Fengting Li, Weiwei Chen, Yanfang Fan, Chenguang Liang and Yang Li
College of Electrical Engineering of Xinjiang University, Xinjiang, China

Li Huang, Yongbiao Yang and Hongjuan Zheng
NARI Technology Development CO., Ltd, Nanjing, Jiangsu Province 211106, China

Honglei Zhao and Xudong Wang
State Grid Tianjin Electric Power Company, Tianjin 300010, China

Yu Cheng and Chengwei Zhang
School of Electrical and Electronic Engineering, North China Electric Power University, Beijing 102206, China

Magdi El-Saadawi and Ahmed Hatata
Mansoura University Faculty of Engineering, Mansoura, Egypt

Aida Asadi Majd, Haidar Samet and Teymoor Ghanbari
School of Electrical and Computer Engineering, Shiraz University, Shiraz, Iran

Index